IMAGE PROCESSING
and MATHEMATICAL
MORPHOLOGY
Fundamentals and Applications

IMAGE PROCESSING
and MATHEMATICAL
MORPHOLOGY
Fundamentals and Applications

Frank Y. Shih

CRC Press
Taylor & Francis Group
Boca Raton London New York

CRC Press is an imprint of the
Taylor & Francis Group, an **Informa** business

CRC Press
Taylor & Francis Group
6000 Broken Sound Parkway NW, Suite 300
Boca Raton, FL 33487-2742

First issued in paperback 2017

ISBN 13: 978-1-138-11228-5 (pbk)
ISBN 13: 978-1-4200-8943-1 (hbk)

Library of Congress Cataloging-in-Publication Data

Shih, Frank Y.
   Image processing and mathematical morphology fundamentals and applications / author, Frank Y. Shih.
       p. cm.
       "A CRC title."
       Includes bibliographical references and index.
       ISBN 978-1-4200-8943-1 (hardcover : alk. paper)
       1. Image processing--Mathematics. 2. Shapes--Mathematical models. 3. Digital filters (Mathematics) I. Title.

   TA1637.S474 2009
   621.36'7--dc22   2009002783

**Visit the Taylor & Francis Web site at
http://www.taylorandfrancis.com**

**and the CRC Press Web site at
http://www.crcpress.com**

*To my loving wife and children, to my parents
who have encouraged me through the years, and
to all those who have helped me in the process of writing this book.*

Contents

Preface

Image processing and mathematical morphology have become increasingly important because digital multimedia, including automated vision detection and inspection and object recognition, are widely used today. For the purposes of image analysis and pattern recognition, one often transforms an image into another better represented form. Image-processing techniques have been developed tremendously during the past five decades and among them, mathematical morphology has always received a great deal of attention because it provides a quantitative description of geometric structure and shape as well as a mathematical description of algebra, topology, probability, and integral geometry.

Mathematical morphology denotes a branch of biology that deals with the forms and structures of animals and plants. It aims at analyzing the shapes and forms of objects. In computer imaging, it has been used as a tool for extracting image components that are useful in the representation and description of object shapes. It is mathematical in the sense that the analysis is based on set theory, topology, lattice algebra, function, and so on.

This book provides a comprehensive overview of the different aspects of morphological mechanisms and techniques. It is written for students, researchers, and professionals who take related courses, want to improve their knowledge, and understand the role of image processing and mathematical morphology. This book provides much technical information and, placing it in a fundamental theoretical framework, helps students understand the many advanced techniques of these processes. By comprehensively considering the essential principles of image morphological algorithms and architectures, readers cannot only obtain novel ideas in implementing these advanced algorithms, but can also discover new problems.

This book introduces basic and advanced techniques in image processing and mathematical morphology. It includes image enhancement, edge detection and linking, order statistics morphology, regulated morphology, alternating sequential morphology, recursive morphology, soft morphology, fuzzy morphology, and sweep morphology. It discusses basic and advanced algorithms and architectures as well as the decomposition methods of structuring elements. It provides practical applications, such as distance transformation, feature extraction, object representation and recognition, shape description, and shortest path planning. The principles of image processing and mathematical morphology are illustrated with numerous graphs and examples to simplify problems so that readers can easily understand even the most complicated theories.

Overview of the Book

In Chapter 1, the basic concept of digital image processing, a brief history of mathematical morphology, the essential morphological approach to image analysis, and the scope that this book covers, are introduced. The Appendix provides a list of recent book publications.

Chapter 2 presents the binary morphological operations on binary images. It introduces set operations and logical operations on binary images. Next, binary dilation and erosion operations are presented, and the extended opening and closing operations as well as hit-or-miss transformations are described. Chapter 3 presents grayscale morphological operations on grayscale images. It introduces grayscale dilation and erosion. The grayscale dilation erosion duality theorem and opening and closing are described. In Chapter 4 several basic morphological algorithms are introduced, including boundary extraction, region filling, connected components extraction, convex hull, thinning, thickening, skeletonization, and pruning. Morphological edge operators are explained.

In Chapter 5 several variations of morphological filters are described, including alternating sequential filters, recursive morphological filters, soft morphological filters, order-statistic soft morphological filters, recursive soft morphological filters, recursive order-statistic soft morphological filters, regulated morphological filters, and fuzzy morphological filters. In Chapter 6 variant distance transformation algorithms are discussed. These include the distance transformation by iterative operations, by mathematical morphology, the approximation of Euclidean distances, the decomposition of distance structuring elements, the iterative erosion algorithm, the two scan–based algorithm, the three-dimensional Euclidean distances, the acquiring approaches, and the deriving approaches. Chapter 7 discusses feature extraction based on morphological filters. It includes edge linking by mathematical morphology, corner detection by regulated morphology, shape database with hierarchical features, the corner and circle detection, and size histogram.

In Chapter 8 object representation and tolerances by mathematical morphology, skeletonization or medial axis transformation, and the morphological shape description called geometric spectrum are introduced. Chapter 9 presents the decomposition of geometric-shaped structuring elements, the decomposition of binary structuring elements, and the decomposition of grayscale structuring elements.

Chapter 10 introduces architectures for mathematical morphology. It includes threshold decomposition architecture of grayscale morphology into binary morphology, implementation of morphological operations using programmable neural networks, multilayer perceptrons as morphological processing modules, systolic array architecture, and their implementation on multicomputers. Chapter 11 introduces general sweep mathematical morphology. It includes its theoretical development and applications, such as the blending of sweep surfaces with deformations, image enhancement and edge linking, geometric modeling, the scheme of its formal language and representation, and its grammar and parsing algorithm. Chapter 12 introduces the morphological approach to shortest path planning. It includes the relationships between the process, the shortest path and mathematical morphology, rotational mathematical morphology, the algorithm to find the shortest path, dynamically rotational mathematical morphology, and the rule of distance functions for achieving shortest path planning.

Features of the Book

This book emphasizes the procedure of gestating an idea by carefully considering the essential requirements of a problem and describes its complexity in a simple way. Several new algorithms are discussed in great detail so that readers can adapt these techniques to derive their own approaches. This book includes the following features:

1. New state-of-the-art techniques for image morphological processing

2. Numerous practical examples

3. More intuitive development of the subject and a clear tutorial to complex technology

4. An updated bibliography

5. Extensive discussion of image morphological methods

6. Inclusion of many different morphological techniques and their examples

Feedback on the Book

It is my wish to provide an opportunity for readers to correct any errors that they may find in this book. Therefore, a clear description of any errors that you may find, along with any suggestions for improvements, will be most welcome. Please use either email (shih@njit.edu) or the regular mail: Frank Y. Shih, College of Computing Sciences, New Jersey Institute of Technology, University Heights, Newark, NJ 07102-1982.

Frank Y. Shih
New Jersey Institute of Technology

Acknowledgments

Portions of the book appeared in earlier forms as conference papers, journal papers, or theses by my students here at New Jersey Institute of Technology. Therefore, these parts of the text are sometimes a combination of my words and those of my students.

I would like to gratefully acknowledge the Institute of Electrical and Electronic Engineers (IEEE) and Elsevier Publishers, for giving me permission to re-use texts and figures that have appeared in some earlier publications.

Author

Frank Y. Shih received his BS degree from the National Cheng-Kung University, Taiwan, in 1980, an MS degree from the State University of New York at Stony Brook in 1984, and a PhD degree from Purdue University, West Lafayette, Indiana, in 1987; all in electrical and computer engineering. His first PhD advisor was Professor King-Sun Fu, a pioneer in syntactic pattern recognition and the founding president of the International Association for Pattern Recognition. After Professor Fu passed away in 1985, Professor Owen R. Mitchell became his advisor.

In January 1988, Dr. Shih joined the faculty of New Jersey Institute of Technology (NJIT), Newark. He was a visiting professor at the US Air Force Rome Laboratory in 1995, National Taiwan University in 1996, and Columbia University in 2006. Currently, he is a professor jointly appointed in the Department of Computer Science, the Department of Electrical and Computer Engineering, and the Department of Biomedical Engineering at NJIT. Professor Shih has served as the departmental acting chair, associate chair, PhD director, and graduate program director. He is currently the director of the Computer Vision Laboratory.

Dr. Shih is also on the Editorial Board of the *International Journal of Pattern Recognition*, the *International Journal of Pattern Recognition Letters*, the *International Journal of Pattern Recognition and Artificial Intelligence*, the *International Journal of Recent Patents on Engineering*, the *International Journal of Recent Patents on Computer Science*, the *International Journal of Internet Protocol Technology*, the *Open Nanoscience Journal*, and the *Journal of Internet Technology*. He was previously on the editorial board of the *International Journal of Information Sciences* and the *International Journal of Systems Integration*. He was a guest editor for the *International Journal of Intelligent Automation and Soft Computing* and the *Journal of Wireless Communication and Mobile Computing*.

Dr. Shih has contributed as a steering member, committee member, and session chair for numerous professional conferences and workshops. He is presently on the Technical Committee of the IEEE Circuits and Systems Society in Visual Signal Processing & Communications and Multimedia Systems & Applications. He was the recipient of the Research Initiation Award from the National Science Foundation in 1991. He won the Honorable Mention Award from the International Pattern Recognition Society for an outstanding paper and also won the Best Paper Award at the International Symposium on Multimedia Information Processing. Dr. Shih has received numerous grants from the National Science Foundation, the navy and air force, and industry. He has received several research merit awards at the New Jersey Institute of Technology. He has served several times on the Proposal Review Panel of the National Science Foundation.

Professor Shih was one of the researchers who initiated mathematical morphology research with applications to image processing, feature extraction, and object representation. His *IEEE PAMI* article "Threshold decomposition of grayscale morphology into binary morphology" is a breakthrough paper that solves the bottleneck problem in grayscale morphological processing. His several *IEEE Image Processing* and *IEEE Signal Processing* articles detail innovations that achieve fast exact Euclidean distance transformation, robust image enhancement, edge linking, and image segmentation using the recursive soft morphological operators, general sweep morphological operators, alternating sequential morphological filters, and regulated morphological filters that he developed.

In the field of digital document processing, Professor Shih developed a fuzzy model for unsupervised character classification, fuzzy typographical analysis for character classification, an adaptive system for block segmentation and classification, and a rule-based character and logo recognition system. In the field of face recognition, Professor Shih developed automatic extraction of head and face boundaries and facial features, a hybrid two-phase algorithm for face recognition, multiview face identification and pose estimation using B-spline interpolation, recognizing facial action units using independent component analysis and a support vector machine, and facial expression recognition.

In the field of pattern recognition, Professor Shih developed improved feature reduction in input and feature spaces, an improved incremental training algorithm for support vector machines using active query, support vector machine networks for multiclass classification, improved adaptive resonance theory networks, model-based partial shape recognition using contour curvature and affine transformation, and a distance-based separator representation for pattern classification. In the field of image segmentation, Professor Shih developed automatic seeded region growing for color image segmentation, locating object contours in complex backgrounds using an improved snake model, a top-down region-dividing approach for image segmentation, and edge linking by adaptive mathematical morphology.

Professor Shih further advanced the field of solar image processing. Collaborating with solar physics researchers, he has made incredible contributions to fill the gaps between solar physics and computer science. They have used innovative computation and information technologies for real-time space weather monitoring and forecasting, and have been granted over $1 million by the National Science Foundation. They have developed several methods to automatically detect and characterize filament/prominence eruptions, flares, and coronal mass ejections. These techniques are currently in use at the Big Bear Observatory in California as well as by NASA.

Professor Shih has also made significant contributions to information hiding, focusing on the security and robustness of digital steganography and watermarking. He has developed several novel methods to increase embedding capacity, enhance robustness, integrate different watermarking platforms, and break steganalytic systems. His recent article, published in *IEEE SMC,* is the first one applying a genetic algorithm-based methodology to break steganalytic systems. He has authored a book entitled *Digital Watermarking and Steganography: Fundamentals and Techniques,* published by CRC Press, 2008.

Dr. Shih has published 95 journal papers, 90 conference papers, 2 books, and 8 book chapters. The journals he has published are top-ranked in the professional societies. He has overcome many difficult research problems in multimedia signal processing, pattern recognition, features extraction, and information security. Some examples are robust information hiding, automatic solar features classification, optimum features reduction, fast accurate Euclidean distance transform, and fully parallel thinning algorithms.

Dr. Shih is a fellow of the American Biographical Institute and a senior member of the IEEE. His current research interests include image forensics, watermarking and steganography, image processing, mathematical morphology, computer vision, pattern recognition, sensor networks, bioinformatics, information security, robotics, fuzzy logic, and neural networks.

1

Introduction to Mathematical Morphology

For the purposes of image analysis and pattern recognition, one often transforms an image into another better represented form. Image-processing techniques have been developed tremendously during the past five decades and mathematical morphology in particular has been continuously receiving a great deal of attention because it provides a quantitative description of geometric structure and shape as well as a mathematical description of algebra, topology, probability, and integral geometry. Mathematical morphology has been proven to be extremely useful in many image processing and analysis applications.

Mathematical morphology denotes a branch of biology that deals with the forms and structures of animals and plants. It analyzes the shapes and forms of objects. In computer vision, it is used as a tool to extract image components that are useful in the representation and description of object shape. It is mathematical in the sense that the analysis is based on set theory, topology, lattice algebra, function, and so on.

The primary applications of multidimensional signal processing are image processing and multichannel time series analyses, which require techniques that can quantify the geometric structure of signals or patterns. The approaches offered by mathematical morphology have been developed to quantitatively describe the morphology (i.e., shape and size) of images. The concepts and operations of mathematical morphology are formal and general enough that they can be applied to the analysis of signals and geometric patterns of any dimensionality.

In this chapter, we first introduce the basic concept in digital image processing, then provide a brief history of mathematical morphology, followed by the essential morphological approach to image analysis. We finally take a brief look at the scope that this book covers.

1.1 Basic Concept in Digital Image Processing

Images are produced by a variety of physical devices, including still and video cameras, scanners, X-ray devices, electron microscopes, radar, and by ultrasound. They can be used for a variety of purposes, including entertainment, medical imaging, business and industry, military, civil, security, and scientific analyses. This ever-widening interest in digital image processing stems from the improvement in the quality of pictorial information available for human interpretation and the processing of scene data for autonomous machine perception.

The Webster's dictionary defines an image as "a representation, likeness or imitation of an object or thing, a vivid or graphic description, something introduced to represent something else." The word "picture" is a restricted type of image. The Webster's dictionary defines a picture as "a representation made by painting, drawing, or photography; a vivid, graphic, accurate description of an object or thing so as to suggest a mental image or give an accurate idea of the thing itself." In image processing, the word "picture" is sometimes equivalent to "image."

Digital image processing starts with one image and produces a modified version of that image. Digital image analysis is a process that transforms a digital image into something other than a digital image, such as a set of measurement data, alphabet text, or a decision. Image digitization is a process that converts a pictorial form to numerical data. A digital image is an image $f(x, y)$ that has been discretized in both spatial coordinates and brightness (intensity). The image is divided into small regions called picture elements, or pixels (see Figure 1.1).

Image digitization includes image sampling [i.e., digitization of spatial coordinates (x, y)] and gray-level quantization (i.e., digitization of brightness amplitude). An image is represented by a rectangular array of integers. The image sizes and the number of gray levels are usually integer powers of 2. The number at each pixel represents the brightness or darkness (generally called the intensity) of the image at that point. For example, Figure 1.2 shows a digital image of 8×8 with one byte (i.e., 8 bits = 256 gray levels) per pixel.

The quality of an image strongly depends upon two parameters: the number of samples and the number of gray levels. The larger these two numbers, the better the quality of an image. However, this will result in a large amount of storage space because the size of an image is the product of its dimensions and the number of bits required to store gray levels. At a lower resolution, an

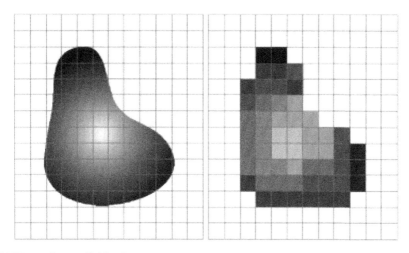

FIGURE 1.1   Image digitization.

image can produce a checkerboard effect or graininess. When an image of 1024×1024 is reduced to 512×512, it may not show much deterioration, but when further reduced to 256×256 and then rescaled back to 1024×1024, it may show discernible graininess.

The visual quality requirement of an image depends upon its applications. To achieve the highest visual quality and at the same time the lowest memory requirement, one may perform fine sampling of an image at the sharp gray-level transition areas and coarse sampling in the smooth areas. This is known as sampling based on the characteristics of an image [Damera-Venkata et al. 2000]. Another method known as tapered quantization [Heckbert 1982] can be used for the distribution of gray levels by computing the occurrence frequency of all allowable levels. The quantization level

1	8	219	51	69	171	81	41
94	108	20	121	17	214	15	74
233	93	197	83	177	215	183	78
41	84	118	62	210	71	122	38
222	73	197	248	125	226	210	5
35	36	127	5	151	2	197	165
196	180	142	52	173	151	243	164
254	62	172	75	21	196	126	224

FIGURE 1.2   A digital image and its numerical representation.

is finely spaced in the regions where gray levels occur frequently, and is coarsely spaced in other regions where the gray levels rarely occur. Images with large amounts of detail can sometimes still have a satisfactory appearance despite a relatively small number of gray levels. This can be seen by examining isopreference curves using a set of subjective tests for images in the Nk-plane, where N is the number of samples and k is the number of gray levels [Huang 1965].

In general, image-processing operations can be categorized into four types:

1. **Pixel operations**. The output at a pixel depends only on the input at that pixel, independent of all other pixels in that image. For example, thresholding, a process of making input pixels above a certain threshold level white, and others black, is simply a pixel operation. Other examples include brightness addition/subtraction, contrast stretching, image inverting, log transformation, and power-law transformation.

2. **Local (neighborhood) operations**. The output at a pixel depends not only on the pixel at the same location in the input image, but also on the input pixels immediately surrounding it (called its neighborhood). The size of the neighborhood is related to the desired transformation. However, as the size increases, so does its computational time. Note that in these operations, the output image must be created separately from the input image, so that the neighborhood pixels in calculation are all from the input image. Some examples are morphological filters, convolution, edge detection, smoothing filters (e.g., the averaging and the median filters), and sharpening filters (e.g., the Laplacian and the gradient filters). These operations can be adaptive because the results depend on the particular pixel values encountered in each image region.

3. **Geometric operations**. The output at a pixel depends only on the input levels at some other pixels defined by geometric transformations. Geometric operations are different from global operations in that the input is only from some allocated pixels calculated by geometric transformation. It is not required to use the input from all the pixels in calculation for this image transformation.

4. **Global operations**. The output at a pixel depends on all the pixels in an image. A global operation may be designed to reflect statistical information calculated from all the pixels in an image, but not from a local subset of pixels. As expected, global operations tend to be very slow. For example, a distance transformation of an image, which assigns to each object pixel the minimum distance from it to all the background pixels, belongs to a global operation. Other examples include histogram equalization/specification, image warping, Hough transform, spatial-frequency domain transforms, and connected components.

1.2 Brief History of Mathematical Morphology

A short summary of the historical development of mathematical morphology is introduced in this section. Early work in this discipline includes the work of Minkowski [1903], Dineen [1955], Kirsch [1957], Preston [1961], Moore [1968], and Golay [1969]. Mathematical morphology has been formalized since the 1960s by Georges Matheron and Jean Serra at the Centre de Morphologie Mathematique on the campus of the Paris School of Mines at Fontainebleau, France, for studying geometric and milling properties of ores. Matheron [1975] wrote a book on mathematical morphology, entitled *Random Sets and Integral Geometry*. Two volumes containing mathematical morphology theory were written by Serra [1982, 1988]. Other books describing fundamental applications include those of Giardina and Dougherty [1988], Dougherty [1993], and Goutsias and Heijmans [2000]. Haralick, Sternberg, and Zhuang [1987] presented a tutorial providing a quick understanding for a beginner. Shih and Mitchell [1989, 1991] created the technique of threshold decomposition of grayscale morphology into binary morphology, which provides a new insight into grayscale morphological processing.

Mathematical morphology has been a very active research field since 1985. Its related theoretical and practical literature has appeared as the main theme in professional journals, such as *IEEE Transactions on Pattern Analysis and Machine Intelligence*; *IEEE Transactions on Image Processing*; *IEEE Transactions on Signal Processing*; *Pattern Recognition*; *Pattern Recognition Letters*; *Computer Vision, Graphics, and Image Processing*; *Image and Vision Computing*; and *Journal of Mathematical Imaging and Vision*; in a series of professional conferences, such as the International Symposium on Mathematical Morphology, IEEE International Conference on Image Processing, and International Conference on Pattern Recognition; and professional societies, such as the International Society for Mathematical Morphology (ISMM) founded in 1993 in Barcelona, the Institute of Electrical and Electronics Engineers (IEEE), Association for Computing Machinery (ACM), the International Association for Pattern Recognition (IAPR), and the International Society for Optical Engineering (SPIE). *The Morphology Digest*, edited by Henk Heijmans and Pierre Soille since 1993, provides a communication channel among people interested in mathematical morphology and morphological image analysis.

1.3 Essential Morphological Approach to Image Analysis

From the Webster's dictionary we find that the word morphology refers to any scientific study of form and structure. This term has been widely used in biology, linguistics, and geography. In image processing, a well-known general approach is provided by mathematical morphology, where the images being analyzed are considered as sets of points, and the set theory is applied to morphological operations. This approach is based upon logical, rather than arithmetic, relations between pixels and can extract geometric features by choosing a suitable structuring shape as a probe.

Haralick, Sternberg, and Zhuang [1987] stated that "as the identification and decomposition of objects, object features, object surface defects, and assembly defects correlate directly with shape, mathematical morphology clearly has an essential structural role to play in machine vision." It has also become "increasingly important in image processing and analysis applications for object recognition and defect inspection" [Shih and Mitchell 1989]. Numerous morphological architectures and algorithms have been developed during the last decades. The advances in morphological image processing have followed a series of developments in the areas of mathematics, computer architectures, and algorithms.

Mathematical morphology provides nonlinear, lossy theories for image processing and pattern analysis. Researchers have developed numerous sophisticated and efficient morphological architectures, algorithms, and applications. One may be interested in morphological techniques such as filtering, thinning, and pruning for image pre- and postprocessing. Morphological operations can simplify image data while preserving their essential shape characteristics and eliminating irrelevancies. It can extract shape features such as edges, fillets, holes, corners, wedges, and cracks by operating with structuring elements of varied sizes and shapes [Serra 1986; Maragos and Schafer 1987].

In industrial vision applications, mathematical morphology can be used to implement fast object recognition, image enhancement, compression of structural properties, enhancement, segmentation [Nomura et al. 2005], restoration [Schonfeld and Goutsias 1991], texture analysis [Sivakumar and Goutsias 1999; Xia, Feng, and Zhao 2006], particle analysis, quantitative description, skeletonization, and defect inspection. Mathematical morphology has been successfully applied to geoscience and remote sensing by analyzing the spatial relationships between groups of pixels [Soille and Pesaresi 2002]. It can also be applied to teeth segmentation [Said et al. 2006] and QRS complex detection [Trahanias 1993]. Note that the QRS complex is the deflections in the tracing of the electrocardiogram (ECG), comprising the Q, R, and S waves that represent the ventricular activity of the heart. Mathematical morphology can also be applied to automatic analysis of DNA microarray images [Angulo and Serra 2003], and Oliveira and Leite [2008] have presented a multiscale directional operator and morphological tools for reconnecting broken ridges in fingerprint images.

1.4 Scope of This Book

This book provides a comprehensive overview of morphological image processing and analysis. It presents the necessary fundamentals, advanced techniques, and practical applications for researchers, scientists, and engineers who work in image processing, machine vision, and pattern recognition disciplines. As there have not been many books published in morphological image processing and analysis in the past few years, this book is novel and up-to-date in providing many of the newest research advances in this discipline.

This book provides students, researchers, and professionals with technical information pertaining to image processing and mathematical morphology, placing it within a fundamental theoretical framework, and leading to the development of many advanced techniques. By comprehensively considering the essential principles of morphological algorithms and architectures, readers can not only obtain novel ideas in implementing advanced algorithms and architectures, but also discover new techniques.

These advanced techniques include image enhancement, edge detection and linking, alternating sequential morphology, recursive morphology, soft morphology, order-statistic soft morphology, recursive soft morphology, recursive order-statistic soft morphology, regulated morphology, fuzzy morphology, and sweep morphology. The book also examines practical applications such as distance transformation, decomposition of structuring elements, feature extraction, skeletonization, object representation and recognition, shape description, and shortest path planning. It covers morphological architectures such as threshold decomposition of grayscale morphology into binary morphology, programmable neural networks, multilayer perceptron, systolic array, and multi computers. The principles of image processing and mathematical morphology in this book are illustrated with abundant graphs, pictures, and examples in order for readers to simplify problems and understand the underlying theoretical concept easily, even though the theories and algorithms are themselves complicated.

References

Angulo, J. and Serra, J., "Automatic analysis of DNA microarray images using mathematical morphology," *Bioinformatics*, vol. 19, no. 5, pp. 553–562, Mar. 2003.

Damera-Venkata, N., Kite, T. D., Geisler, W. S., Evans, B. L., and Bovik, A. C., "Image quality assessment based on a degradation model," *IEEE Trans. Image Processing*, vol. 9, no. 4, pp. 636–650, Apr. 2000.

Dineen, G. P., "Programming pattern recognition," Proc. Western Joint Computer Conference, Los Angeles, CA, pp. 94–100, Mar. 1–3, 1955.

Dougherty, E. R., *Mathematical Morphology in Image Processing*, Miami: CRC Press, 1993.

Giardina, C. and Dougherty, E., *Morphological Methods in Image and Signal Processing*, Prentice-Hall, Englewood Cliffs, NJ, 1988.

Golay, M. J. E., 1969, "Hexagonal parallel pattern transformations," *IEEE Trans. Computers*, vol. 18, no. 8, pp. 733–740, Aug. 1969.

Goutsias, J. and Heijmans, H. J., *Mathematical Morphology*, IOS Press, Amsterdam, the Netherlands, 2000.

Haralick, R. M., Sternberg, S. R., and Zhuang, X., "Image analysis using mathematical morphology," *IEEE Trans. Pattern Analysis and Machine Intelligence*, vol. 9, no. 4, pp. 532–550, July 1987.

Heckbert, P., "Color image quantization for frame buffer display," *ACM SIGGRAPH Computer Graphics*, vol. 16, no. 3, pp. 297–307, July 1982.

Huang, T. S., "PCM picture transmission," *IEEE Spectrum*, vol. 2, no. 12, pp. 57–63, Dec. 1965.

Kirsch, R. A."Experiments in processing life motion with a digital computer," *Proc. Eastern Joint Computer Conference*, pp. 221–229, 1957.

Maragos, P. and Schafer, R., "Morphological filters—Part II: Their relations to median, order-statistics, and stack filters," *IEEE Trans. Acoustics, Speech, Signal Processing*, vol. 35, no. 8, pp. 1170–1184, Aug. 1987.

Matheron, G., *Random Sets and Integral Geometry*, John Wiley, New York, 1975.

Minkowski, H., "Volumen und oberflache," *Mathematical Annals*, vol. 57, pp. 447–495, 1903.

Moore, G. A., "Automatic sensing and computer process for the quantitative analysis of micrographs and equivalued subjects," in C. George, R. S. Ledley, D. K. Pollock, and A. Rosenfeld (Eds.), *Pictorial Pattern Recognition*, Thompson Book Co., Durank, OK, pp. 275–326, 1968.

Nomura, S., Yamanaka, K., Katai, O., Kawakami, H., and Shiose, T., "A novel adaptive morphological approach for degraded character image segmentation," *Pattern Recognition*, vol. 38, no. 11, pp. 1961–1975, Nov. 2005.

Oliveira, M. A. and Leite, N. J., "A multiscale directional operator and morphological tools for reconnecting broken ridges in fingerprint images," *Pattern Recognition*, vol. 41, no. 1, pp. 367–377, Jan. 2008.

Preston, K. Jr., "Machine techniques for automatic identification of binucleate lymphocyte," Proc. Int. Conf. Medical Electronics, D.C., 1961.

Said, E. H., Nassar, D. E. M., Fahmy, G., and Ammar, H. H., "Teeth segmentation in digitized dental X-ray films using mathematical morphology," *IEEE Trans. Information Forensics and Security*, vol. 1, no. 2, pp. 178–189, June 2006.

Schonfeld, D. and Goutsias, J., "Optimal morphological pattern restoration from noisy binary images," *IEEE Trans. Pattern Analysis and Machine Intelligence*, vol. 13, no. 1, pp. 14–29, Jan. 1991.

Serra, J., *Image Analysis and Mathematical Morphology*, Academic Press, New York, 1982.

Serra, J., "Introduction to mathematical morphology," *Computer Vision, Graphics, and Image Processing*, vol. 35, no. 3, pp. 283–305, Sep. 1986.

Serra, J., *Image Analysis and Mathematical Morphology, Vol. 2: Theoretical Advances*, Academic Press, New York, 1988.

Shih, F. Y. and Mitchell, O. R., "Threshold decomposition of grayscale morphology into binary morphology," *IEEE Trans. Pattern Analysis and Machine Intelligence*, vol. 11, no. 1, pp. 31–42, Jan. 1989.

Shih, F. Y. and Mitchell, O. R., "Decomposition of gray scale morphological structuring elements," *Pattern Recognition*, vol. 24, no. 3, pp. 195–203, March 1991.

Sivakumar, K. and Goutsias, J., "Morphologically constrained GRFs: applications to texture synthesis and analysis," *IEEE Trans. Pattern Analysis and Machine Intelligence*, vol. 21, no. 2, pp. 99–113, Feb. 1999.

Soille, P. and Pesaresi, M., "Advances in mathematical morphology applied to geoscience and remote sensing," *IEEE Trans. Geoscience and Remote Sensing*, vol. 40, no. 9, pp. 2042–2055, Sep. 2002.

Trahanias, P. E., "An approach to QRS complex detection using mathematical morphology," *IEEE Trans. Biomedical Engineering*, vol. 40, no. 2, pp. 201–205, Feb. 1993.

Xia, Y., D. Feng, D., and Zhao, R., "Morphology-based multifractal estimation for texture segmentation," *IEEE Trans. Image Processing*, vol. 15, no. 3, pp. 614–623, Mar. 2006.

Appendix: Selected List of Books on Image Processing and Mathematical Morphology

(*Note*: Books that appear in the reference list are not listed here.)

Dougherty, E. R. and R. A. Lotufo, *Hands-on Morphological Image Processing*, Bellingham, WA: SPIE Press, 2003.

Heijmans, H., *Morphological Image Operators*, Boston: Academic Press, 1994.

Heijmans, H. and J. Roerdink, eds., *Mathematical Morphology and its Applications to Image and Signal Processing*, Boston: Kluwer Academic, 1998.

Jonker, P. P., *Morphological Image Processing: Architecture and VLSI Design*, Springer, the Netherlands, 1992.

Maragos, P, R. Schafer, and M. Butt, eds., *Mathematical Morphology and Its Applications to Image and Signal Processing*, Boston: Kluwer Academic, 1996.

Ronse, C., L. Najman, and E. Decenciere, eds., *Mathematical Morphology: 40 Years On*, Proceedings of the 7th International Symposium on Mathematical Morphology, Springer, 2005.

Serra J. and P. Salembier, eds., *Mathematical Morphology and Its Applications to Signal Processing*, Barcelona: Universitat Politecnica Catalunya, 1993.

Serra, J. and P. Soille, eds., *Mathematical Morphology and Its Applications to Image Processing*, Dordrecht: Kluwer Academic, 1994.

Soille, P., *Morphological Image Analysis: Principles and Applications*, 2nd ed., Berlin: Springer, 2003.

Vogt, R., *Automatic Generation of Morphological Set Recognition Algorithms*, New York: Springer, 1989.

2

Binary Morphology

Image analysis aims at extracting information from an image. Image transformations are of three types: unary (one image in and one image out), dyadic (two images in and one image out), and information extraction (image in and information out). The language of mathematical morphology is set theory, topology, lattice algebra, and function. In binary images, each picture element (pixel) is a tuple or a 2-dimensional vector of the (x, y)-plane. The binary image is a matrix containing object (or foreground) pixels of value 1 and nonobject (or background) pixels of value 0. We treat the binary image as a set whose elements are composed of all the pixels of value 1. In grayscale images, each element is a triple (or a 3-dimensional vector); that is, (x, y) plus the third dimension corresponding to the pixel's intensity value. Sets in higher dimensional spaces can contain other image attributes.

Morphological operations on binary images are presented in this chapter; on grayscale images in the next. This chapter is organized as follows. Section 2.1 introduces set operations and Section 2.2 describes logical operations on binary images. Binary morphological dilation and erosion operations are presented in Sections 2.3 and 2.4, respectively. Section 2.5 describes morphological opening and closing operations. Finally, Section 2.6 presents the morphological hit-or-miss transformation.

2.1 Set Operations on Binary Images

In mathematical morphology, an image is defined as a set of coordinate vectors in the Euclidean space. Let E^N denote the set of all points $p = (x_1, x_2, \ldots, x_N)$ in an N-dimensional Euclidean space. Each set A corresponds to a binary image; that is, an N-dimensional composition in black and white, where the point p is black in the binary image if and only if $p \in A$; otherwise, p is white. A binary image in E^2 is a silhouette, a set representing foreground regions (or black pixels). A binary image in E^3 is a solid, a set representing the surface and interior of objects. The notion of a binary image correlates the notion of black and white pixels to a Cartesian coordinate system for the binary image.

Let A denote a set (or a binary image) in E^2. If a set contains no elements, it is called an empty set or a null set, denoted ϕ. Let $a \in A$ denote an element $a = (a_1, a_2)$ in A. The complement (or inverse) of the image A is the binary image that exchanges black and white, as given by

$$\overline{A} = (A)^c = \{a \mid a \notin A\}. \tag{2.1}$$

The reflection of an image A is the reflected image of A across the origin (i.e., the version of A rotated $180°$ on the plane), as given by

$$\hat{A} = \{w \mid w = -a \quad \text{for } a \in A\}, \tag{2.2}$$

in which the elements with (a_1, a_2) are negated. An example of reflection is shown in Figure 2.1, where the origin of coordinates is located at the center pixel.

The union of two images A and B is a binary image in which the pixels are black if the corresponding input pixels are black in A or black in B, as given by

$$A \cup B = \{p \mid p \in A \quad \text{or} \quad p \in B\}. \tag{2.3}$$

The intersection of two images A and B is a binary image where the pixels are black if the corresponding input pixels are black in both A and B, as

$$A \cap B = \{p \mid p \in A \quad \text{and} \quad p \in B\}. \tag{2.4}$$

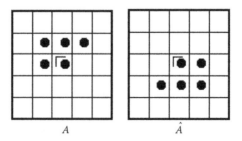

$$A \qquad\qquad\qquad \hat{A}$$

FIGURE 2.1   Reflection.

The union and intersection can be exchanged by logical duality (DeMorgan's Law) as

$$(A \cap B)^c = A^c \cup B^c, \tag{2.5}$$

and

$$(A \cup B)^c = A^c \cap B^c. \tag{2.6}$$

The difference operation between A and B is denoted as A–B or $A \backslash B$, which indicates the pixels in A that are not in B. Therefore, we can rewrite the difference as

$$A - B = A \cap \overline{B}. \tag{2.7}$$

The cardinality of A, indicating the area of the object or the total number of elements in the object set, is denoted as $\#A$, $|A|$, or $Card(A)$. If a set A has n distinct elements for some natural number n, the cardinality of A is n.

2.2 Logical Operations on Binary Images

Logical operations are derived from *Boolean algebra*, which is a mathematical approach to dealing with the truth-values of concepts in an abstract way. A truth-value in Boolean algebra can have just one of two possible values: true or false. Boolean logic was developed by George Boole (1815–1864) and is often used to refine the determination of system status. Boolean logic is a useful tool to implement image-processing algorithms based on mathematical morphology. It simply compares individual bits of input data in different ways.

The primary operators in Boolean logic are AND, OR, and NOT. The AND operation produces the output 1 (truth) if and only if all inputs are 1s. In other words, the output will be 0 (false) if any of the inputs are 0s. The OR operation produces the output 1 if any input is 1. The NOT (inverse) operation simply produces the opposite in state of the input. Similar to AND and OR operations, there are two operations using the same logic but with inverted outputs. The NAND operation produces the output 0 if and only if all inputs are 1, and the NOR operation produces the output 0 if any input is 1. Another variation of OR is called XOR (exclusive OR), which produces the output 1 if the two inputs are different. In other words, XOR yields the output 1 when either one or other input is 1, but not if both are 1. The truth

AND		
Input A	Input B	Output
0	0	0
0	1	0
1	0	0
1	1	1

OR		
Input A	Input B	Output
0	0	0
0	1	1
1	0	1
1	1	1

NOT	
A	$\bar{A}$
0	1
1	0

NAND		
Input A	Input B	Output
0	0	1
0	1	1
1	0	1
1	1	0

NOR		
Input A	Input B	Output
0	0	1
0	1	0
1	0	0
1	1	0

XOR		
Input A	Input B	Output
0	0	0
0	1	1
1	0	1
1	1	0

FIGURE 2.2 The truth tables of AND, OR, NOT, NAND, NOR, and XOR operations.

tables of each logical operation illustrating all of the input combinations are shown in Figure 2.2. Their schematic symbols used in circuit diagrams are shown in Figure 2.3.

A binary image consists of object pixels 1 and background pixels 0, which can be interpreted as true and false, respectively, in the truth tables. By applying this concept, one can conduct logic operations on binary images by simply

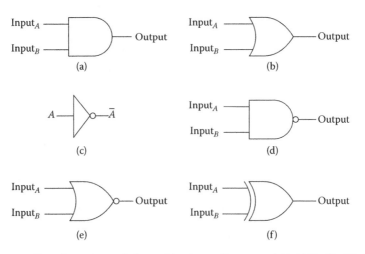

FIGURE 2.3 The schematic symbols used in circuit diagrams of (a) AND, (b) OR, (c) NOT, (d) NAND, (e) NOR, and (f) XOR operations.

applying truth-table combination rules to pixel values from a pair of input images (or a single input image when using the NOT operation). In other words, logic operations compare the corresponding input pixels of two binary images of the same size and generate the output image of the same size.

Furthermore, the logic operations can be extended to process grayscale images. Logic operations are conducted in a bitwise fashion on the binary representation of pixels, comparing corresponding bits to yield the output pixel value. For example, a pixel in an image of 256 gray levels has 8 bits. Let two input pixels be 13 and 203 in decimal. Their respective binary representations are 00001101 and 11001011. When the AND operation is applied, the resulting binary representation is 00001001, which is equal to 9 in decimal.

2.3 Binary Dilation

Mathematical morphology involves geometric analysis of shapes and textures in images. An image can be represented by a set of pixels. Morphological operators work with two images. The image being processed is referred to as the *active image*, and the other image, being a kernel, is referred to as the *structuring element*. Each structuring element has a designed shape, which can be thought of as a probe or a filter of the active image. The active image can be modified by probing it with the structuring elements of different sizes and shapes. The elementary operations in mathematical morphology are dilation and erosion, which can be combined in sequence to produce other operations, such as opening and closing.

The definitions and properties of dilation and erosion from a tutorial on mathematical morphology by Haralick, Sternberg, and Zhuang [1987] have been adopted here, which are slightly different from Matheron's [1975] and Serra's [1982] definitions. Binary dilation combines 2 sets using vector addition of set elements. It was first introduced by Minkowski and is also called *Minkowski addition*. Let A and B denote 2 sets in E^N with elements a and b, respectively, where $a = (a_1, a_2, \ldots, a_N)$ and $b = (b_1, b_2, \ldots, b_N)$ being N-tuples of element coordinates. The binary dilation of A by B is the set of all possible vector sums of pairs of elements, one coming from A and the other from B. Since A and B are both binary, the morphological operation applied on the two sets is called *binary morphology*.

Definition 2.1: Let $A \subset E^N$ and $b \in E^N$. The *translation* of A by b, denoted by $(A)_b$, is defined as

$$(A)_b = \{c \in E^N \mid c = a + b \quad \text{for some } a \in A\} \tag{2.8}$$

Let the coordinates (r, c) in sets A and B denote (row number, column number).

Example 2.1:

$A = \{(0, 2), (1, 1), (1, 2), (2, 0), (2, 2), (3, 1)\}$
$b = (0, 1)$
$(A)_b = \{(0, 3), (1, 2), (1, 3), (2, 1), (2, 3), (3, 2)\}$

	0	1	2	3
0	0	0	1	0
1	0	1	1	0
2	1	0	1	0
3	0	1	0	0

A

	0	1	2	3
0	0	0	0	1
1	0	0	1	1
2	0	1	0	1
3	0	0	1	0

$A_{(0,1)}$

Definition 2.2: Let $A, B \subset E^N$. The *binary dilation* of A by B, denoted by $A \oplus_b B$, is defined as

$$A \oplus_b B = \{c \in E^N \mid c = a + b \quad \text{for some } a \in A \quad \text{and } b \in B\}. \qquad (2.9)$$

The subscript of dilation "b" indicates binary. Equivalently, we may write

$$A \oplus_b B = \bigcup_{b \in B} (A)_b = \bigcup_{a \in A} (B)_a. \qquad (2.10)$$

The representation, $A \oplus_b B = \bigcup_{b \in B} (A)_b$, states that the dilation of A by B can be implemented by delaying the raster scan of A by the amounts corresponding to the points in B and then ORing the delayed raster scans. That is

$$(A \oplus_b B)_{(i,j)} = \underset{m,n}{\text{OR}}[\text{AND}(B_{(m,n)}, A_{(i-m, j-n)})]. \qquad (2.11)$$

Example 2.2:

$A = \{(0, 1), (1, 1), (2, 1), (3, 1)\}$
$B = \{(0, 0), (0, 2)$
$A \oplus_b B = \{(0, 1), (1, 1), (2, 1), (3, 1), (0, 3), (1, 3), (2, 3), (3, 3)\}.$

	0	1	2	3
0	0	1	0	0
1	0	1	0	0
2	0	1	0	0
3	0	1	0	0

A

	0	1	2
0	1	0	1

B

	0	1	2	3
0	0	1	0	1
1	0	1	0	1
2	0	1	0	1
3	0	1	0	1

$A \oplus_b B$

In dilation, the roles of sets A and B are symmetric. Dilation has a local interpretation: The dilation $A \oplus_b B$ is the locus of all centers c, such that the translation $(B)_c$ (by placing the origin of B at c) hits the set A. That is, if we think of each point, $a \in A$, as a seed that grows the flower $(B)_a$, then the union of all the flowers is the dilation of A by B. Note that the dilation by a disk-structuring element corresponds to the isotropic expansion algorithm in image processing. Dilation by a small square (3×3) is an eight-neighborhood operation easily implemented by adjacently connected array architectures and is the one known by the names "fill," "expand," or "grow."

2.3.1 Properties of Dilation

We now list the other properties of dilation.

1. If B contains the origin, that is, if $0 \in B$, then $A \oplus_b B \supseteq A$.
2. $A \oplus_b B = B \oplus_b A$ (commutative).
3. $(A \oplus_b B) \oplus_b C = A \oplus_b (B \oplus_b C)$ (associative).
4. $(A)_x \oplus_b B = (A \oplus_b B)_x$ (translation invariance).
5. $(A)_x \oplus_b (B)_{-x} = A \oplus_b B$.
6. If $A \subseteq B$, then $A \oplus_b C \subseteq B \oplus_b C$ (increasing).
7. $(A \cap B) \oplus_b C \subseteq (A \oplus_b C) \cap (B \oplus_b C)$.
8. $(A \cup B) \oplus_b C = (A \oplus_b C) \cup (B \oplus_b C)$ (distributive with respect to union).

2.4 Binary Erosion

Erosion is the morphological dual to dilation. It combines two sets using vector subtraction of set elements. If A and B denote two sets in E^N with elements a and b, respectively, then the *binary erosion* of A by B is the set of all elements x, for which $x + b \subset A$, for every $b \in B$.

Definition 2.3: The *binary erosion* of A by B, denoted by $A \ominus_b B$, is defined as

$$A \ominus_b B = \{x \in E^N \mid x + b \in A \quad \text{for every } b \in B\}. \tag{2.12}$$

Equivalently, we may write

$$A \ominus_b B = \bigcap_{b \in B} (A)_{-b}. \tag{2.13}$$

Note that binary dilation is the same as the Minkowski addition, but binary erosion is slightly different from *Minkowski subtraction*, in which it is defined as $\bigcap_{b \in B} (A)_b$. The above equation indicates that the implementation of binary erosion is similar to dilation except for a change of the OR function to an AND function and use of the image translated by the negated points of B. That is

$$(A \ominus_b B)_{(i,j)} = \underset{m,n}{\text{AND}}\{\text{OR}[\text{AND}(B_{(m,n)}, A_{(i+m,j+n)}), \bar{B}_{(m,n)}]\} \tag{2.14}$$

where $\bar{B}$ denotes an inverse function. Equivalently, we may simplify

$$(A \ominus_b B)_{(i,j)} = \underset{m,n}{\text{AND}}[\text{OR}(A_{(i+m,j+n)}), \bar{B}_{(m,n)}] \tag{2.15}$$

Erosion $A \ominus_b B$ can be interpreted as the locus of all centers c, such that the translation $(B)_c$ is entirely contained within the set A

$$A \ominus_b B = \{c \in E^N \mid (B)_c \subseteq A\}. \tag{2.16}$$

Example 2.3:

$AA = \{(0, 1), (1, 1), (2, 1), (2, 2), (3, 0)\}$
$B = \{(0, 0), (0, 1)\}$
$A \ominus_b B = \{(2, 1)\}$

	0	1	2	3
0	0	1	0	0
1	0	1	0	0
2	0	1	1	0
3	1	0	0	0

A

	0	1	2
0	1	1	0

B

	0	1	2	3
0	0	0	0	0
1	0	0	0	0
2	0	1	0	0
3	0	0	0	0

$A \ominus_b B$

Erosion does not possess the commutative property. Some erosion equivalent terms are "shrink" and "reduce." It should be noted that the erosion by B defined here is the same as in [Haralick et al. 1987]. However, this definition is equivalent to the erosion in [Serra 1982] by $\hat{B}$ (the reflection of the structuring element).

2.4.1 Properties of Erosion

Other properties of erosion are listed below:

1. If B contains the origin, that is, $0 \in B$, then $A \ominus_b B \subseteq A$.
2. $(A \ominus_b B) \ominus_b C = A \ominus_b (B \oplus_b C)$.

3. $A \oplus_b (B \ominus_b C) \subseteq (A \oplus_b B) \ominus_b C$.

4. $A_x \ominus_b B = (A \ominus_b B)_x$ (translation invariance).

5. If $A \subseteq B$, then $A \ominus_b C \subseteq B \ominus_b C$ (increasing).

6. $(A \cap B) \ominus_b C = (A \ominus_b C) \cap (B \ominus_b C)$ (distributive with respect to inter-section).

7. $(A \cup B) \ominus_b C \supseteq (A \ominus_b C) \cup (B \ominus_b C)$.

8. $A \ominus_b (B \cup C) = (A \ominus_b B) \cap (A \ominus_b C)$.

9. $A \ominus_b (B \cap C) \supseteq (A \ominus_b B) \cup (A \ominus_b C)$.

10. Erosion dilation duality: $(A \ominus_b B)^c = A^c \oplus_b \hat{B}$ and $(A \oplus_b B)^c = A^c \ominus_b \hat{B}$.

Heijmans and Ronse [1990] presented the algebraic basis of morphological dilation and erosion. Although dilation and erosion are dual, this does not imply that the equality can hold for any morphological cancellation. For example, if $C = A \oplus_b B$, then eroding both sides of the expression by B results in $C \ominus_b B = (A \oplus_b B) \ominus_b B \neq A$. Instead of equality, a containment relationship does hold: $(A \oplus B) \ominus B \supseteq A$. Note that if the dilation and erosion are clearly used in the binary image A and binary structuring element B, then the binary "b" subscript will be skipped. An example of binary dilation and erosion is shown in Figure 2.4. Figure 2.5 shows that when a square is dilated by a circular structuring element, the result is an enlarged square with four

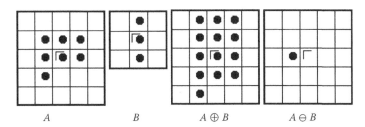

| A | B | $A \oplus B$ | $A \ominus B$ |

FIGURE 2.4 Binary dilation and erosion.

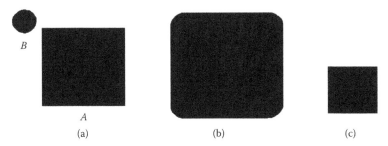

(a)   (b)   (c)

FIGURE 2.5 Binary morphological operations. (a) A and B, (b) $A \oplus B$, (c) $A \ominus B$.

corners rounded; however, when erosion is applied, the result is exactly a shrunken version of the square.

Given a Cartesian grid, the two commonly used structuring elements are the 4-connected (or cross-shaped) and 8-connected (or square-shaped) sets, N_4 and N_8. They are given by

$$B_{N_4} = \begin{bmatrix} 0 & 1 & 0 \\ 1 & 1 & 1 \\ 0 & 1 & 0 \end{bmatrix}, \quad B_{N_8} = \begin{bmatrix} 1 & 1 & 1 \\ 1 & 1 & 1 \\ 1 & 1 & 1 \end{bmatrix}.$$

2.5 Opening and Closing

In practical applications, dilation and erosion pairs are combined in sequence: either dilation of an image followed by erosion of the dilated result, or vice versa. In either case, the result of iteratively applying dilations and erosions is an elimination of specific image details whose sizes are smaller than the structuring element without the global geometric distortion of unsuppressed features. These properties were first explored by Matheron [1975] and Serra [1982]. Their definitions for both opening and closing are identical to the ones given here, but their formulas appear different because they use the symbol $\ominus$ to mean Minkowski subtraction rather than erosion. The algebraic basis of morphological opening and closing has been further developed by Heijmans and Ronse [1991].

Definition 2.4: The *opening* of an image A by a structuring element B, denoted by $A \circ B$, is defined as

$$A \circ B = (A \ominus B) \oplus B. \tag{2.17}$$

Definition 2.5: The *closing* of an image A by a structuring element B, denoted by $A \bullet B$, is defined as

$$A \bullet B = (A \oplus B) \ominus B. \tag{2.18}$$

Note that in opening and closing, the symbols $\oplus$ and $\ominus$ can respectively represent either binary or grayscale dilation and erosion. Grayscale morphological operations will be presented in the next chapter. Equivalently, opening can be expressed as the union of all translations of B that are contained in A and is formulated as

$$A \circ B = \bigcup_{B_y \subseteq A} B_y. \tag{2.19}$$

In set-theoretical terms, closing is defined as

$$A \bullet B = \bigcap_{\{y|\hat{B}_y \cap A = \phi\}} \hat{B}_y^c,$$

(2.20)

where ϕ denotes an empty set. Equivalently, closing can be expressed as the complement of the union of all translations of $\hat{B}$ that are contained in A^c and is formulated as

$$A \bullet B = \left(\bigcup_{\hat{B}_y \subseteq A^c} \hat{B}_y \right)^c.$$

(2.21)

Note that the following two relationships are equivalent:

$$\hat{B}_y \subseteq A^c \Leftrightarrow \hat{B}_y \cap A = \phi.$$

(2.22)

The opening and closing can be interpreted as follows. The opening will remove all of the pixels in the regions that are too small to contain the probe (structuring element). It tends to smooth outward bumps, break narrow sections, and eliminate thin protrusions. The opposite sequence, closing, will fill in holes and concavities smaller than the probe. It tends to eliminate small holes and remove inward bumps. Such filters can be used to suppress spatial features or discriminate against objects based on their size distribution. As an example, if a disk-shaped structuring element of diameter h is used, the opening of an image is equivalent to a low-pass filter. The opening residue is a high-pass filter. When an image is opened with two elements of non-equal diameters, the difference between the openings is a band-pass filter. The symbolic expressions are:

- Low pass $= A \circ B^h$
- High pass $= A - (A \circ B^h)$
- Band pass $= (A \circ B^{h_1}) - (A \circ B^{h_2})$, where, for diameters of B, $h_1 < h_2$.

2.5.1 Properties of Opening and Closing

1. $(A \circ B) \subseteq A \subseteq (A \bullet B)$.
2. $A \oplus B = (A \oplus B) \circ B = (A \bullet B) \oplus B$.
3. $A \ominus B = (A \ominus B) \bullet B = (A \circ B) \ominus B$.
4. If $X \subseteq Y$, then $X \circ B \subseteq Y \circ B$ (increasing).
5. If $X \subseteq Y$, then $X \bullet B \subseteq Y \bullet B$ (increasing).
6. $(A \circ B) \circ B = A \circ B$ (idempotent).
7. $(A \bullet B) \bullet B = A \bullet B$ (idempotent).
8. Opening closing duality: $(A \bullet B)^c = A^c \circ \hat{B}$.

Example 2.4: Let A and B be as shown below. Draw diagrams for $A \circ B$ and $A \bullet B$.

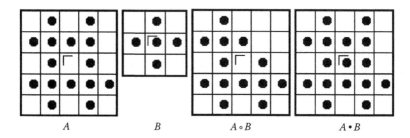

    A B $A \circ B$ $A \bullet B$

2.6 Hit-or-Miss Transformation

The hit-or-miss transformation (HMT) on a binary image A is defined as follows. The structuring element B is a pair of binary images B_1 and B_2, where $B_1 \subseteq A$ and $B_2 \subseteq \bar{A}$ (complement of A). The HMT of A by (B_1, B_2), denoted by $A \circledast (B_1, B_2)$, is defined as

$$A \circledast (B_1, B_2) = (A \ominus B_1) \cap (\bar{A} \ominus B_2) \tag{2.23}$$

Note that erosion is a special case of HMT, where B_2 is an empty set. Because $\bar{A} \ominus B_2 = \overline{A \oplus \hat{B}_2}$, we have

$$A \circledast (B_1, B_2) = (A \ominus B_1) \cap \overline{(A \oplus \hat{B}_2)} \tag{2.24}$$

It is equivalent to set difference:

$$A \circledast (B_1, B_2) = (A \ominus B_1) - (A \oplus \hat{B}_2). \tag{2.25}$$

  Morphological HMT is a natural and powerful morphological tool for shape recognition and the processing of binary images. The generalization of HMT to process grayscale images has been proposed based on grayscale erosion [Bloomberg and Maragos 1990; Naegel et al. 2007a]. Naegel et al. [2007b] applied gray-level HMTs to angiographic image processing. Using grayscale HMT, Khosravi and Schafer [1996] developed a class of rank-based template-matching criteria that are multiplier-free and independent of the dc variations of the image. Schaefer and Casasent [1995] presented a modified version of morphological HMT for object detection. Raducanu and Grana [2000] proposed that HMT be based on level sets, calling it the level set HMT (LSHMT), to obtain a translation invariant recognition tool, with some robustness regarding small deformations and variations in illumination. It can be applied to localization on grayscale images based on a set of face patterns.

Example 2.5: Let A be

0	0	0	0	0	0	0	0
0	0	1	0	1	0	0	0
0	1	0	1	1	1	1	0
0	0	1	0	1	0	0	0
0	0	0	0	1	1	1	0
0	1	1	0	1	0	1	0
0	1	1	0	0	0	1	0
0	0	0	0	0	0	0	0

Let B_1 and B_2 be

B_1:

0	1	0
0	1	1
0	1	0

B_2:

1	0	1
1	0	0
1	0	1

Then $A \circledast (B_1, B_2)$ is

0	0	0	0	0	0	0	0
0	0	0	0	0	0	0	0
0	0	0	0	0	0	0	0
0	0	0	0	0	0	0	0
0	0	0	0	1	0	0	0
0	0	0	0	0	0	0	0
0	0	0	0	0	0	0	0
0	0	0	0	0	0	0	0

References

Bloomberg, D. and Maragos, P., "Generalized hit-miss operations," in P. Gader (Ed.), *Image Algebra and Morphological Image Processing 1998, Proceedings*, vol. 1350, SPIE, San Diego, CA, USA, 1990, pp. 116–128.

Haralick, R. M., Sternberg, S. R., and Zhuang, X., "Image analysis using mathematical morphology," *IEEE Trans. Pattern Analysis and Machine Intelligence*, vol. 9, no. 4, pp. 532–550, July 1987.

Heijmans, H. J. A. M. and Ronse, C., "The algebraic basis of mathematical morphology—I: dilations and erosions," *Computer Vision, Graphics and Image Processing: Image Understanding*, vol. 50, no. 3, pp. 245–295, June 1990.

Heijmans, H. J. A. M. and Ronse, C., "The algebraic basis of mathematical morphology—II: openings and closings," *Computer Vision, Graphics and Image Processing: Image Understanding*, vol. 54, no. 1, pp. 74–97, July 1991.

Khosravi, M. and Schafer, R. W., "Template matching based on a grayscale hit-or-miss transform," *IEEE Trans. Image Processing*, vol. 5, no. 6, pp. 1060–1066, June 1996.

Matheron, G., *Random Sets and Integral Geometry*, John Wiley, New York, 1975.

Naegel, B., Passat, N., and Ronse, C., "Grey-level hit-or-miss transforms—Part I: Unified theory," *Pattern Recognition*, vol. 40, no. 2, pp. 635–647, Feb. 2007a.

Naegel, B., Passat, N., and Ronse, C., "Grey-level hit-or-miss transforms—Part II: Application to angiographic image processing," *Pattern Recognition*, vol. 40, no. 2, pp. 648–658, Feb. 2007b.

Raducanu, B. and Grana, M., "A grayscale hit-or-miss transform based on level sets," *Proc. IEEE Intl. Conf. Image Processing*, Vancouver, BC, Canada, pp. 931–933, 2000.

Schaefer, R. and Casasent, D., "Nonlinear optical hit-miss transform for detection," *Applied Optics*, vol. 34, no. 20, pp. 3869–3882, July 1995.

Serra, J., *Image Analysis and Mathematical Morphology*, Academic Press, New York, 1982.

3

Grayscale Morphology

Mathematical morphology represents image objects as sets in a Euclidean space. In morphological analysis, the set is the primary notion and a function is viewed as a particular case of a set [e.g., an N-dimensional, multivalued function can be viewed as a set in an $(N + 1)$-dimensional space]. From this viewpoint then, any function- or set-processing system is viewed as a set mapping, a transformation, from one class of sets into another. The extension of the morphological transformations from binary to grayscale processing by Serra [1982], Sternberg [1986], and Haralick, Sternberg, and Zhuang [1987], introduced a natural morphological generalization of the dilation and erosion operations. Heijmans [1991] further showed how to use binary morphological operators and thresholding techniques to build a large class of grayscale morphological operators.

The implementation of grayscale morphological operations in terms of minimum and maximum selections is impossible in some systems using optical data processors. Furthermore, grayscale morphological operations are difficult to run in real time. Fitch, Coyle, and Gallagher [1985] developed a superposition property called threshold decomposition and another property called stacking for rank order operations. Shih and Mitchell [1989] invented the threshold decomposition algorithm of grayscale morphology into binary morphology. This algorithm allows grayscale signals and structuring elements to be, respectively, decomposed into multiple binary ones, which are processed using binary morphological operators. These operations are processed in parallel and the results are combined to produce the same output as using the grayscale morphological operators. The architecture of applying threshold decomposition makes grayscale morphological operations easier to run in real time.

This chapter is organized as follows. Section 3.1 introduces grayscale dilation and erosion. Section 3.2 presents the grayscale dilation erosion duality theorem. Section 3.3 describes grayscale opening and closing.

3.1 Grayscale Dilation and Erosion

Grayscale images can be represented as binary images in a three-dimensional space, with the third dimension representing brightness. Grayscale images can then be viewed as three-dimensional surfaces, with the height at each point equal to the brightness value. We begin with definitions of the top surface of a set and the umbra of a surface. Given a set A in E^N, the top surface of A is a function defined on the projection of A onto its first $(N-1)$ coordinates, and the highest value of the N-tuple is the function value (or gray value). We assume that for every $x \in F$, $\{y \mid (x, y) \in A\}$ is topologically closed, and also that the sets E and K are finite.

Definition 3.1: Let $A \subseteq E^N$ and $F = \{x \in E^{N-1} \mid \text{for some } y \in E, (x, y) \in A\}$. The *top* or *top surface* of A, denoted by $T[A]: F \to E$, is defined as

$$T[A](x) = \max\{y \mid (x, y) \in A\}. \tag{3.1}$$

Definition 3.2: Let $F \subseteq E^{N-1}$ and $f: F \to E$. The *umbra* of f, denoted by $U[f]$, $U[f] \subseteq F \times E$, is defined as

$$U[f] = \{(x, y) \in F \times E \mid y \le f(x)\}. \tag{3.2}$$

Example 3.1: Let $f = \{2, 1, 0, 2\}$. Show the umbra $U[f]$.

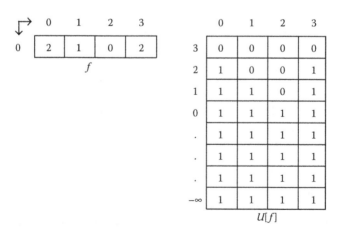

$U[f]$

The top surface of a set and the umbra of a function are essentially inverses of each other. In the above example, the top surface $T[A]$ is equal to the input function f where $A = U[f]$. That is, the one-dimensional grayscale input function f maps the one-dimensional coordinates to its function values. The umbra extends the value "1" from the function value (as coordinates) of the top surface to $-\infty$. It is convenient that the rows below the zeroth row are not shown in the following examples whenever they are all ones. A function is the top surface of its umbra. The umbra can also be illustrated by a graph in R^2, as shown by Serra [1982, Chapter XII, e.g., Fig. XII.1].

The top surface and the umbra operations possess the following properties [Haralick, Sternberg, and Zhuang 1987]:

1. $T[U[f]] = f$.
2. $U[T[U[f]]] = U[f]$.
3. $A \subseteq U[T[A]]$.
4. If A is an umbra, then $A = U[T[A]]$.
5. If A and B are umbras, then $A \oplus B$ and $A \ominus B$ are umbras.
6. If A and B are umbras and $A \subseteq B$, then $T[A](x) \le T[B](x)$.

Definition 3.3: Let $F, K \subseteq E^{N-1}$ and $f: F \to E$ and $k: K \to E$. The *grayscale dilation* of f by k, denoted by $f \oplus_g k$, is defined as

$$f \oplus_g k = T[U[f] \oplus_b U[k]]. \tag{3.3}$$

Note that the subscripts g and b, respectively, denote grayscale and binary. An alternate definition for $f \oplus_g k$ is

$$U[f] \oplus_b U[k] = U[f \oplus_g k]. \tag{3.4}$$

Example 3.2: Let $f = \{2, 1, 0, 2\}$ and $k = \{2, 1, 0\}$. Show $U[f] \oplus_b U[k]$.

	0	1	2
0	2	1	0

k

	0	1	2
4	0	0	0
3	0	0	0
2	1	0	0
1	1	1	0
0	1	1	1

$U[k]$

	0	1	2	3
4	1	0	0	1
3	1	1	0	1
2	1	1	1	1
1	1	1	1	1
0	1	1	1	1

$U[f] \oplus_b U[k]$

Definition 3.4: Let $F, K \subseteq E^{N-1}$ and $f : F \to E$ and $k : K \to E$. The *grayscale erosion* of f by k, denoted by $f \ominus_g k$, is defined as

$$f \ominus_g k = T[U[f] \ominus_b U[k]]. \tag{3.5}$$

An alternate definition for $f \ominus_g k$ is

$$U[f] \ominus_b U[k] = U[f \ominus_g k]. \tag{3.6}$$

Example 3.3: Let f and k be $\{2, 1, 0, 2\}$ and $\{2, 1, 0\}$, respectively. Show $U[f] \ominus_b U[k]$.

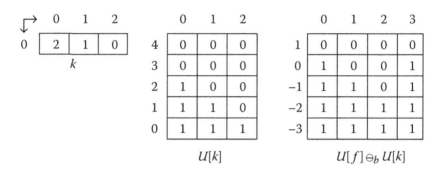

$$U[k] \qquad\qquad U[f] \ominus_b U[k]$$

The following propositions give an alternative method of computing grayscale dilation and erosion (for a proof, see Haralick, Sternberg, and Zhuang [1987]). Let F and K be domains of the grayscale image $f(x,y)$ and the grayscale structuring element $k(m,n)$, respectively. When f or k is grayscale, the morphological operators applied on the two functions are called *grayscale morphology*.

Proposition 3.1: Let $f(x,y): F \to E$ and $k(m,n): K \to E$. Then $(f \oplus_g k)(x,y):$ $F \oplus_g K \to E$ can be computed by

$$(f \oplus_g k)(x, y) = \max\{f(x - m, y - n) + k(m, n)\}, \tag{3.7}$$

for all $(m, n) \in K(x - m, y - n) \in F$.

Grayscale dilation is done using the maximum selection of a set of sums. It is similar to convolution except that the sum replaces the product and the maximum replaces the summation. The convolution in image processing is represented as

$$\text{convolution}(f, k) = \sum_m \sum_n f(x - m, y - n)k(m, n). \tag{3.8}$$

The integrand, called *convolution*, is a product of two functions, f and k, with the latter rotated by 180° and shifted by x and y along the x- and y-directions, respectively. The following two examples illustrate one- and two-dimensional grayscale dilation operations.

Example 3.4: Compute $f \oplus_g k$.

	0	1	2	3
0	2	1	0	2

f

	0	1	2
0	2	1	0

k

	0	1	2	3
0	4	3	2	4

$f \oplus_g k$

Example 3.5: Perform grayscale dilation of f and k shown below. Use the upper-left pixel of k as the origin (as circled), and do not consider the computation on the pixels that are outside the boundary of f.

f:

2	3	0	4
1	3	5	7

k:

⊝1	1
2	3

Answer: Grayscale dilation is equivalent to reflecting k about the origin, pointwise adding it to f and taking the maximum. The reflected k is therefore:

$\hat{k}$:

3	2
1	⊝1

where the origin is now located at the lower-right pixel, and the values produced can be represented by

$2 + (-1)$	$2 + 1, 3 + (-1)$	$3 + 1, 0 + (-1)$	$0 + 1, 4 + (-1)$
$2 + 2,$ $1 + (-1)$	$2 + 3, 3 + 2, 1 + 1,$ $3 + (-1)$	$3 + 3, 0 + 2, 3 + 1,$ $5 + (-1)$	$0 + 3, 4 + 2, 5 + 1,$ $7 + (-1)$

where the maximum is selected among those values. The result of grayscale dilation is

1	3	4	3
4	5	6	6

Proposition 3.2: Let $f(x, y): F \rightarrow E$ and $k(m, n): K \rightarrow E$. Then $(f \ominus_g k)(x, y)$: $F \ominus_g K \rightarrow E$ can be computed by

$$(f \ominus_g k)(x, y) = \min\{f(x + m, y + n) - k(m, n)\}, \tag{3.9}$$

for all $(m, n) \in K$ and $(x + m, y + n) \in F$.

Grayscale erosion is done using the minimum selection of a set of differences. It is similar to correlation except that the subtraction replaces the product and the minimum replaces the summation. The correlation in image processing is represented as

$$\text{correlation}(f, k) = \sum_m \sum_n f^*(x + m, y + n)k(m, n), \tag{3.10}$$

where f^* is the complex conjugate of f. The integrand, called *correlation*, is a product of two functions, f and k. In image processing, since $f^* = f$, the difference of correlation (or grayscale erosion) as compared to convolution (or grayscale dilation) is that the function k is not folded across the origin. The following two examples illustrate one- and two-dimensional grayscale erosion operations.

Example 3.6: Compute $f \ominus_g k$.

	0	1	2	3
0	2	1	0	2

f

	0	1	2
0	2	1	0

k

	0	1	2	3
0	0	−1	−2	0

$f \ominus_g k$

Example 3.7: Perform grayscale erosion of f and k shown below. Use the upper-left pixel of k as the origin (as circled) and do not consider computation of pixels outside the boundary of f.

f:

2	3	0	4
1	3	5	7

k:

⊝1	1
2	3

Answer: Grayscale erosion is equivalent to pointwise subtracting k from f and taking the minimum. In this case, the values can be represented by

$2 - (-1), 3 - 1,$	$3 - (-1), 0 - 1,$	$0 - (-1), 4 - 1,$	$4 - (-1),$
$1 - 2, 3 - 3$	$3 - 2, 5 - 3$	$5 - 2, 7 - 3$	$7 - 2$
$1 - (-1), 3 - 1$	$3 - (-1), 5 - 1$	$5 - (-1), 7 - 1$	$7 - (-1)$

Taking the minimum, the result of grayscale erosion is

-1	-1	1	5
2	4	6	8

Properties of Dilation

1. $f \oplus_g k = k \oplus_g f$
2. $(f \oplus_g k_1) \oplus_g k_2 = f \oplus_g (k_1 \oplus_g k_2)$
3. If $\vec{\mathbf{P}}_c$ is a constant vector, then $(f + \vec{\mathbf{P}}_c) \oplus_g k = (f \oplus_g k) + \vec{\mathbf{P}}_c$
4. $(f + \vec{\mathbf{P}}_c) \oplus_g (k - \vec{\mathbf{P}}_c) = f \oplus_g k$
5. If $f_1 \le f_2$, then $f_1 \oplus_g k \le f_2 \oplus_g k$

Properties of Erosion

1. $(f \ominus_g k_1) \ominus_g k_2 = f \ominus_g (k_1 \oplus_g k_2)$
2. If $\vec{\mathbf{P}}_c$ is a constant vector, then $(f + \vec{\mathbf{P}}_c) \ominus_g k = (f \ominus_g k) + \vec{\mathbf{P}}_c$
3. $(f + \vec{\mathbf{P}}_c) \ominus_g (k - \vec{\mathbf{P}}_c) = f \ominus_g k$
4. If $f_1 \le f_2$, then $f_1 \ominus_g k \le f_2 \ominus_g k$

In analyzing the computational complexity of grayscale dilation, Equation 3.3 requires a three-dimensional binary dilation (for two-dimensional images) followed by conversion back to a two-dimensional image by the top surface operation. Equation 3.7 requires many grayscale additions and a local maximum operation for each pixel in an input image. A new method was developed by Shih and Mitchell [1989], based on threshold decomposition of the grayscale image and structuring element, which results in calculations that require binary dilations of one dimension less than that required by Equation 3.3.

The *threshold decomposition* of an M-level sequence $\vec{f}$ is the set of M binary sequences, called *thresholded sequences*, $\vec{f0}_b, \vec{f1}_b, \ldots, \vec{f(M-1)}_b$, whose elements are defined as

$$fi_b(j) = \begin{cases} 1 & \text{if } f(j) \ge i \\ 0 & \text{if } f(j) < i \end{cases} \tag{3.11}$$

Assume that the largest gray value in the structuring element k is Q. Let $\vec{Q}_c$ be a constant vector of value Q. Grayscale dilation can be derived by using threshold decomposition and binary dilation operations as follows. A similar derivation can be obtained for grayscale erosion (see Shih and Mitchell [1989]).

$$f \oplus_g k = \vec{Q}_c + \max \left\{ \vec{f1}_b \oplus_b \vec{kQ}_b, \sum_{i=Q-1}^{Q} (\vec{f2}_b \oplus_b \vec{ki}_b), \sum_{i=Q-2}^{Q} (\vec{f3}_b \oplus_b \vec{ki}_b), \ldots \right\} \tag{3.12}$$

FIGURE 3.1 An example of grayscale morphological operations by a 3 × 3 square with all 0s: (a) the Lena image, (b) grayscale dilation, and (c) grayscale erosion.

Note that convolution and correlation are linear operations, but grayscale morphological dilation and erosion are nonlinear. Let a structuring element be a 3 × 3 square with all 0s. An example of grayscale dilation and erosion on the Lena image is shown in Figure 3.1.

3.2 Grayscale Dilation Erosion Duality Theorem

Let $f: F \rightarrow E$ and $k: K \rightarrow E$. Let $x \in (F \oplus K) \cap (F \ominus \hat{K})$ be given. Then

$$-(f \oplus_g k) = (-f) \ominus_g \hat{k}, \tag{3.13}$$

where $-f$ is the inverted image (i.e., gray levels are negated) and $\hat{k}(x) = k(-x)$ denotes the reflection of k. Hence, grayscale erosion can be obtained by computing grayscale dilation as

$$f \ominus_g k = -((-f) \oplus_g \hat{k}). \tag{3.14}$$

Proof: According to Equation 3.13, $(-f') \ominus_g \hat{k}' = -(f' \oplus_g k')$.

Let $f' = -f$. We get $f \ominus_g \hat{k}' = -((-f) \oplus_g k')$.

Let $k' = \hat{k}$. Because $\hat{\hat{k}}(x) = \hat{k}(-x) = k(x)$, we get $f \ominus_g k = -((-f) \oplus_g \hat{k})$. □

If the structuring element is symmetric about the origin (i.e., $\hat{k} = k$), then grayscale erosion of an image by a structuring element can be obtained by negating the image, then performing grayscale dilation with the same structuring element, and finally negating the result.

3.3 Grayscale Opening and Closing

Grayscale opening of an image f by a structuring element k, denoted by $f \circ_g k$, is defined as

$$f \circ_g k = (f \ominus_g k) \oplus_g k \qquad (3.15)$$

Grayscale closing of an image f by a structuring element k, denoted by $f \bullet_g k$, is defined as

$$f \bullet_g k = (f \oplus_g k) \ominus_g k \qquad (3.16)$$

In the one-dimensional signal shown in Figure 3.2, the grayscale opening can be interpreted as rolling a ball as the structuring element under the signal's surface, and the highest values of the ball that can reach the top at each location constitute the opening result, as shown by the bold surface in Figure 3.2b. The opening tends to remove bright objects that are small in size and break narrow connections between two bright objects. Conversely, grayscale closing can be interpreted as rolling the ball above the signal's surface, and the lowest values of the ball that can reach the bottom at each location constitute the closing result, as shown by the bold surface in Figure 3.3b. The closing tends to preserve small objects that are brighter than the background and connect bright objects with small gaps in between.

Grayscale opening and closing operations possess idempotence, translation invariance, and increasing properties as follows:

1. $(f \circ k) \leq f \leq (f \bullet k)$
2. $f \oplus k = (f \oplus k) \circ k = (f \bullet k) \oplus k$
3. $f \ominus k = (f \ominus k) \bullet k = (f \circ k) \ominus k$
4. If $f \leq g$, then $f \circ k \leq g \circ k$ (increasing)
5. If $f \leq g$, then $f \bullet k \leq g \bullet k$ (increasing)
6. $(f \circ k) \circ k = f \circ k$ (idempotent)
7. $(f \bullet k) \bullet k = f \bullet k$ (idempotent)
8. $f \circ k = -((-f) \bullet \hat{k})$ and $f \bullet k = -((-f) \circ \hat{k})$ (duality)

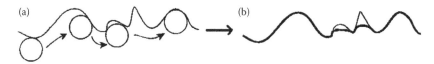

FIGURE 3.2   The grayscale opening using a ball-structuring element.

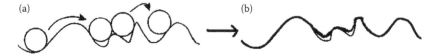

FIGURE 3.3 The grayscale closing using a ball-structuring element.

In many cases, grayscale morphological processing adopts symmetrical structuring elements (i.e., $\hat{k} = k$) so as to reduce computational complexity. Furthermore, some cases use a flat-top structuring element (i.e., k is a constant vector and can be set to a zero vector). In these cases, grayscale morphological operations are significantly reduced to maximum and minimum selections as

$$(f \oplus_g k)(x, y) = \max_{(m,n)\in k} \{f(x-m, y-n)\} = \max_k f, \tag{3.17}$$

$$(f \ominus_g k)(x, y) = \min_{(m,n)\in k} \{f(x+m, y+n)\} = \min_k f, \tag{3.18}$$

$$f \circ_g k = \max_k (\min_k f) \tag{3.19}$$

$$f \bullet_g k = \min_k (\max_k f) \tag{3.20}$$

Let a structuring element be a 3×3 square with all 0s. An example of grayscale opening and closing on the Lena image is shown in Figure 3.4. Another example of grayscale opening and closing on the Cameraman image using a structuring element of a 5×5 square with all 0s is shown in Figure 3.5.

The successful applications of univariate morphological operators along with the increasing need to process the plethora of available multivalued images have led to the extension of the mathematical morphology framework to multivariate data. Aptoula and Lefèvre [2007] presented a comprehensive

FIGURE 3.4 An example of grayscale morphological operations by a 3×3 square with all 0s: (a) the Lena image, (b) the grayscale opening, and (c) the grayscale closing.

FIGURE 3.5 An example of grayscale morphological operations by a 5 × 5 square with all 0s: (a) the Cameraman image, (b) the grayscale opening, and (c) the grayscale closing.

review of the multivariate morphological framework. Burgeth et al. [2007] extended fundamental morphological operations to the setting of matrices, sometimes referred to in the literature as tensors despite the fact that matrices are only rank-two tensors. They explored two approaches to mathematical morphology for matrix-valued data: one is based on a partial ordering and the other uses non-linear partial differential equations.

References

Aptoula, E. and Lefèvre, S., "A comparative study on multivariate mathematical morphology," *Pattern Recognition*, vol. 40, no. 11, pp. 2914–2929, Nov. 2007.

Burgeth, B., Bruhn, A., Didas, S., Weickert, J., and Welk, M., "Morphology for matrix data: Ordering versus PDE-based approach," *Image and Vision Computing*, vol. 25, no. 4, pp. 496–511, Apr. 2007.

Fitch, J. P., Coyle, E. J., and Gallagher, N. C., "Threshold decomposition of multi-dimensional ranked order operations," *IEEE Trans. Circuits and Systems*, vol. 32, no. 5, pp. 445–450, May 1985.

Haralick, R. M., Sternberg, S. R., and Zhuang, X., "Image analysis using mathematical morphology," *IEEE Trans. Pattern Analysis and Machine Intelligence*, vol. 9, no. 4, pp. 532–550, July 1987.

Heijmans, H. J. A. M., "Theoretical aspects of gray-level morphology," *IEEE Trans. Pattern Analysis and Machine Intelligence*, vol. 13, no. 6, pp. 568–582, June 1991.

Serra, J., *Image Analysis and Mathematical Morphology*, Academic Press, New York, 1982.

Shih, F. Y. and Mitchell, O. R., "Threshold decomposition of grayscale morphology into binary morphology," *IEEE Trans. Pattern Analysis and Machine Intelligence*, vol. 11, no. 1, pp. 31–42, Jan. 1989.

Sternberg, S. R., "Grayscale morphology," *Computer Vision, Graphics, and Image Processing*, vol. 35, no. 3, pp. 333–355, Sep. 1986.

4

Basic Morphological Algorithms

In image processing and analysis, it is important to extract features of objects, describe shapes, and recognize patterns. Such tasks often refer to geometric concepts, such as size, shape, and orientation. Mathematical morphology takes this concept from set theory, geometry, and topology and analyzes geometric structures in an image. Most essential image-processing algorithms can be represented in the form of morphological operations.

The operations, described in morphological notations, can easily be implemented in a parallel processor designed for cellular transformations. Each pixel can be thought of as a cell in a given state. If we define a neighborhood relationship and a pixel transition function, the application of the transition function in parallel across the space will cause the configuration of pixel states forming the image to be modified or transformed into new configurations. The implementation of these transforms in neighborhood processing stages allows parallel processing of images in real-time by conceptually high-level operations. This allows the development of extremely powerful image-processing algorithms in a short time.

In this chapter we introduce several basic morphological algorithms, including boundary extraction, region filling, connected components extraction, convex hull, thinning, thickening, skeletonization, pruning, and morphological edge operators.

4.1 Boundary Extraction

When an input image is grayscale, one can perform the segmentation process by thresholding the image. This involves the selection of a gray-level threshold between 0 and 255 via a histogram. Any pixels greater than this threshold are assigned 255 and otherwise 0. This produces a simple binary image in preparation for boundary extraction, in which 255 is replaced by 1 as the foreground pixels.

In a binary image, an object set is a connected component having pixels of value 1. The boundary pixels are those object pixels whose 8-neighbors have at least one with value 0. Boundary extraction of a set A requires first the eroding of A by a structuring element B and then taking the set difference between A and its erosion. The structuring element must be isotropic; ideally, it is a circle, but in digital image processing, a 3×3 matrix of 1s is often used. That is, the boundary of a set A is obtained by

$$\partial A = A - (A \ominus B). \tag{4.1}$$

For example, let A be

0	0	0	0	0	0	0	0	0
0	1	1	1	0	0	0	0	0
0	1	1	1	0	0	0	0	0
0	1	1	1	0	1	1	1	0
0	1	1	1	0	1	1	1	0
0	1	1	1	1	1	1	1	0
0	1	1	1	1	1	1	1	0
0	1	1	1	1	1	1	1	0
0	0	0	0	0	0	0	0	0

The boundary of A is extracted as

0	0	0	0	0	0	0	0	0
0	1	1	1	0	0	0	0	0
0	1	0	1	0	0	0	0	0
0	1	0	1	0	1	1	1	0
0	1	0	1	0	1	0	1	0
0	1	0	1	1	1	0	1	0
0	1	0	0	0	0	0	1	0
0	1	1	1	1	1	1	1	0
0	0	0	0	0	0	0	0	0

Note that the size of structuring element determines the thickness of the object contour. For instance, a 3×3 structuring element will generate a thickness of 1 and a 5×5 structuring element will generate a thickness of 3.

4.2 Region Filling

The objective of region filling is to fill value 1 into the entire object region. A binary image set A contains all boundary pixels labeled 1 and non-boundary pixels labeled 0. Region filling starts by assigning 1 to a pixel p inside the object boundary, and then grows by performing iterative dilations under a limited condition restricted by A^c. If the restriction is not placed, the growing process would flood the entire image area. Let the initial step $X_0 = p$ and B be a cross structuring element as shown in Figure 4.1. The iterative process is conducted as in the kth step [Gonzalez and Woods 2007]:

$$X_k = (X_{k-1} \oplus B) \cap A^c, \quad k = 1, 2, 3, \ldots \tag{4.2}$$

The iteration terminates at step k if $X_k = X_{k-1}$. Therefore, X_{k-1} is the interior region of the set A. The union of X_{k-1} and A will yield the filled region and the boundary. An example of region filling is shown in Figure 4.1, where the region is entirely painted after seven iterations.

4.3 Extraction of Connected Components

Similar to region filling, the iterative process can be applied also to extraction of connected components. Assume that a binary image set A contains several connected components, in which all object pixels are labeled 1 and background pixels are labeled 0. One can extract a desired connected component by picking up a pixel in the component and growing it by performing iterative dilations under a limited condition restricted by A. Let the initial step $X_0 = p$ and B be a 3×3 structuring element of 1s. The iterative process is conducted as in the kth step

$$X_k = (X_{k-1} \oplus B) \cap A, \quad k = 1, 2, 3, \ldots \tag{4.3}$$

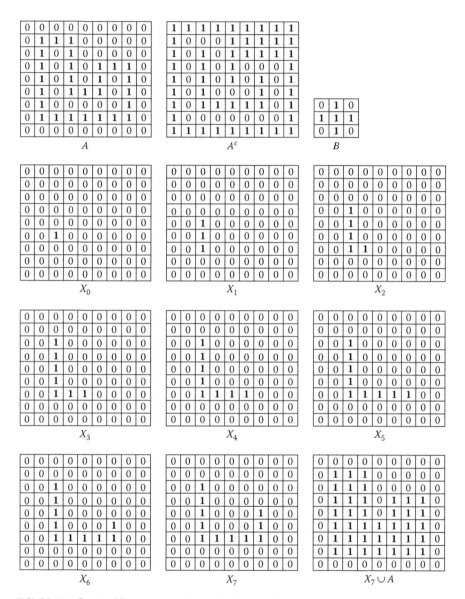

FIGURE 4.1   Region filling using mathematical morphology.

The iteration terminates at step k if $X_k = X_{k-1}$. Therefore, X_{k-1} is the extracted connected component. An example of connected component extraction is shown in Figure 4.2.

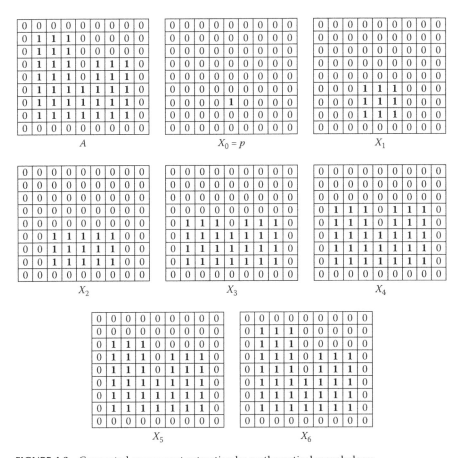

FIGURE 4.2 Connected component extraction by mathematical morphology.

4.4 Convex Hull

A *convex* set is one in which the straight line joining any two points in the set consists of points that are also in the set. The convex hull of a set A is defined as the smallest convex set that contains A. One can visualize the convex hull of an object as using an elastic band surrounding the object, so that the elastic band crosses over the concave parts and follows the convex contours of the object. This results in a convex set informing the object's shape without concavities. In digital images, we use an approximated 45° convex hull set, which contours in orientations of multiples of 45°.

The convex hull can be easily obtained by hit-or-miss operations with four 45°-rotated structuring elements. Let A be a binary image set and let B^i, $i = 1, 2, 3, 4$, represent four structuring elements as

$$B^1 = \begin{bmatrix} 1 & \times & \times \\ 1 & 0 & \times \\ 1 & \times & \times \end{bmatrix}, \quad B^2 = \begin{bmatrix} 1 & 1 & 1 \\ \times & 0 & \times \\ \times & \times & \times \end{bmatrix}, \quad B^3 = \begin{bmatrix} \times & \times & 1 \\ \times & 0 & 1 \\ \times & \times & 1 \end{bmatrix}, \quad B^4 = \begin{bmatrix} \times & \times & \times \\ \times & 0 & \times \\ 1 & 1 & 1 \end{bmatrix}, \quad (4.4)$$

where the symbol $\times$ denotes don't care. Let the initial step be $X_0^i = A$. The iterative process is conducted as in the kth step

$$X_k^i = (X \circledast B^i) \cup A, \quad i = 1, 2, 3, 4 \text{ and } k = 1, 2, 3, \dots \quad (4.5)$$

If $X_k^i = X_{k-1}^i$, then it converges. Let $D^i = X_{conv}^i$. Finally, the convex hull of A is

$$C(A) = \bigcup_{i=1}^{4} D^i \quad (4.6)$$

An example of a convex hull by mathematical morphology is shown in Figure 4.3. Note that in this example, we obtain $D^1 = X_1^1$, $D^2 = X_1^2$, $D^3 = X_1^3$, and $D^4 = X_4^4$.

$X_0^i = A$

0	0	0	0	0	0	0	0	0
0	0	0	0	0	0	0	0	0
0	0	1	1	0	0	0	0	0
0	1	1	1	0	1	0	0	0
0	1	1	1	0	1	1	0	0
0	1	1	1	1	1	1	1	0
0	0	1	1	1	1	1	0	0
0	0	0	0	1	1	0	0	0
0	0	0	0	1	0	0	0	0

X_1^1

0	0	0	0	0	0	0	0	0
0	0	0	0	0	0	0	0	0
0	0	1	1	0	0	0	0	0
0	1	1	1	1	1	0	0	0
0	1	1	1	1	1	1	0	0
0	1	1	1	1	1	1	1	0
0	0	1	1	1	1	1	0	0
0	0	0	0	1	1	0	0	0
0	0	0	0	1	0	0	0	0

X_1^2

0	0	0	0	0	0	0	0	0
0	0	0	0	0	0	0	0	0
0	0	1	1	0	0	0	0	0
0	1	1	1	0	1	0	0	0
0	1	1	1	0	1	1	0	0
0	1	1	1	1	1	1	1	0
0	0	1	1	1	1	1	0	0
0	0	0	1	1	1	0	0	0
0	0	0	0	1	0	0	0	0

X_1^3

0	0	0	0	0	0	0	0	0
0	0	0	0	0	0	0	0	0
0	0	1	1	0	0	0	0	0
0	1	1	1	0	1	0	0	0
1	1	1	1	1	1	1	0	0
0	1	1	1	1	1	1	1	0
0	0	1	1	1	1	1	0	0
0	0	0	1	1	1	0	0	0
0	0	0	0	1	0	0	0	0

X_4^4

0	0	0	0	0	0	0	0	0
0	0	0	1	0	0	0	0	0
0	0	1	1	1	0	0	0	0
0	1	1	1	1	1	0	0	0
0	1	1	1	1	1	1	0	0
0	1	1	1	1	1	1	1	0
0	0	1	1	1	1	1	0	0
0	0	0	0	1	1	0	0	0
0	0	0	0	1	0	0	0	0

$C(A)$

0	0	0	0	0	0	0	0	0
0	0	0	1	0	0	0	0	0
0	0	1	1	1	0	0	0	0
0	1	1	1	1	1	0	0	0
1	1	1	1	1	1	1	0	0
0	1	1	1	1	1	1	1	0
0	0	1	1	1	1	1	0	0
0	0	0	1	1	1	0	0	0
0	0	0	0	1	0	0	0	0

FIGURE 4.3 Convex hull by mathematical morphology.

4.5 Thinning

Thinning is similar to erosion, but it does not cause disappearance of the components of the object. It reduces objects to the thickness of one pixel, generating a minimally connected axis that is equidistant from the object's edges. Thinning is an early, fundamental processing step in representing the structural shape of a pattern as a graph. It can be applied to inspection of industrial parts, fingerprint recognition, optical character recognition, and biomedical diagnosis.

Digital skeletons, generated by thinning algorithms, are often used to represent objects in a binary image for shape analysis and classification. Jang and Chin [1990] presented a precise definition of digital skeletons and a mathematical framework for the analysis of a class of thinning algorithms, based on morphological set transformation. The thinning of a binary image set can be represented in terms of hit-or-miss operation as in Serra [1982] and Meyer [1988]:

$$A \wedge B = A - (A \circledast B) = A \cap (A \circledast B)^c. \tag{4.7}$$

This can be thought of as a search-and-delete process. The operation $A \circledast B$ locates all occurrences of B in A, and set subtraction removes those pixels that have been located from A. For thinning A symmetrically, we use a sequence of eight structuring elements:

$$\{B\} = \{B^1, B^2, \dots, B^8\}, \tag{4.8}$$

where B^i is a rotated version of B^{i-1}. In digital images, this sequence contains eight structuring elements as

$$B^1 = \begin{bmatrix} 0 & 0 & 0 \\ \times & 1 & \times \\ 1 & 1 & 1 \end{bmatrix}, \quad B^2 = \begin{bmatrix} \times & 0 & 0 \\ 1 & 1 & 0 \\ 1 & 1 & \times \end{bmatrix}, \quad B^3 = \begin{bmatrix} 1 & \times & 0 \\ 1 & 1 & 0 \\ 1 & \times & 0 \end{bmatrix}, \quad B^4 = \begin{bmatrix} 1 & 1 & \times \\ 1 & 1 & 0 \\ \times & 0 & 0 \end{bmatrix},$$

$$B^5 = \begin{bmatrix} 1 & 1 & 1 \\ \times & 1 & \times \\ 0 & 0 & 0 \end{bmatrix}, \quad B^6 = \begin{bmatrix} \times & 1 & 1 \\ 0 & 1 & 1 \\ 0 & 0 & \times \end{bmatrix}, \quad B^7 = \begin{bmatrix} 0 & \times & 1 \\ 0 & 1 & 1 \\ 0 & \times & 1 \end{bmatrix}, \quad B^8 = \begin{bmatrix} 0 & 0 & \times \\ 0 & 1 & 1 \\ \times & 1 & 1 \end{bmatrix}. \tag{4.9}$$

Therefore, thinning a sequence of eight structuring elements is represented as

$$A \wedge \{B\} = (\ \dots\ ((A \wedge B^1) \wedge B^2)\ \dots\ ) \wedge B^8 \tag{4.10}$$

The thinning process removes pixels from the outside edges of an object. The structuring elements are designed to find those edge pixels whose removal will not change the object's connectivity. After thinning for the first pass with these eight structuring elements is completed, the entire process for the second pass is repeated until no further changes occur. An example of thinning is shown in Figure 4.4, where convergence is achieved

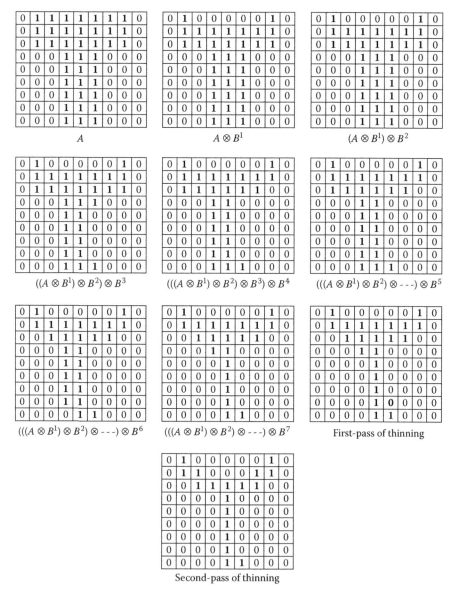

A $A \otimes B^1$ $(A \otimes B^1) \otimes B^2$

$((A \otimes B^1) \otimes B^2) \otimes B^3$ $(((A \otimes B^1) \otimes B^2) \otimes B^3) \otimes B^4$ $(((A \otimes B^1) \otimes B^2) \otimes \text{- - -}) \otimes B^5$

$(((A \otimes B^1) \otimes B^2) \otimes \text{- - -}) \otimes B^6$ $(((A \otimes B^1) \otimes B^2) \otimes \text{- - -}) \otimes B^7$ First-pass of thinning

Second-pass of thinning

FIGURE 4.4 Morphological thinning.

in two passes. Note that the skeleton after this thinning process often contains undesirable short spurs as small irregularities in the object boundary. These spurs can be removed by a process called *pruning*, which will be introduced in Section 4.8.

4.6 Thickening

Thickening is similar to dilation, but it does not cause merging of disconnected objects. Thickening is the morphological dual to thinning. It is used to grow some concavities in an object and can be represented in terms of a hit-or-miss operation and union as

$$A \odot B = A \cup (A \circledast B). \tag{4.11}$$

The thickened image contains the original set plus some filled-in pixels determined by the hit-or-miss operation. The sequence of structuring elements used in thickening is the same as that in thinning. The sequence of thickening operations can be represented as

$$A \odot \{B\} = (\cdots((A \odot B^1) \odot B^2)\cdots) \odot B^n. \tag{4.12}$$

4.7 Skeletonization

Skeletonization is similar to thinning, but it explores in greater detail the structure of an object. The skeleton emphasizes certain properties of images: for instance, curvatures of the contour correspond to topological properties of the skeleton. The concept of skeleton was first proposed by Blum [1967]. Skeleton, medial axis, or symmetrical axis, has been extensively used for characterizing objects satisfactorily using the structures composed of line or arc patterns. This has the advantage of reducing the memory space required for storage of essential structural information and simplifying the data structure required in pattern analysis. Applications include the representation and recognition of hand-written characters, fingerprint ridge patterns, biological cell structures, circuit diagrams, engineering drawings, robot path planning, and the like.

Let us visualize a connected object region as a field of grass and place a fire starting from its contour. Assume that this fire spreads uniformly in all directions. The skeleton is where waves collide with each other in a frontal or circular manner. It is called medial axis because the pixels are located at midpoints or along local symmetrical axes of the region. The skeleton of an object pattern is a line representation of the object. It must preserve the topology of the object, be one-pixel thick, and be located in the middle of the object. However, in digital images, some objects are not always realizable for skeletonization, as shown in Figure 4.5. In one case, Figure 4.5a, it is impossible to generate a one-pixel thick skeleton from a two-pixel thick line. In case Figure 4.5b, it is impossible to produce the skeleton from an alternative-pixel pattern in order to preserve the topology.

In this section we introduce the skeletonization of an object by morphological erosion and opening [Serra 1982]. A study on the use of morphological set operations to represent and encode a discrete binary image by parts of its skeleton can be found in Maragos and Schafer [1986]. The morphological skeleton was shown to unify many skeletonization algorithms, and fast algorithms were developed for skeleton decomposition and reconstruction. Fast homotopy-preserving skeletons mathematical morphology was proposed to preserve homotopy [Ji and Piper 1992]. Another approach to obtaining the skeleton by Euclidean distance transformation and maximum value tracking can be found in the work of Shih and Pu [1995] and Shih and Wu [2004].

Let A denote a binary image set containing object pixels of 1s and background pixels of 0s. Let B be a 3×3 structuring element of 1s. The skeleton of A is obtained by

$$S(A) = \bigcup_{k=0}^{K} S_k(A), \tag{4.13}$$

where $S_k(A) = \bigcup_{k=0}^{K} \{(A \ominus kB) - [(A \ominus kB) \circ B]\}$. Note that $(A \ominus kB)$ denotes k successive erosions of $(A \ominus B)$, and K is the final step before A is eroded to an empty set. An example of a morphological skeleton is shown in Figure 4.6.

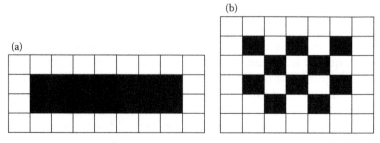

(a)

(b)

FIGURE 4.5 Two cases of nonrealizable skeletonization.

0	1	1	1	1	1	1	1	0
0	1	1	1	1	1	1	1	0
0	1	1	1	1	1	1	1	0
0	0	0	1	1	1	0	0	0
0	0	0	1	1	1	0	0	0
0	0	0	1	1	1	0	0	0
0	0	0	1	1	1	0	0	0
0	0	0	1	1	1	0	0	0
0	0	0	1	1	1	0	0	0

A

0	0	0	0	0	0	0	0	0
0	0	1	1	1	1	1	0	0
0	0	0	0	1	0	0	0	0
0	0	0	0	1	0	0	0	0
0	0	0	0	1	0	0	0	0
0	0	0	0	1	0	0	0	0
0	0	0	0	1	0	0	0	0
0	0	0	0	1	0	0	0	0
0	0	0	0	0	0	0	0	0

$A \ominus B$

0	0	0	0	0	0	0	0	0
0	0	0	0	0	0	0	0	0
0	0	0	0	0	0	0	0	0
0	0	0	0	0	0	0	0	0
0	0	0	0	0	0	0	0	0
0	0	0	0	0	0	0	0	0
0	0	0	0	0	0	0	0	0
0	0	0	0	0	0	0	0	0
0	0	0	0	0	0	0	0	0

$(A \ominus B) \circ B$

0	0	0	0	0	0	0	0	0
0	0	1	1	1	1	1	0	0
0	0	0	0	1	0	0	0	0
0	0	0	0	1	0	0	0	0
0	0	0	0	1	0	0	0	0
0	0	0	0	1	0	0	0	0
0	0	0	0	1	0	0	0	0
0	0	0	0	1	0	0	0	0
0	0	0	0	0	0	0	0	0

$S_1(A) = (A \ominus B) - (A \ominus B) \circ B$

0	0	0	0	0	0	0	0	0
0	0	0	0	0	0	0	0	0
0	0	0	0	0	0	0	0	0
0	0	0	0	0	0	0	0	0
0	0	0	0	0	0	0	0	0
0	0	0	0	0	0	0	0	0
0	0	0	0	0	0	0	0	0
0	0	0	0	0	0	0	0	0
0	0	0	0	0	0	0	0	0

$S_2(A)$

0	0	0	0	0	0	0	0	0
0	0	1	1	1	1	1	0	0
0	0	0	0	1	0	0	0	0
0	0	0	0	1	0	0	0	0
0	0	0	0	1	0	0	0	0
0	0	0	0	1	0	0	0	0
0	0	0	0	1	0	0	0	0
0	0	0	0	1	0	0	0	0
0	0	0	0	0	0	0	0	0

$S(A)$

FIGURE 4.6   A morphological skeleton.

After obtaining the skeleton, one can easily reconstruct the original set A from these skeleton subsets using successive dilations as

$$A = \bigcup_{k=0}^{K} (S_k(A) \,\%\, kB) \tag{4.14}$$

4.8 Pruning

The skeleton of a pattern after thinning usually appears as extra short noisy branches that need to be cleaned up in post processing. This process is called *pruning*. For instance, in automated recognition of hand-printed characters, the skeleton is often characterized by "spurs" caused during erosion by non-uniformities in the strokes composing the characters. The short noisy branch can be located by finding end points that have only one neighboring pixel in the set, and moving along the path until another branch is reached within a very small number of pixels. The length of a noisy branch is related to object characteristics and image size; it is often given as at most three pixels. A branch junction is considered as the pixel having at least three neighboring pixels. The pruning process can become complicated if an object has a complex shape. A detailed pruning process can be found in the work of Serra [1982, Chapter 11].

0	1	0	0	1	1	0	0	0
0	0	1	1	0	0	0	0	0
0	0	0	1	0	0	0	0	0
0	0	0	1	0	0	0	0	0
0	0	0	1	0	1	1	0	0
0	0	0	1	1	0	0	1	0
0	0	0	1	0	0	0	1	0
0	0	0	1	1	0	0	1	0
0	0	0	1	0	1	1	0	1

A

0	0	0	0	0	0	0	0	0
0	0	0	1	0	0	0	0	0
0	0	0	1	0	0	0	0	0
0	0	0	1	0	0	0	0	0
0	0	0	1	0	1	1	0	0
0	0	0	1	1	0	0	1	0
0	0	0	1	0	0	0	1	0
0	0	0	1	1	0	0	1	0
0	0	0	1	0	1	1	0	0

X

FIGURE 4.7 Morphological pruning of character "b".

Pruning is represented in terms of thinning as $X = A \otimes \{B\}$, where $\{B\} = \{B^1, B^2, \ldots, B^8\}$, as given below.

$$B^1 = \begin{bmatrix} \times & 0 & 0 \\ 1 & 1 & 0 \\ \times & 0 & 0 \end{bmatrix}, \quad B^2 = \begin{bmatrix} \times & 1 & \times \\ 0 & 1 & 0 \\ 0 & 0 & 0 \end{bmatrix}, \quad B^3 = \begin{bmatrix} 0 & 0 & \times \\ 0 & 1 & 1 \\ 0 & 0 & \times \end{bmatrix}, \quad B^4 = \begin{bmatrix} 0 & 0 & 0 \\ 0 & 1 & 0 \\ \times & 1 & \times \end{bmatrix},$$

$$B^5 = \begin{bmatrix} 1 & 0 & 0 \\ 0 & 1 & 0 \\ 0 & 0 & 0 \end{bmatrix}, \quad B^6 = \begin{bmatrix} 0 & 0 & 1 \\ 0 & 1 & 0 \\ 0 & 0 & 0 \end{bmatrix}, \quad B^7 = \begin{bmatrix} 0 & 0 & 0 \\ 0 & 1 & 0 \\ 0 & 0 & 1 \end{bmatrix}, \quad B^8 = \begin{bmatrix} 0 & 0 & 0 \\ 0 & 1 & 0 \\ 1 & 0 & 0 \end{bmatrix}.$$

(4.15)

An example of pruning a character "b" after two passes is shown in Figure 4.7.

4.9 Morphological Edge Operator

Edge operators based on grayscale morphological operations were introduced in Lee, Haralick, and Shapiro [1987]. These operators can be efficiently implemented in real-time machine vision systems that have special hardware architecture for grayscale morphological operations. The simplest morphological edge detectors are the dilation residue and erosion residue operators. Another grayscale gradient is defined as the difference between the dilated result and the eroded result. Figure 4.8 shows an example of grayscale gradient. A blur-minimum morphological edge operator is also introduced.

4.9.1 Simple Morphological Edge Operators

A simple method of performing grayscale edge detection in a morphology-based vision system is to take the difference between an image and its erosion by a small structuring element. The difference image is the image of edge

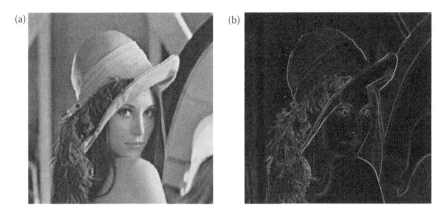

FIGURE 4.8 (a) Lena image, (b) the grayscale gradient.

strength. We can then select an appropriate threshold value to threshold the edge strength image into a binary edge image.

Let f denote a grayscale image, and let k denote a structuring element as below, where $\times$ means don't care.

$$
k = \begin{array}{|c|c|c|}
\hline
\times & 0 & \times \\
\hline
0 & 0 & 0 \\
\hline
\times & 0 & \times \\
\hline
\end{array}
$$

The erosion of f by k is given by $e(x, y) = \min[f(x + i, y + j) - k(x, y)]$ for $(i, j) \in N_4(x, y)$. Since k contains 4-connected neighbors of position $(0, 0)$, the erosion residue edge operator produces the edge strength image G_e as:

$$
G_e(x, y) = f(x, y) - e(x, y) = \max[f(x, y) - f(i, j)], \quad \text{for } (i, j) \in N_4(x, y), \quad (4.16)
$$

where $N_4(x, y)$ is the set of 4-connected neighbors of pixel (x, y).

A natural nonmorphological variation of this operator takes the summation instead of maximization. This is the familiar linear digital Laplacian operator, $\nabla^2 f(x, y)$, which is the digital convolution of $f(x, y)$ with the kernel

$$
\begin{array}{|c|c|c|}
\hline
0 & -1 & 0 \\
\hline
-1 & 4 & -1 \\
\hline
0 & -1 & 0 \\
\hline
\end{array}
$$

It is also possible to increase the neighborhood size of the morphological edge operator by increasing the size of the structuring element used for the erosion. For example, we can have an 8-connected neighborhood edge operator by changing the structuring element to the following:

$$k = \begin{array}{|c|c|c|} \hline 0 & 0 & 0 \\ \hline 0 & 0 & 0 \\ \hline 0 & 0 & 0 \\ \hline \end{array}$$

The erosion residue edge operator produces the edge strength image G_e defined by:

$$G_e(x, y) = \max[f(x, y) - f(i, j)], \quad \text{for } (i, j) \in N_8(x, y), \qquad (4.17)$$

where $N_8(x, y)$ is the set of 8-connected neighbors of pixel (x, y).

The corresponding linear Laplacian operator, which has an 8-connected neighborhood support, can be implemented as the digital convolution of $f(x, y)$ with

$$\begin{array}{|c|c|c|} \hline -1 & -1 & -1 \\ \hline -1 & 8 & -1 \\ \hline -1 & -1 & -1 \\ \hline \end{array}$$

The performance of the edge operator can be easily evaluated. This operator performs perfectly on ideal step edge patterns. However, it is noise sensitive and position-biased. To correct this bias and give both inside and outside boundaries of the checkers their corresponding edge strengths, the erosion residue morphological edge detector can be used in conjunction with the dilation residue operator. The dilation residue operator takes the difference between a dilated image and its original image. The dilation residue edge strength image is

$$G_d(x, y) = \max[f(i, j) - f(x, y)], \quad \text{for } (i, j) \in N_8(x, y) \qquad (4.18)$$

A position-unbiased edge operator can be obtained by a combination of the operators $G_e(x, y)$ and $G_d(x, y)$ using the pixel-wise minimum, maximum, or summation [Lee, Haralick, and Shapiro 1987]. For example, considering summation we have the edge operator strength $E(x, y) = G_e(x, y) + G_d(x, y)$. The summation and maximum versions perform perfectly on ideal step edge patterns, but they are also noise sensitive. The minimum version is noise insensitive, but has no response when applied to a single noise-point. Unfortunately, it is unable to detect ideal step edge patterns. This creates a new edge operator, which first performs a blur operation to convert all the ideal step edges into ideal ramp edges and then applies the minimum version of the edge operator on the output.

4.9.2 Blur-Minimum Morphological Edge Operators

This blur-minimum morphological edge operator is defined by

$$I_{\text{edge–strength}} = \min\{\tilde{I} - Erosion(\tilde{I}), Dilation(\tilde{I}) - \tilde{I}\}, \tag{4.19}$$

where $\tilde{I} = \text{Blur}(I)$, which is the result after a blurring operation [Lee, Haralick, and Shapiro 1987]. An example of the blur-minimum morphological edge operator is shown in Figure 4.9. The blur-minimum morphological edge operator is noise insensitive. For the ideal step edge, it produces a result that has nonzero edge strength on both the edge pixels. However, due to the effect of blurring, the edge strength assigned to the edge pixels is one-third of the edge contrast. For ideal ramp edges of larger spatial extent, a nonzero edge strength will be assigned to more than one pixel.

(a)

(b)

FIGURE 4.9 (a) The original image, (b) the edge image by the blur-minimum morphological edge operator.

However, the edge strength of the true edge pixel is usually higher than that of its neighbors.

The performance of the blur-minimum morphological edge operator may be poor in the presence of noise. To overcome this problem, Feehs and Arce [1987] proposed a twofold modification that uses a so-called *alpha-trimmed mean* (ATM) filter to smooth noise models other than Gaussian distribution. They define the morphological edge operator as

$$I_{atm} = \min\{[(f_{atm} \circ k) - (f_{atm} \ominus k)], \ [(f_{atm} \oplus k) - (f_{atm} \bullet k)]\}, \qquad (4.20)$$

where f_{atm} is the image f smoothed by an ATM filter. Since opening and closing are smoothing and noise-suppressing operations, this result will be less sensitive to noise. Furthermore, Moran [1990] developed a local neighborhood operator to sharpen edges using selective erosion followed by selective dilation.

References

Blum, H., "A transformation for extracting new descriptors of shape," in Walthen Dunn (Ed.), *Models for the Perception of Speech and Visual Forms*, Proceedings of Meeting held on Nov. 1964, MIT Press, MA, 1967, pp. 362–380.

Feehs, R. J. and Arce, G. R., "Multidimensional morphological edge detection," *Proc. SPIE Visual Comm. and Image Processing II*, vol. 845, pp. 285–292, 1987.

Gonzalez, R. and Woods, R., *Digital Image Processing*, Prentice-Hall, third edition, 2007.

Jang, B. and Chin, R. T., "Analysis of thinning algorithms using mathematical morphology," *IEEE Trans. Pattern Analysis and Machine Intelligence*, vol. 12, no. 6, pp. 541–551, June 1990.

Ji, L. and Piper J., "Fast homotopy-preserving skeletons using mathematical morphology," *IEEE Trans. Pattern Analysis and Machine Intelligence*, vol. 14, no. 6, pp. 653–664, June 1992.

Lee, J. S., Haralick, R. M., and Shapiro, L. G., "Morphologic edge detection," *IEEE Trans. Robotics and Automation*, vol. 3, no. 2, pp. 142–156, Apr. 1987.

Maragos, P. A. and Schafer, R. W., "Morphological skeleton representation and coding of binary images," *IEEE Trans. Acoustics, Speech, and Signal Processing*, vol. 34, no. 5, pp. 1228–1244, Oct. 1986.

Meyer, F. "Skeletons in digital spaces," in J. Serra (Ed.), *Image Analysis and Mathematical Morphology*, vol. 2, Academic Press, New York, pp. 257–296, 1988.

Moran, C. J., "A morphological transformation for sharpening edges of features before segmentation," *Compute Vision, Graphics and Image Processing*, vol. 49, no. 1, pp. 85–94, Jan. 1990.

Serra, J., *Image Analysis and Mathematical Morphology*, Academic Press, New York, 1982.

Shih, F. Y. and Pu, C. C., "A skeletonization algorithm by maxima tracking on Euclidean distance transform," *Pattern Recognition*, vol. 28, no. 3, pp. 331–341, Mar. 1995.

Shih, F. Y. and Wu, Y., "The efficient algorithms for achieving Euclidean distance transformation," *IEEE Trans. Image Processing*, vol. 13, no. 8, pp. 1078–1091, Aug. 2004.

5

Basic Morphological Filters

Image filters are specifically designed to remove unwanted image components and/or enhance wanted ones. Mathematical morphology, applying set-theoretic concept, provides an approach to digital image processing and analysis based on the geometric shape of objects. Using an *a priori* determined structuring element, features in an image can be extracted, suppressed, or preserved by applying morphological operators. Morphological filters, using the basic operations such as dilation, erosion, opening, and closing, are suited for many purposes including digitization, enhancement, compression, restoration, segmentation, and description.

Linear filtering techniques have serious limitations in dealing with the images that have been created or processed by a system exhibiting some degree of nonlinearity. A nonlinear filter produces the output, which is not a linear function of its input. Nonlinear filtering techniques have shown clear superiority over linear filters in image processing. For example, an image often contains meaningful high frequency components, such as edges and fine details. A linear lowpass filter would blur sharp edges thus producing unacceptable results, so nonlinear filters must be used.

Morphological filters belong to nonlinear operations. Morphological dilation is used to smooth small dark regions. If all the values in the structuring element are positive, the output image tends to be brighter than the input. Dark elements are reduced or eliminated depending on how their shapes and sizes relate to the structuring element used. Morphological erosion is used to smooth small light regions as opposed to dilation. A morphological opening removes bright objects that are small in size and breaks narrow connections between two bright objects. A morphological closing preserves small objects that are brighter than the background and connects bright objects with small gaps between. A morphological top-hat operation, that is, the difference between the original image and the opened image, is used to enhance low-contrast, high frequency details within an image.

In this chapter, we will introduce several variations of morphological filters, including alternating sequential filters, recursive morphological filters, soft morphological filters, order-statistic soft morphological (OSSM) filters, recursive soft morphological (RSM) filters, recursive order-statistic soft morphological (ROSSM) filters, regulated morphological filters, and fuzzy morphological filters.

5.1 Alternating Sequential Filters

Alternating sequential filters (ASFs) in mathematical morphology are a combination of iterative morphological filters with increasing size of structuring elements. They offer a hierarchical structure for extracting geometric characteristics of objects. In addition, they provide less distortion in feature extraction than those filters, which directly process the images with the largest structuring element [Serra 1982]. The alternating filter is composed of morphological openings and closings whose primitive morphological operations are dilation and erosion.

The class of alternating filters (AFs) has been demonstrated to be useful in image analysis applications. Sternberg [1986] introduced a new class of morphological filters called ASFs, which consist of iterative operations of openings and closings with structuring elements of increasing sizes. Schonfeld and Goutsias [1991] have shown that ASFs are the best in preserving crucial structures of binary images in the least differentiation sense. The ASFs have been successfully used in a variety of applications, such as remote sensing and medical imaging [Destival 1986; Preteux et al. 1985]. Morales, Acharya, and Ko [1995] presented a relationship between ASFs and the morphological sampling theorem (MST) developed by Haralick et al. [1989]. The motivation is to take advantage of the computational efficiency offered by the MST to implement morphological operations.

Let X denote a binary image and let B denote a binary structuring element. The AF is defined as an opening followed by a closing or a closing followed by an opening, and is represented as

$$AF_B(X) = (X \circ B) \bullet B \tag{5.1}$$

or

$$AF_B(X) = (X \bullet B) \circ B. \tag{5.2}$$

Another type of AF is defined as

$$AF_B(X) = ((X \circ B) \bullet B) \circ B \tag{5.3}$$

or

$$AF_B(X) = ((X \bullet B) \circ B) \bullet B. \tag{5.4}$$

An ASF is an iterative application of $AF_B(X)$ with increasing size of structuring elements, denoted as

$$ASF(X) = AF_{B_N}(X)AF_{B_{N-1}}(X) \cdots AF_{B_1}(X), \tag{5.5}$$

where N is an integer and B_N, $B_{N-1}, \ldots, B_1$ are structuring elements with decreasing sizes. The B_N is constructed by

$$B_N = B_{N-1} \oplus B_1 \quad \text{for } N \geq 2. \tag{5.6}$$

The ASF offers a method of extracting image features hierarchically. The features can be divided into different layers according to their corresponding structuring element sizes. The features, such as size distribution [Giardina and Dougherty 1988], of each layer can be used in many applications; for example, feature classification and recognition.

5.1.1 Morphological Adjunction

A noticeable problem of applying ASFs to image analysis is their high computational complexity. For a structuring element of size n, the computational complexity is the fourth power of n when the opening-closing operation is used. This leads to a huge number of operations. An efficient class of ASFs based on the adjunctional property of morphology was proposed to reduce the computational complexity, and the applications to image filtering and texture classification were given in Pei, Lai, and Shih [1997].

A pair of morphological operators (ε, δ) on Z^2 is called *an adjunction* [Heijmans 1995] if

$$\delta(Y) \subseteq X \Leftrightarrow Y \subseteq \varepsilon(X), \quad \text{where } X, Y \subseteq Z^2. \tag{5.7}$$

That is, if (ε, δ) is an adjunction, then ε is an erosion and δ is a dilation. For each erosion there exists a unique dilation such that this pair constitutes an adjunction. Heijmans [1997] extended the class of ASFs by composing over filters and under filters. It is shown that any composition consisting of an equal number of dilations and erosions from an adjunction is a filter.

Let a function f be dilated n times by a structuring element B, denoted as $f(\delta^n)B$. That is equivalent to $f \oplus nB$. This notation is also applied for erosions n times as ε^n. If (ε, δ) is an adjunction, it has the following property.

Proposition 5.1: Let a filter be of the form $\psi = \varepsilon^{e_1} \delta^{d_1} \varepsilon^{e_2}$. If $e_1 = d_1 = e_2$, this filter can be reduced to

$$\psi = \varepsilon^{e_1} \delta^{e_1} \varepsilon^{e_1} = \varepsilon^{e_1}. \tag{5.8}$$

Similarly, a filter of the form $\phi = \delta^{d_1}\varepsilon^{d_1}\delta^{d_1}$ can be reduced to

$$\phi = \delta^{d_1}\varepsilon^{d_1}\delta^{d_1} = \delta^{d_1}. \tag{5.9}$$

Filters of the form

$$\psi = \varepsilon^{e_n}\delta^{d_n} \cdots \varepsilon^{e_2}\delta^{d_2}\,\varepsilon^{e_1}\delta^{d_1} \tag{5.10}$$

are called *adjunctional filters*. There exists a large amount of redundancy in adjunctional filters, which can be removed.

5.1.2 Redundancy Removal in ASFs

To simplify the notations of different ASFs, several definitions are given to represent different types of ASFs as follows.

Definition 5.1: An ASF with $AF_B(X)$ having the form of Equation 5.1 is called a TYPE-I ASF.

Definition 5.2: An ASF with $AF_B(X)$ having the form of Equation 5.2 is called a TYPE-II ASF.

Definition 5.3: An ASF with $AF_B(X)$ having the form of Equation 5.3 is called a TYPE-III ASF.

Definition 5.4: An ASF with $AF_B(X)$ having the form of Equation 5.4 is called a TYPE-IV ASF.

For convenience, a type-i ASF with a structuring element B is denoted as $ASF_i^B(X)$. It is observed that the type-III and -IV ASFs have computational redundancy in their structure. If a type-III ASF has n levels, it can be rewritten as

$$
\begin{aligned}
ASF_{III}^B(X) &= (((((((X \circ B) \bullet B) \circ B) \circ 2B) \bullet 2B) \circ 2B) \cdots \circ nB) \bullet nB) \circ n_i \\
&= X(\varepsilon\delta\delta\varepsilon\varepsilon\delta\varepsilon^2\delta^2\delta^2\varepsilon^2\varepsilon^2\delta^2 \cdots \varepsilon^n\delta^n\delta^n\varepsilon^n\varepsilon^n\delta^n)B \\
&= X(\varepsilon\delta^2\underline{\varepsilon^2\delta\varepsilon}\delta^4\varepsilon^4\delta^2 \cdots \varepsilon^n\delta^{2n}\varepsilon^{2n}\delta^n)B \\
&= X(\varepsilon\delta^2\underline{\varepsilon^3\delta^4}\varepsilon^4\delta^2 \cdots \varepsilon^n\delta^{2n}\varepsilon^{2n}\delta^n)B \\
&\;\;\vdots \\
&= X(\varepsilon\delta^2\varepsilon^3\delta^4 \cdots \delta^{2n}\varepsilon^{2n}\delta^n)B \\
&= ((((((X \circ B) \bullet B) \circ 2B) \bullet 2B) \cdots \circ nB) \bullet nB) \circ nB \\
&= ASF_I^B(X) \circ nB \tag{5.11}
\end{aligned}
$$

Equation 5.11 states that a type-III ASF can be reduced to a type-I ASF but with an additional opening operation padded to the end. Similarly, an n level type-IV ASF can be rewritten as

$$
\begin{aligned}
ASF_{IV}^B(X) &= (((((((((X \bullet B) \circ B) \bullet B) \bullet 2B) \circ 2B) \bullet 2B) \cdots \bullet nB) \circ nB) \bullet nB \\
&= ((((((X \bullet B) \circ B) \bullet 2B) \circ 2B) \cdots \bullet nB) \circ nB) \bullet nB \\
&= ASF_{II}^B(X) \bullet nB
\end{aligned}
\tag{5.12}
$$

Thus, a type-IV ASF can be reduced to a type-II ASF, but with an additional closing operation padded to the end. That is, by using the adjunction property, a type-III (-IV) ASF can be implemented by a type-I (-II) ASF, adding an additional opening (closing) operation. Compared with the original implementation of type-III and -IV ASFs, the computational complexity of this new method is reduced to about 2/3 of their original defined operations.

5.1.3 Definitions of the New Class of ASFs

The definitions of the new class of ASFs are given as follows.

Definition 5.5: An ASF of the following form is called a TYPE-V ASF.

$$
ASF_V^B(X) = ((((((X \circ B) \bullet B) \bullet 2B) \circ 2B) \circ 3B) \bullet 3B) \cdots
\tag{5.13}
$$

Definition 5.6: An ASF of the following form is called a TYPE-VI ASF.

$$
ASF_{VI}^B(X) = ((((((X \bullet B) \circ B) \circ 2B) \bullet 2B) \bullet 3B) \circ 3B) \cdots
\tag{5.14}
$$

Definition 5.7: An ASF of the following form is called a TYPE-VII ASF.

$$
ASF_{VIII}^B(X) = ((((((((X \circ B) \bullet B) \circ B) \bullet 2B) \circ 2B) \bullet 2B) \circ 3B) \bullet 3B) \circ 3B) \cdots
\tag{5.15}
$$

Definition 5.8: An ASF of the following form is called a TYPE-VIII ASF.

$$
ASF_{VIII}^B(X) = (((((((((X \bullet B) \circ B) \bullet B) \circ 2B) \bullet 2B) \circ 2B) \bullet 3B) \circ 3B) \bullet 3B) \cdots
\tag{5.16}
$$

It is noted that ASF_V, ASF_{VI}, ASF_{VII}, and ASF_{VIII} are the modifications of ASF_I, ASF_{II}, ASF_{III}, and ASF_{IV}, respectively. Definitions of the eight different types of ASFs are listed in Table 5.1.

TABLE 5.1

Definitions of Eight Different Types of ASF

ASF Type	Definitions
ASF_I	$X \circ B \bullet B \circ 2B \bullet 2B \cdots$
ASF_{II}	$X \bullet B \circ B \bullet 2B \circ 2B \cdots$
ASF_{III}	$X \circ B \bullet B \circ B \circ 2B \bullet 2B \circ 2B \cdots$
ASF_{IV}	$X \bullet B \circ B \bullet B \bullet 2B \circ 2B \bullet 2B \cdots$
ASF_V	$X \circ B \bullet B \bullet 2B \circ 2B \cdots$
ASF_{VI}	$X \bullet B \circ B \circ 2B \bullet 2B \cdots$
ASF_{VII}	$X \circ B \bullet B \circ B \bullet 2B \circ 2B \bullet 2B \cdots$
ASF_{VIII}	$X \bullet B \circ B \bullet B \circ 2B \bullet 2B \circ 2B \cdots$

To achieve the computational efficiency of the new ASFs, they are rewritten as follows. Let n be an odd number throughout the following derivations. A type-V ASF can be expressed as

$$
\begin{aligned}
ASF_V^B(X) &= (((((X \circ B) \bullet B) \bullet 2B) \circ 2B) \cdots \circ nB) \bullet nB \\
&= X(\varepsilon\delta\delta\varepsilon\delta^2\varepsilon^2\varepsilon^2\delta^2 \cdots \varepsilon^n\delta^n\delta^n\varepsilon^n)B \\
&= X(\varepsilon\delta^2\varepsilon\delta^2\varepsilon^4\delta^2 \cdots \varepsilon^n\delta^{2n}\varepsilon^n)B \\
&= X(\varepsilon\delta^2\varepsilon^5 \cdots \delta^{2n-3}\varepsilon^{2n-1}\delta^{2n}\varepsilon^n)B \\
&= ((((X \circ B) \bullet 2B) \circ 3B) \cdots \circ nB) \bullet nB
\end{aligned}
\tag{5.17}
$$

Similarly, a type-VI ASF can be expressed as

$$
\begin{aligned}
ASF_{VI}^B(X) &= (((((X \bullet B) \circ B) \circ 2B) \bullet 2B) \cdots \bullet nB) \circ nB \\
&= ((((X \bullet B) \circ 2B) \bullet 3B) \cdots \bullet nB) \circ nB
\end{aligned}
\tag{5.18}
$$

It is clear that, for each $AF_B(X) = (X \circ B) \bullet B$ layer, since either a closing or an opening operation can be saved in ASF_V and ASF_{VI}, their complexities are nearly one half the amount of those in ASF_I and ASF_{II}. On the other hand, rewriting the type-VII and -VIII ASFs by the same method produces the similar results as follows.

$$
\begin{aligned}
ASF_{VII}^B(X) &= ((((((((X \circ B) \bullet B) \circ B) \bullet 2B) \circ 2B) \bullet 2B) \cdots \circ nB) \bullet nB) \circ nB \\
&= ((((((X \circ B) \bullet 2B) \circ 3B) \cdots \bullet (n-1)B) \circ nB) \bullet nB) \circ nB \\
&= ASF_V^B \circ nB
\end{aligned}
\tag{5.19}
$$

and

$$ASF_{VIII}^{B}(X) = ((((((((X \bullet B) \circ B) \bullet B) \circ 2B) \bullet 2B) \circ 2B) \cdots \bullet nB) \circ nB) \bullet nB$$
$$= ((((((X \bullet B) \circ 2B) \bullet 3B) \cdots \circ (n-1)B) \bullet nB) \circ nB) \bullet nB$$
$$= ASF_{VI}^{B} \bullet nB \qquad (5.20)$$

Therefore, an additional advantage in reducing complexity can be achieved if we rewrite the four new defined ASFs using the adjunction property. However, although additional advantages are obtained, there are some important questions to be addressed. What are the properties of the new ASFs? Do they have a performance boundary? These questions will be answered in the next section.

5.1.4 Properties of the New Class of ASFs

Property 5.1: The four operations of types V–VIII are increasing.

Property 5.2: The four operations of types V–VIII are idempotent.

The proofs of property 5.1 and 5.2 are trivial since they are composed of increasing and idempotent operators. These two properties are used to ensure that types V–VIII operators are morphological filters. In the following properties their upper and lower bounds are derived. To ease the analysis, the padded operations of types V–VIII are ignored. Thus, an odd number n level type-V ASF can be written as: $X \circ B \bullet 2B \cdots \circ nB$.

Property 5.3: A type-V ASF is nearly bounded by a type-II ASF (upper bound) and exactly bounded by a type-I ASF (lower bound) when n is odd, and is nearly bounded by a type-I ASF (lower bound) and exactly bounded by a type-II ASF (upper bound) when n is even.

Proof: By using the anti-extensive and extensive properties of the opening and closing filters, it is observed that

$$X \circ B \subset (X \bullet B) \circ B \subset X \bullet B \qquad (5.21)$$

Since closing is increasing, if a closing operator is applied at each side, Equation 5.21 can be rewritten as

$$(X \circ B) \bullet 2B \subset ((X \bullet B) \circ B) \bullet 2B \subset (X \bullet B) \bullet 2B$$

where the present equation is obtained by applying the operation inside the bracket to its previous equation. By continuously utilizing the same method, the above equation implies:

$$X \circ B \bullet B \circ 2B \bullet 2B \circ 3B \bullet 3B \subset X \circ B \bullet 2B \circ 3B \bullet 3B$$
$$\subset X \bullet B \circ B \bullet 2B \circ 2B \bullet 3B \circ 3B \bullet 3B$$

when n is odd. It is noted that the left-hand side of the above equation is exactly a type-I ASF, while the right-hand side is a type-II ASF with an additional closing padded to its end. On the other hand, if n is even, Equation 5.21 implies

$$X \circ B \bullet B \circ 2B \bullet 2B \circ 2B \subset X \circ B \bullet 2B \circ 2B \subset X \bullet B \circ B \bullet 2B \circ 2B$$

It is observed that the left-hand side of the above equation is a type-I ASF with an additional opening padded to its end, while the right-hand side is an exact type-II ASF. That is, the type-V ASFs are bounded by the type-I and -II ASFs.   □

Property 5.4: A type-VI ASF is nearly bounded by a type-I ASF (lower bound) and exactly bounded by a type-II ASF (upper bound) when n is odd, and is nearly bounded by a type-II ASF (upper bound) and exactly bounded by a type-I ASF (lower bound) when n is even.

Proof: By using the same method in the proof of Property 5.3, the following equation

$$X \circ B \subset (X \circ B) \bullet B \subset X \bullet B \tag{5.22}$$

implies

$$X \circ B \bullet B \circ 2B \bullet 2B \circ 3B \bullet 3B \circ 3B \subset X \bullet B \circ 2B \bullet 3B \circ 3B$$
$$\subset X \bullet B \circ B \bullet 2B \circ 2B \bullet 3B \circ 3B$$

in case when n is an odd number. It is noted that the left-hand side and the right-hand side are nearly the type-I and exactly the type-II ASFs, respectively. If n is an even number, Equation 5.22 implies

$$X \circ B \bullet B \circ 2B \bullet 2B \subset X \bullet B \circ 2B \bullet 2B \subset X \bullet B \circ B \bullet 2B \circ 2B \bullet 2B$$

Thus, the left-hand side and right-hand side are exactly the type-I and nearly the type-II ASFs, respectively.   □

Property 5.5: The type-VII and -VIII ASFs are bounded by the type-V and -VI ASFs if the padded operations are ignored.

Proof: It is trivial that this property can be obtained by observing Equations 5.17 through 5.20. That is, the type-VII ASF is equivalent to the type-V ASF, and the type-VIII ASF is equivalent to the type-VI ASF if the padded operations are ignored.   □

Properties 5.3 through 5.5 suggest the performance boundary of the new class of ASFs. If they are applied to one of image processing applications, it is interesting to note that we can predict the performance in advance. Thus, the proposed ASFs have much better computational efficiency, while their filtered results remain comparable to the traditional ASFs.

Two kinds of applications are adopted as computer simulations to illustrate the analysis. One is for image filtering and the other for texture classification [Pei, Lai, and Shih 1997]. Both new and conventional types of ASFs are used for comparison of their advantages and disadvantages.

Figures 5.1a and b, respectively, show a binary image and its corrupted version by adding pepper-salt noise with 0.1 occurrence probability. The structuring elements are square blocks of sizes 3×3, 5×5, and 7×7. Figures 5.1c–f show the filtered results of type-I, -II, -V, and -VI ASFs, respectively. Similarly, the grayscale image "Lena," distorted version, and its filtered results are, respectively, shown in Figures 5.2a–h. From experimental results it is observed that the new types of ASFs perform better results than the present ASFs. Phenomenon of the bound properties can be easily observed in Figures 5.1a–d. That is, the dark part of the filtering results by the type-V and -VI ASFs is contained in that of the type-I filtering result, while the bright part of the filtering results by the type-V and -VI ASFs is contained in that of the type-II filtering result. The bound properties are useful in image filtering because we can predict the range of the results from their bounds in advance. However, it is a little hard to view these properties in Figure 5.2 since grayscale is used. From Figures 5.2c and d, it is observed that the ASFs are better than the morphological filters, which directly apply the largest structuring element.

For performance comparisons, an application of texture classification is adopted. We use 20 texture images of size 512×512 from Brodatz [1966] in classification. Each texture image is decomposed into five component images by choosing a squared structuring element with increasing sizes of 1, 9, 25, 49, and 81 pixels. The decomposition method is similar to Li and Coat [1995], and is described briefly as follows. Let B_i denote a flat structuring element of size $i \times i$ and $B_0 = (0, 0)$ in E^2. The texture image $f(x, y)$ is processed first by an AF with the largest structuring element B_9. By sequentially applying this procedure through decreasing the size of the structuring element, the image primitives of different sizes can be decomposed into different image layers s_i. This decomposition procedure can be described by

$$\begin{cases} f_0(x,y) = f(x,y) \\ s_i(x,y) = (AF)f_i(x,y), \quad j = 0, 1, \ldots, n \\ f_{i+1}(x,y) = f_i(x,y) - s_i(x,y) \end{cases}$$

For each component image, it is first divided into 16 blocks; then the mean value of the training blocks (diagonal blocks) is calculated and stored. The test images are randomly selected from these 20 textures and corrupted by the zero mean Gaussian noise with variance equals 100. They are also decomposed into five components and the rest 12 blocks, not the training

FIGURE 5.1   (a) A binary image, (b) a version corrupted by pepper-salt noise (occurrence probability 0.1), and filtered results from (c) type-I, (d) type-II, (e) type-V, and (f) type-VI ASFs.

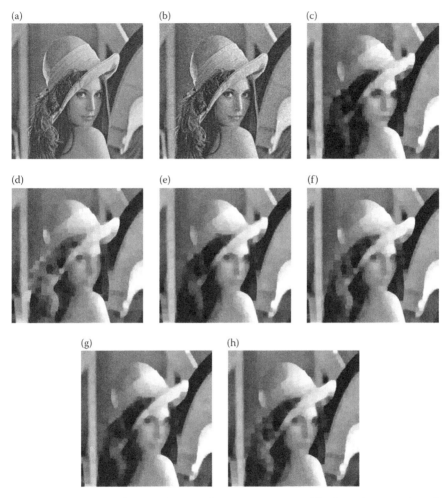

FIGURE 5.2 (a) Grayscale "Lena" image, (b) a version corrupted by Gaussian noise ($\sigma = 10$), and filtered results from (c) opening-closing by 7×7 structuring element, (d) closing-opening by 7×7 structuring element, (e) type-I, (f) type-II, (g) type-V, and (h) type-VI ASFs.

blocks, of each component are used to calculate their mean value. The five-dimensional mean vector is then used as the difference measurement. For each type of ASF, 300 trials are used. Figure 5.3 shows the 20 texture images. The classification results by using type-I, -II, -V, and -VI ASFs are shown in Table 5.2. It is important to note that properties 5.3, 5.4, and 5.5 are not available here since the decomposition methods of this application are not exactly the forms of type-V and -VI ASFs. That is, we cannot ensure that the correct classification rates of type-V and -VI ASFs are between the rates of type-I and -II ASFs. Indeed, it is observed in Table 5.2 that the correct classification rates of type-V and -VI ASFs are somewhat lower than

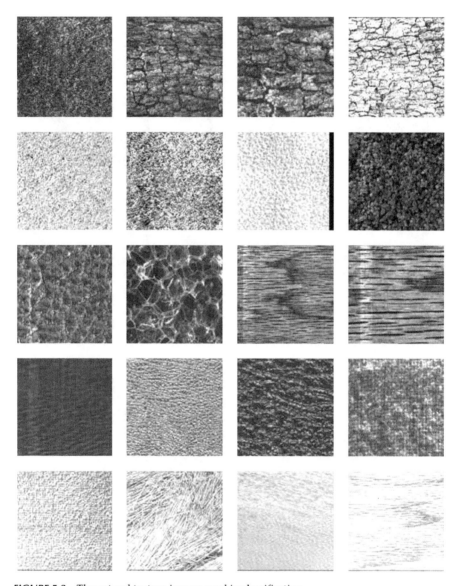

FIGURE 5.3   The natural texture images used in classification.

TABLE 5.2

Classification Results of Type-I, -II, -V, and -VI ASFs

	Type-I	Type-II	Type-V	Type-VI
Correct classification rate	69%	76%	64%	67%

those of type-I and -II ASFs. However, only half of the computational complexity is needed for the new ASFs.

From experimental results it is concluded that the new types of ASFs not only offer efficient implementation structures in ASFs, but also maintain comparable results in image decomposition and filtering. Moreover, the bounding properties are also illustrated in the experimental results.

5.2 Recursive Morphological Filters

Recursive morphological filters can be used to avoid the iterative processing in the distance transform and skeletonization [Shih and Mitchell 1992; Shih, King, and Pu 1995]. The intent of the recursive morphological operations is to feed back the output at the current scanning pixel to overwrite its corresponding input pixel to be considered into computation at the following scanning pixels. Therefore, each output value obtained depends on all the updated input values at positions preceding the current location. The resulting output image in recursive mathematical morphology inherently varies with different image scanning sequences.

Let $N(O)$ be the set of the pixel O plus the neighbors prior to it in a scanning sequence of the picture within the domain of a defined structuring element. Suppose that a set A in the Euclidean N-space (E^N) is given. Let F and K be $\{x \in E^{N-1} \mid$ for some $y \in E, (x, y) \in A\}$, and let the domain of the grayscale image be f and of the grayscale structuring element be k. The *recursive dilation* of f by k, which is denoted by $f \oplus k$, is defined as follows:

$$(f \oplus k)(x, y) = \max\{[(f \oplus k)(x - m, y - n)] + k(m, n)\} \tag{5.23}$$

for all $(m, n) \in K$ and $(x - m, y - n) \in (N \cap F)$. The *recursive erosion* of f by k, which is denoted by $f \ominus k$, is defined as follows:

$$(f \ominus k)(x, y) = \min\{[(f \ominus k)(x + m, y + n)] - k(m, n)\} \tag{5.24}$$

for all $(m, n) \in K$ and $(x + m, y + n) \in (N \cap F)$.

The scanning order will affect the output image in recursive morphology. Starting with a corner pixel, the scanning sequences in a 2-D digital image are in general divided into:

(i) "LT" denotes left-to-right and top-to-bottom. (Note that this is a usual television raster scan.)

(ii) "RB" denotes right-to-left and bottom-to-top.

(iii) "LB" denotes left-to-right and bottom-to-top.

(iv) "RT" denotes right-to-left and top-to-bottom.

Assume that a 3×3 structuring element k is denoted by

$$k = \begin{bmatrix} A_1 & A_2 & A_3 \\ A_4 & A_5 & A_6 \\ A_7 & A_8 & A_9 \end{bmatrix}.$$

Because they are associated with different scanning directions, not all of the nine elements $A_1, \ldots, A_9$ are used in computation. The redefined structuring element in recursive mathematical morphology must satisfy the following criterion: Wherever a pixel is being dealt with, all its neighbors defined within the structuring element k must have already been visited before by the specified scanning sequence. Thus, the k is redefined as:

$$k_{LT} = \begin{bmatrix} A_1 & A_2 & A_3 \\ A_4 & A_5 & \times \\ \times & \times & \times \end{bmatrix}, \quad k_{RB} = \begin{bmatrix} \times & \times & \times \\ \times & A_5 & A_6 \\ A_7 & A_8 & A_9 \end{bmatrix},$$

$$k_{LB} = \begin{bmatrix} \times & \times & \times \\ A_4 & A_5 & \times \\ A_7 & A_8 & A_9 \end{bmatrix}, \quad k_{RT} = \begin{bmatrix} A_1 & A_2 & A_3 \\ \times & A_5 & A_6 \\ \times & \times & \times \end{bmatrix}$$

(5.25)

where $\times$ means don't care. A complete scanning in the image space must contain the dual directions, that is, the first one being any of the four scans *LT, RB, LB, RT* working with its associated structuring element and the second one being the opposite scanning sequence of the first scan and its associated structuring element. For example, the opposite scanning of *LT* is *RB*.

The advantage of recursive mathematical morphology can be illustrated by applying distance transform. The distance transform is to convert a binary image, which consists of object (foreground) and nonobject (background) pixels into an image in which every object pixel has a value corresponding to the minimum distance from the background. Three types of distance measures in digital image processing are usually used: Euclidean, city-block, and chessboard. The city-block and chessboard distances are our concern since their structuring elements can be decomposed into iterative dilations of small structuring components. The distance computation is in principle a global operation. By applying the well-developed decomposition properties of mathematical morphology, we can significantly reduce the tremendous cost of global operations to that of small neighborhood operations, which is suited for parallel-pipelined computers.

The iteration in the distance transform results in a bottleneck in real-time implementation. To avoid iteration, the recursive morphological operation is used. Only two scannings of the image are required: one is left-to-right top-to-bottom, and the other is right-to-left bottom-to-top. Thus, the distance transform becomes independent of the object size.

By setting the origin point as P, we can represent all of the points by their distances from P (so that P is represented by 0). The values of the structuring element are selected to be negative of the associated distance measures because the erosion is the minimum selection after the subtraction is applied. Let f denote a binary image, which consists of two classes, object pixels (foreground) and nonobject pixels (background). Let the object pixels be the value $+\infty$ (or any number larger than a half of the object's diameter) and the nonobject pixels be the value 0. Let k_1 and k_2 be the structuring elements associated with the LT and RB scannings, respectively; that is, in the chessboard distance

$$k_1 = \begin{bmatrix} -1 & -1 & -1 \\ -1 & 0 & \times \\ \times & \times & \times \end{bmatrix}, \quad k_2 = \begin{bmatrix} \times & \times & \times \\ \times & 0 & -1 \\ -1 & -1 & -1 \end{bmatrix}. \tag{5.26}$$

In the city-block distance

$$k_1 = \begin{bmatrix} -2 & -1 & -2 \\ -1 & 0 & \times \\ \times & \times & \times \end{bmatrix}, \quad k_2 = \begin{bmatrix} \times & \times & \times \\ \times & 0 & -1 \\ -2 & -1 & -2 \end{bmatrix}. \tag{5.27}$$

Distance Transform using Recursive Morphology:

1. $d_1 = f \ominus k_1$.
2. $d_2 = d_1 \ominus k_2$.

The distance transform algorithm has widely-used applications in image analysis. One of the applications is to compute a shape factor that is a measure of the compactness of the shape based on the ratio of the total number of object pixels to the summation of all distance measures [Danielsson 1978]. Another is to obtain the medial axis (or skeleton) that can be used for features extraction.

Example 5.1: Apply the two-step distance transformation using recursive morphology to calculate the chessboard distance transform on the following image.

0	0	0	0	0	0	0	0	0	0	0	0	0
0	7	7	7	7	7	0	0	0	7	7	7	0
0	7	7	7	7	7	0	0	0	7	7	7	0
0	7	7	7	7	7	0	0	0	7	7	7	0
0	7	7	7	7	7	7	7	7	7	7	7	0
0	7	7	7	7	7	7	7	7	7	7	7	0
0	7	7	7	7	7	7	7	7	7	7	7	0
0	7	7	7	7	7	7	7	7	7	7	7	0
0	7	7	7	7	7	7	7	7	7	7	7	0
0	0	0	0	0	0	0	0	0	0	0	0	0

Answer: First, we apply $d_1 = f \circledcirc k_1$, where $k_1 = \begin{bmatrix} -1 & -1 & -1 \\ -1 & 0 & \times \\ \times & \times & \times \end{bmatrix}$, and obtain the following:

0	0	0	0	0	0	0	0	0	0	0	0	0
0	1	1	1	1	1	0	0	0	1	1	1	0
0	1	2	2	2	1	0	0	0	1	2	1	0
0	1	2	3	2	1	0	0	0	1	2	1	0
0	1	2	3	2	1	1	1	1	1	2	1	0
0	1	2	3	2	2	2	2	2	2	2	1	0
0	1	2	3	3	3	3	3	3	3	2	1	0
0	1	2	3	4	4	4	4	4	3	2	1	0
0	1	2	3	4	5	5	5	4	3	2	1	0
0	0	0	0	0	0	0	0	0	0	0	0	0

Second, we apply $d_2 = d_1 \circledcirc k_2$, where $k_2 = \begin{bmatrix} \times & \times & \times \\ \times & 0 & -1 \\ -1 & -1 & -1 \end{bmatrix}$, and obtain the chessboard distance transform as

0	0	0	0	0	0	0	0	0	0	0	0	0
0	1	1	1	1	1	0	0	0	1	1	1	0
0	1	2	2	2	1	0	0	0	1	2	1	0
0	1	2	3	2	1	0	0	0	1	2	1	0
0	1	2	3	2	1	1	1	1	1	2	1	0
0	1	2	3	2	2	2	2	2	2	2	1	0
0	1	2	3	3	3	3	3	3	3	2	1	0
0	1	2	2	2	2	2	2	2	2	2	1	0
0	1	1	1	1	1	1	1	1	1	1	1	0
0	0	0	0	0	0	0	0	0	0	0	0	0

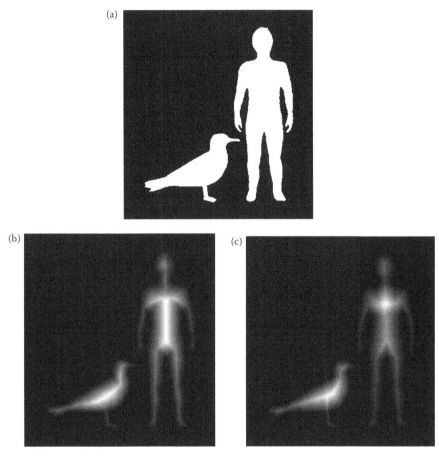

FIGURE 5.4 (a) A binary image of human and bird figures, (b) its chessboard distance transformation, (c) its city-block distance transformation.

Figure 5.4 shows a binary image of human and bird figures and its chessboard and city-block distance transformations. Note that after distance transformations, the distances are linearly rescaled to be in the range of [0, 255] for clear display.

Note that the recursive morphological filters defined here are different from the commonly-used recursive filters in signal processing, which reuse one or more of the outputs as an input and can efficiently achieve a long impulse response, without having to perform a long convolution [Smith 1997]. Other recursive algorithms may use different definitions. For example, the recursive algorithm based on observation of the basis matrix and block basis matrix of grayscale structuring elements was used to avoid redundant steps in computing overlapping local maximum or minimum operations [Ko, Morales, and Lee 1996]. Another recursive two-pass algorithm that runs at a constant time for simultaneously obtaining binary dilations and erosions with all possible length line structuring elements are referred to Nadadur and Haralick [2000].

5.3 Soft Morphological Filters

Soft morphological filters adopt the concept of weighted order-statistic filters and morphological filters. The primary difference from traditional morphological filters is that the maximum and minimum operations are replaced by general weighted order statistics and the "soft" boundary is added to the structuring element. In general, soft morphological filters are less sensitive to additive noise and to small variations in object shape, and can preserve most of the desirable properties of traditional morphological filters [Koskinen, Astola, and Neuvo 1991; Kuosmanen and Astola 1995; Shih and Pu 1995; Shih and Puttagunta 1995]. Hamid, Harvey, and Marshall [2003] presented a technique for the optimization of multidimensional grayscale soft morphological filters for applications in automatic film archive restoration, specific to the problem of film dirt removal.

The basic idea of soft morphological operations is that the structuring element B is split into two subsets: the core subset A (i.e., $A \subseteq B$) and the soft boundary subset $B \backslash A$, where "$\backslash$" denotes the *set difference*. Soft morphological dilation (erosion) of a function with respect to these two finite sets A and B using the order index k is a function whose value at location x is obtained by sorting in a descendent (ascendent) order the total $\text{Card}(B \backslash A) + k \times \text{Card}(A)$ values of the input function, including the elements within $(B \backslash A)_x$ and repetition k times of the elements within $(A)_x$, and then selecting the kth order from the sorted list. Let $\{k \Diamond f(a)\}$ denote the repetition k times of $f(a)$, which means $\{k \Diamond f(a)\} = \{f(a), f(a), \ldots, f(a)\}$ (k times). The soft morphological operations of function-by-set are defined as follows, where f is a grayscale image and A and B are flat structuring elements.

Definition 5.9: The *soft morphological dilation* of f by $[B, A, k]$ is defined as

$$(f \oplus [B, A, k])(x) = k\text{th largest of } (\{k \Diamond f(a) \mid a \in A_x\} \cup \{f(b) \mid b \in (B \backslash A)_x\}). \quad (5.28)$$

Definition 5.10: The *soft morphological erosion* of f by $[B, A, k]$ is defined as

$$(f \ominus [B, A, k])(x) = k\text{th smallest of } (\{k \Diamond f(a) \mid a \in A_x\} \cup \{f(b) \mid b \in (B \backslash A)_x\}). \quad (5.29)$$

The soft morphological operations of set-by-set can be simplified as in the definition below, where $n = \text{Card}(A)$ and $N = \text{Card}(B)$.

Definition 5.11: The *soft morphological dilation* of a set X by $[B, A, k]$ is defined as

$$X \oplus [B, A, k] = \{x \mid k \times \text{Card}(X \cup A_x) + \text{Card}(X \cap (B \backslash A)_x) \le k\}. \quad (5.30)$$

Definition 5.12: The *soft morphological erosion* of a set X by $[B, A, k]$ is defined as

$$X \ominus [B, A, k] = \{x \mid k \times \text{Card}(X \cap A_x) + \text{Card}(X \cap (B \backslash A)_x)$$
$$\ge N + (k - 1) \times n - k + 1\}. \quad (5.31)$$

The logic-gate implementation of soft morphological dilation and erosion is illustrated in Figure 5.5, where the parallel counter counts the number of 1s of the input signal and the comparator outputs 1 if the input is greater than or equal to the index k.

The cross-section $X_t(f)$ of f at level t is the set obtained by thresholding f at level t:

$$X_t(f) = \{x \mid f(x) \ge t\}, \quad \text{where } -\infty < t < \infty. \quad (5.32)$$

Theorem 5.1: The soft morphological operations of function-by-set commute with thresholding. That is, for any t, we have

$$X_t(f \oplus [B, A, k]) = X_t(f) \oplus [B, A, k]. \quad (5.33)$$

This theorem can be easily derived from Shih and Mitchell [1989] and Shih and Pu [1995]. The implementation and analysis of soft morphological operations of function-by-set can be achieved by using the set-by-set operations, which are much easier to deal with because they are involved in only counting the number of pixels instead of sorting numbers. Consider the thresholding of a binary image:

$$f_a(x) = \begin{cases} 1 & \text{if } f(x) \ge a \\ 0 & \text{otherwise} \end{cases}, \quad (5.34)$$

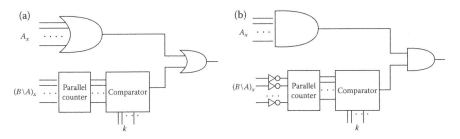

FIGURE 5.5 The logic-gate implementation of soft morphological operations (a) dilation and (b) erosion.

where $0 \leq a \leq L$ and L is the largest value in f. It is simple to show that f can be reconstructed from its thresholded binary images as

$$f(x) = \sum_{a=1}^{L} f_a(x) = \max\{a \mid f_a(x) = 1\}. \qquad (5.35)$$

A transformation Ψ is said to possess threshold-linear superposition if it satisfies

$$\Psi(f) = \sum_{a=1}^{L} \Psi(f_a). \qquad (5.36)$$

Such a transformation Ψ can be realized by decomposing f into all its binary images f_as, processing each thresholded images by Ψ, and creating the output $\Psi(f)$ by summing up the processed f_as. The soft morphological operations of function-by-set can be easily shown to obey the threshold-linear superposition.

Theorem 5.2: The soft morphological operations of functions by sets obey the threshold-linear superposition. That is

$$f \oplus [B, A, k] = \sum_{a=1}^{L} (f_a \oplus [B, A, k]). \qquad (5.37)$$

This theorem is very useful in the sense that the grayscale input image is thresholded at each gray level a to get f_a, followed by the soft morphological operations of sets by sets for f_a, and then summing up all f_as, which is exactly the same as the soft morphological operations of functions by sets. Considering A_1, A_2, and B in Figure 5.6, an example of illustrating the threshold decomposition is given in Figure 5.7.

The grayscale soft morphological operation can be decomposed into binary soft morphological operation [Pu and Shih 1995]. A new hardware structure for implementation of soft morphological filters can be referred to Gasteratos, Andreadis, and Tsalides [1997]. Let the reflected sets of A and B be $\hat{A}(x) = A(-x)$ and $\hat{B}(x) = B(-x)$, respectively.

Definition 5.13: The *soft morphological closing* of f by $[B, A, k]$ is defined as

$$f \bullet [B, A, k] = (f \oplus [B, A, k]) \ominus [\hat{B}, \hat{A}, k]. \qquad (5.38)$$

Definition 5.14: The *soft morphological opening* of f by $[B, A, k]$ is defined as

$$f \circ [B, A, k] = (f \ominus [B, A, k]) \oplus [\hat{B}, \hat{A}, k]. \qquad (5.39)$$

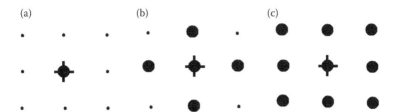

FIGURE 5.6 Two core sets (a) A_1, (b) A_2, and (c) the structuring element B.

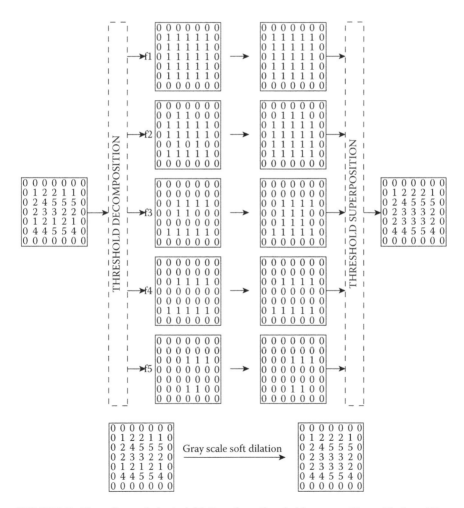

FIGURE 5.7 The soft morphological dilation obeys threshold superposition with A_1 and B as shown in Figure 5.6, and k is equal to 4.

It is interesting to note that if $k = 1$, the soft morphological operations are equivalent to standard morphological operations. If $k > \mathrm{Card}(B \backslash A)$, the soft morphological operations are reduced to consider only the hard set A [i.e., the soft boundary set $(B \backslash A)$ will actually not affect the result]. Therefore, in order to preserve the nature of soft morphological operations, the constraint that $k \leq \min\{\mathrm{Card}(B)/2, \mathrm{Card}(B \backslash A)\}$ is used.

Figure 5.8a shows an original image, and Figure 5.8b shows the same image but with added Gaussian noise being zero mean and standard deviation 20. Figure 5.8c shows the result of applying a standard morphological closing with a structuring element of size 3×3, and Figure 5.8d shows the result of applying a soft morphological closing with $B = \langle (0, -1), (-1, 0), (0, 0), (1, 0), (0, 1) \rangle$, $A = \langle (0, 0) \rangle$, and $k = 3$. The mean-square signal-to-noise ratios for Figures 5.8c and d are 39.78 and 55.42, respectively. One can observe that the result of soft morphological filtering produces a smoother image and a higher signal-to-noise ratio.

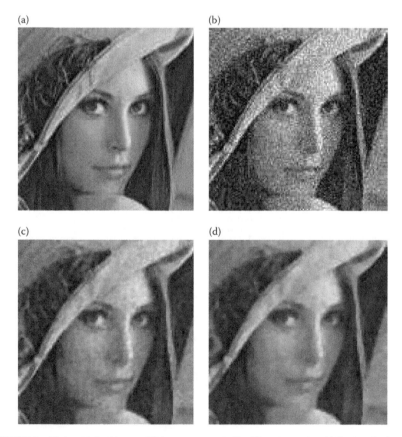

FIGURE 5.8 (a) An original image, (b) image corrupted by Gaussian noise, (c) the result of applying a standard morphological closing, (d) the result of applying a soft morphological closing.

Example 5.2: Perform the soft morphological dilation $f \oplus [B, A, 3]$ and the soft morphological erosion $f \ominus [B, A, 3]$. Use the center (underlined) of A and B as the origin.

$$f: \begin{bmatrix} 0 & 0 & 0 & 1 & 0 \\ 0 & 5 & 5 & 5 & 1 \\ 4 & 5 & 1 & 5 & 0 \\ 3 & 6 & 5 & 6 & 0 \\ 0 & 0 & 0 & 0 & 1 \end{bmatrix}, \; A = \begin{bmatrix} 0 & 0 & 0 \\ 0 & \underline{1} & 0 \\ 0 & 0 & 0 \end{bmatrix}, \; B = \begin{bmatrix} 1 & 1 & 1 \\ 1 & \underline{1} & 1 \\ 1 & 1 & 1 \end{bmatrix}$$

Answer: The soft morphological dilation $f \oplus [B, A, 3]$ repeats the central element of the 3×3 window three times and others once, and then arrange them in an ascending order. Taking out the third largest element we obtain the following result:

$$\begin{matrix} 0 & 0 & 5 & 1 & 1 \\ 4 & 5 & 5 & 5 & 1 \\ 5 & 5 & 5 & 5 & 5 \\ 4 & 6 & 5 & 6 & 1 \\ 0 & 3 & 5 & 1 & 1 \end{matrix}$$

The soft morphological erosion $f \ominus [B, A, 3]$ repeats the central element of the 3×3 window three times and others once, and then arranges in an ascending order. Taking out the third smallest element we obtain the following result:

$$\begin{matrix} 0 & 0 & 0 & 1 & 0 \\ 0 & 0 & 1 & 0 & 1 \\ 4 & 3 & 1 & 1 & 0 \\ 3 & 0 & 0 & 0 & 0 \\ 0 & 0 & 0 & 0 & 1 \end{matrix}$$

5.3.1 Properties of Soft Morphological Operations

Definition 5.15: A transformation Ψ is said to be *idempotent* if $\Psi(\Psi(f)) = \Psi(f)$ for any input signal f.

Definition 5.16: A transformation Ψ is said to be *increasing* if for any two input signals f and g such that $f(x) \leq g(x)$ for every x, the resultant outputs satisfy the relationship $\Psi(f(x)) \leq \Psi(g(x))$.

Definition 5.17: A transformation Ψ is said to be *translation-invariant* if $\Psi_x(f) = \Psi(f_x)$ for any $x \in Z^m$.

In the following, we introduce general properties of soft morphological filters, which are valid for both sets by sets and functions by sets. The proofs are omitted. Interested readers should refer to Shih and Pu [1995].

Proposition 5.2: The soft morphological dilation and erosion are increasing.

Proposition 5.3: The soft morphological closing and opening are increasing.

Proposition 5.4: The soft morphological operations are translation-invariant.

Proposition 5.5: The soft morphological dilation propagates the local maximum of the pixels within B to all the pixels within A and fills up the valley whose area is less than or equal to Card(A).

Proposition 5.6: The soft morphological erosion propagates the local minimum of the pixels within B to all the pixels within A and eliminates the peak whose area is less than or equal to Card(A).

5.3.2 Idempotent Soft Morphological Filters

An idempotent filter maps a class of input signals into an associated set of root sequences. Each of these root signals is invariant to additional filter passes. Since the nature of the soft morphological operations and their fair nonlinearity, idempotency usually does not exist in such operations at the first stage. Koskinen, Astola, and Neuvo [1991] constructed a special class of idempotent soft morphological filters under the constraints that A only contains the origin of B and $k = $ Card(B) $- 1$. In this section, we present a general class of idempotent soft morphological closing. The idempotent soft morphological opening can be similarly derived.

Proposition 5.7: The soft morphological closing fills up the valley whose area is less than or equal to Card(A), when $k = $ Card($B \backslash A$). When the area of valley is equal to Card(A), there must be no other valley in ($B \backslash A$).

Proof: According to Proposition 5.5, the soft morphological dilation fills up the valley whose area is less than or equal to Card(A). The result followed by a soft morphological erosion will not change in those filled areas. However, when the area of valley is equal to Card(A), there must be no other valley in ($B \backslash A$). Otherwise, the value of the kth largest must exist in valleys. That is the valley will not be filled in fully.   □

Proposition 5.8: The soft morphological closing suppresses the local maximum when $k = $ Card($B \backslash A$).

An observation is made that when k is equal to Card($B \backslash A$), the soft morphological closing becomes idempotent. For simplicity, only the one-dimensional case is given. Let $D[X]$ and $E[X]$ denote the positive Boolean functions of binary soft morphological dilation and erosion, respectively, where X is the input binary signal. Let the core set $A = \{-a, -a+1, \dots, -1, 0, 1, \dots, a-1, a\}$ and the soft boundary $B \setminus A = \{-b, -b+1, \dots, -a-1, a+1, \dots, b-1, b\}$. Since $k = \text{Card}(B \setminus A)$, $D[X]$ can be represented as a logical function below.

$$D[X](0) = \sum_{i \in A} X(i) + \prod_{j \in (B \setminus A)} X(j), \qquad (5.40)$$

where $X(i)$ is the value of X at location i, and the symbols of addition and multiplication denote the logic OR and AND, respectively. Similarly, the binary soft morphological erosion can be expressed as

$$E[X](0) = \prod_{i \in A} X(i) \cdot \sum_{j \in (B \setminus A)} X(j). \qquad (5.41)$$

Let $\Psi[X]$ be the soft morphological closing. Replacing X in Equation 5.41 by $D[X]$ yields

$$
\begin{aligned}
\Psi[X](0) &= E[D[X]](0) \\
&= \prod_{i \in A} D[X](i) \cdot \sum_{j \in (B \setminus A)} D[X](j)
\end{aligned}
\qquad (5.42)
$$

Replacing $D[X](i)$ using Equation 5.40 and making some arrangements yield

$$\Psi[X](0) = \prod_{i \in A} \left(\sum_{i' \in A_i} X(i') + \sum_{j' \in (B \setminus A)_i} X(j') \right) \cdot \sum_{j \in (B \setminus A)} \left(\sum_{i' \in A_j} X(i') + \prod_{j' \in (B \setminus A)_j} X(j') \right). \qquad (5.43)$$

Expanding the two terms of $\prod_{i \in A}$ and $\sum_{j \in (B \setminus A)}$ gives

$$
\begin{aligned}
\Psi[X](0) = &\left[\left(\sum_{i' \in A_{-a}} X(i') + \prod_{j' \in (B \setminus A)_{-a}} X(j') \right) \cdots \left(\sum_{i' \in A_a} X(i') + \prod_{j' \in (B \setminus A)_a} X(j') \right) \right] \\
&\times \left[\left(\sum_{i' \in A_{-b}} X(i') + \prod_{j' \in (B \setminus A)_{-b}} X(j') \right) + \cdots + \left(\sum_{i' \in A_b} X(i') + \prod_{j' \in (B \setminus A)_b} X(j') \right) \right]
\end{aligned}
\qquad (5.44)
$$

For increasing filters, the idempotency can be characterized in terms of sum-of-products expressions. The minimal sum-of-products of $\Psi[X]$ containing $X(0)$ is separated from the terms that do not contain $X(0)$. Therefore, we have

$$\Psi[X](0) = X(0)\Psi_0[X] + \Psi_1[X]. \qquad (5.45)$$

When $k = \text{Card}(B \backslash A)$, the valley with the size $\text{Card}(A)$ will not be filled in fully. However, the other valley within $(B \backslash A)$ will be filled up after a soft morphological dilation is applied. After one iteration of a soft morphological closing, the valley with $(B \backslash A)$ will disappear, and the valley with the size $\text{Card}(A)$ remains the same. Applying a soft morphological dilation again, the valley with the size $\text{Card}(A)$ will be filled up according to Proposition 5.7. Thus, the idempotency will hold after two iterations.

Theorem 5.3: The soft morphological closing is idempotent when $k = \text{Card}(B \backslash A)$.

Proof: Applying Ψ a second time to Equation 5.45 yields

$$\Psi[\Psi[X]](0) = \Psi[X](0)\Psi_0[\Psi[X]] + \Psi_1[\Psi[X]]. \tag{5.46}$$

Idempotency takes the forms:

$$\Psi[X](0) = \Psi[X](0)\Psi_0[\Psi[X]] + \Psi_1[\Psi[X]]. \tag{5.47}$$

Let $\Psi[X](0)$ be p, $\Psi_0[\Psi[X]]$ be q, and $\Psi_1[\Psi[X]]$ be r. The logical identity of Equation 5.47 is $p = pq + r$. If we can show $r = p$, then $pq + r = pq + p = p$. When $r = 1$, we have

$$\Psi_1[\Psi[X]] = 1 \rightarrow \prod_{i \in (B \backslash A)_x} \Psi[X](i) = 1 \quad \text{for } x \in A. \tag{5.48}$$

If

$$\Psi[X](0) = 0 \rightarrow D[X](0) = 0, \tag{5.49}$$

and

$$\sum_{i \in A} D[X](i) = 0 \rightarrow \sum_{i \in (A \cup P)} X(i) = 0, \tag{5.50}$$

where $P = \{y \mid \text{some } y \in (B \backslash A)\}$. However, when Equation 5.50 is true, according to Propositions 5.5 and 5.6, there must be some i such that $\Psi[X](i) = 0$, where $i \in (B \backslash A)$, which is contradictory to Equation 5.48. It implies that $\Psi[X](0) = 1$. That is

$$\Psi_1[\Psi[X]] = 1 \rightarrow \Psi[X](0) = 1 \tag{5.51}$$

$$\rightarrow p = 1 \quad \text{and} \quad r = p \tag{5.52}$$

$$\rightarrow pq + r = p \tag{5.53}$$

$$\rightarrow \Psi[\Psi[X]](0) = \Psi[X](0). \tag{5.54}$$

(a)

0	0	0	0	0	0	0
0	1	2	2	1	1	0
0	2	4	5	7	9	0
0	2	3	3	2	2	0
0	1	2	1	2	1	0
0	4	4	5	5	4	0
0	0	0	0	0	0	0

(b)

0	0	0	0	0	0	0
0	1	2	2	1	1	0
0	2	4	5	7	7	0
0	2	3	3	2	2	0
0	1	2	2	2	1	0
0	4	4	5	5	4	0
0	0	0	0	0	0	0

(c)

0	0	1	1	1	0	0
0	1	2	2	1	1	0
1	2	4	5	7	7	1
1	2	3	4	5	2	1
1	1	4	4	4	1	1
0	4	4	5	5	4	0
0	0	4	4	4	0	0

FIGURE 5.9   (a) The image f, (b) $f \bullet [B, A_1, 8]$, and (c) $f \bullet [B, A_2, 4]$.

If $r = 0$ and $q \geq p$, then $pq = p$. When $\Psi_1[\Psi[X]](0)$, according to Proposition 5.8, the local maximum will be suppressed after a closing, we have

$$\sum_{i \in (B \backslash \{0\})} \Psi[X](i) \geq \Psi[X](0) \rightarrow \Psi_0[\Psi[X]] \geq \Psi[X](0) \quad (5.55)$$

$$\rightarrow q \geq p \quad (5.56)$$

$$\rightarrow pq = p \quad (5.57)$$

$$\rightarrow \Psi[\Psi[X]](0) = \Psi[X](0). \quad (5.58)$$

$\square$

Theorem 5.4: The soft morphological opening is idempotent when $k = \text{Card}(B \backslash A)$.

Examples are given to show that the soft morphological closing is idempotent when $k = \text{Card}(B \backslash A)$. The core sets A_1, A_2, and the structuring element set B are shown in Figure 5.6. The input image is shown in Figure 5.9a, and the results are shown in Figures 5.9b and c, which are the idempotent cases; that is, if the operation is applied again, the same results will be obtained. Figure 5.10 shows that when $k \neq \text{Card}(B \backslash A)$, the filter is not idempotent and its results are changed by repeated operations.

(a)

0	0	0	0	0	0	0
0	0	1	2	1	0	0
0	1	2	3	2	1	0
0	2	3	3	3	2	0
0	2	3	3	3	2	0
0	0	2	3	2	0	0
0	0	0	0	0	0	0

(b)

0	0	0	0	0	0	0
0	0	0	1	0	0	0
0	0	2	2	2	0	0
0	1	2	3	2	1	0
0	0	2	3	2	0	0
0	0	0	2	0	0	0
0	0	0	0	0	0	0

FIGURE 5.10   (a) $f \bullet [B, A_1, 4]$ and (b) $(f \bullet [B, A_1, 4]) \bullet [B, A_1, 4]$.

5.4 OSSM Filters

The disadvantage of soft morphological filters is that the repetition number is set up to be the rank order. This constraint limits the flexibility and applicability of the filters in noise removal and edge preservation. In this section we present the OSSM filters (OSSMFs), which provide a free order selection parameter. We also develop the properties and explore the restoration performance when a signal is corrupted by variant noises.

Many nonlinear image-processing techniques have their root in statistics. Their applications in images, image sequences, and color images with order statistics [David 1981] and their closely related morphological filters have by far been very successful [Maragos and Schafer 1987a, b; Stevenson and Arce 1987; Soille 2002]. The median filter (MF) was first introduced in statistics for smoothing economic time series, and was then used in image processing applications to remove noises with retaining sharp edges. The stacking property unifies all subclasses of stack filters, such as ranked order, median, weighted median, and weighted order-statistic filters.

Let f be a function defined on m-dimensional discrete Euclidean space Z^m. Let W be a window of a finite subset in Z^m which contains N points, where $N = \text{Card}(W)$, the *cardinality* of W. The kth order statistics (OS^k) of a function $f(x)$ with respect to the window W is a function whose value at location x is obtained by sorting in a descendent order the N values of $f(x)$ inside the window W whose origin is shifted to location x and picking up the kth number from the sorted list, where k ranges from 1 to N. Let W_x denote the translation of the origin of the set W to location x. The kth OS is represented by

$$OS^k \, (f, W)(x) = k\text{th largest of } \{f(a) \mid a \in W_x\}. \tag{5.59}$$

The OSSM operations, like in the soft morphology, adopt order statistics and use the split subsets, core set A and boundary set $B\backslash A$, as the structuring element. The obvious difference is that a free order selection parameter is introduced to increase flexibility and efficiency of the operations. That is, we select the lth order, rather than the kth order, from the sorted list of the total "$\text{Card}(B\backslash A) + k \times \text{Card}(A)$" values of the input function. The definitions of function by sets are given as follows, where f is a gray scale image and B is a flat structuring element.

Definition 5.18: The *OSSM dilation* of f by $[B, A, k, l]$ is defined as

$$(f \oplus [B, A, k, l])(x) = l\text{th largest of } (\{k \, \lozenge \, f(a) \mid a \in A_x\} \cup \{f(b) \mid b \in (B\backslash A)_x\}).$$

$$\tag{5.60}$$

Definition 5.19: The *OSSM erosion* of f by $[B, A, k, l]$ is defined as

$$(f \ominus [B, A, k, l])(x) = l\text{th smallest of } (\{k \lozenge f(a) \mid a \in A_x\} \cup \{f(b) \mid b \in (B \setminus A)_x\}).$$

(5.61)

One of the advantages of using the OSSM operators over standard morphological operators can be observed in the following example. Assume that we have a flat signal f corrupted by impulsive noise as shown in Figure 5.11a. To remove the impulsive noise, a standard morphological opening and a closing are applied to the distorted signal. Note that the positive impulse is removed by the opening operation and the negative impulse by closing. On the other hand, by applying a single soft opening and closing, the impulse noise can be removed. It is due to that the soft dilation can remove one part (positive or negative) of the noise and the soft erosion can remove the other. Figures 5.11b and c, respectively, show the results of a soft dilation and a soft erosion by $[(1, 1, \underline{1}, 1, 1), (\underline{1}), 3]$, where $B = (1, 1, \underline{1}, 1, 1)$ with underlined one indicating the origin, $A = (\underline{1})$, and $k = 3$. Therefore, four standard morphological operations or two soft morphological operations are needed to eliminate the impulse noise. The same goal of filtering can be achieved by using an OSSM dilation by $[(1, 1, \underline{1}, 1, 1), (\underline{1}), 3, 4]$. This leads to a significant reduction in computational complexity near a half amount.

Another important characteristic of the OSSMFs is the edge-preserving ability. Figures 5.12a and b, respectively, show a signal with upward and downward step edges. The edges of applying a standard morphological dilation or an erosion to the signal are shifted and are preserved if an opening or a closing is used. However, if an OSSM dilation by $[(1, 1, \underline{1}, 1, 1), (\underline{1}), 2, 3]$ is used, the location of edges is preserved.

A question is raised that whether there are constraints on the selection of parameter l. Under what circumstances can the OSSMFs get better performance in filtering than other filters? To answer these questions, a detailed

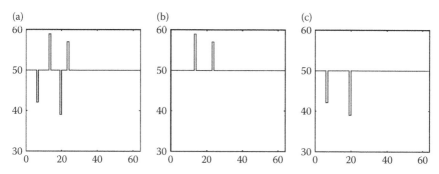

FIGURE 5.11 (a) A flat signal corrupted by positive and negative impulsive noises, and the filtered results by using (b) a soft dilation filter, and (c) a soft erosion filter.

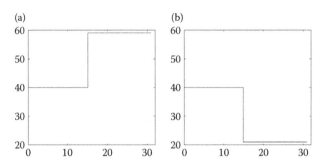

FIGURE 5.12 A signal with (a) upward step edge and (b) downward step edge.

filtering analysis is given below in one- and two-dimensional signals. We use the OSSM dilation in the illustration and the OSSM erosion can be similarly derived.

5.4.1 One-Dimensional Filtering Analysis

In this section, we concentrate on a symmetric structuring element. For non-symmetric structuring elements, the result can be derived accordingly. To simplify the representation, a notation $mBnA$ is used to denote a symmetric structuring element with m pixels in B and n pixels in A. For example, the structuring element $[-2, -1, \underline{0}, 1, 2]$ is denoted as $5B1A$ while $[-2, \underline{-1, 0, 1}, 2]$ is denoted as $5B3A$.

Let the structuring element be denoted as $sBhA = [b, \ldots, b, h, \ldots, \underline{h}, \ldots, h, b, \ldots, b]$, where h denotes the components of the hard center with underlined indicating the origin and b denotes the boundary components. Then there are $s = 2S + 2H + 1$ components in total in the structuring element and $h = 2H + 1$ components in the hard center, where $2S$ and $2H + 1$ denote the total number of components in the soft boundary and the hard center, respectively. Let the repetition value be k. The total number of components after repetition becomes $2S + (2H + 1)k$. We aim to apply an OSSM dilation to remove noises and preserve edges. The constraints for the selection of parameter l are described below.

1. For the positive impulse noise, when h is shifted to the noise location and the repetition of k times is used, l must be greater than k; that is, $l > k$, in order to avoid choosing the noise as output.

2. Similarly, for the negative impulse noise, we cannot choose the last k orders since the noise is repeated k times, that is, $l < 2Hk + 2S + 1$.

3. For the upward step edge, when h is shifted to the left point before edge transition, there are $kH + S$ positive numbers that we cannot select, that is, $l > kH + S$.

4. Similarly for the downward step edge, there are $kH + S$ smallest numbers that we cannot choose, that is, $l < (H + 1)k + S + 1$.

Let Z denote an integer. To satisfy the above four requirements, the range of l should be

$$\min[2(Hk + S) + 1, (H + 1)k + S + 1] > l > \max[k, kH + S], l \in Z \qquad (5.62)$$

Note that for $H > 0$, we always have the result, $kH + S \geq k$ and $2(Hk + S) + 1 \geq (H + 1)k + S + 1$, so Equation 5.62 is reduced to $(H + 1)k + S + 1 > l > kH + S$.

5.4.2 Two-Dimensional Filtering Analysis

The two-dimensional structuring element is divided with an arbitrarily directional symmetric line drawn through the origin into six parts, namely, b_r, b_c, b_l, h_r, h_c, and h_l, where subscripts l, r, and c denote the left side of, the right side of, and on the line, respectively. Let S and H represent the total number of components in the soft boundary and the hard center. There are "$k(H_r + H_c + H_l) + S_r + S_c + S_l$" components after repetition k times. To eliminate the impulse noise and preserve the step edge by an OSSM dilation, the constraints of l value are described below.

1. For the positive impulse noise, l must be greater than k; that is, $l > k$, since the repetition number of the noise is k when it is located at h_c.
2. For the negative impulse noise, we cannot choose the last k orders since the noise is repeated k times; that is, $l < k(H_r + H_c + H_l - 1) + S_r + S_c + S_l + 1$.
3. For the upward step edge, when the symmetric line is located at the left to the edge transition, there are $kH_r + S_r$ positive numbers that we cannot select; that is, $l > kH_r + S_r$.
4. For the downward step edge, there are $kH_r + S_r$ smallest numbers that we cannot choose; that is, $l < k(H_c + H_l) + S_c + S_l + 1$.

According to the above constraints, the range of value l is

$$\min \begin{bmatrix} k(H_r + H_c + H_l - 1) + S_r + S_c + S_l + 1, \\ k(H_c + H_l) + S_c + S_l + 1 \end{bmatrix} > l > \max \begin{bmatrix} kH_r + S_r, \\ k \end{bmatrix} \qquad (5.63)$$

When $k(H_r - 1) + S_r > 0$, it becomes

$$k(H_c + H_l) + S_c + S_l + 1 > l > \max[kH_r + S_r, \ k] \qquad (5.64)$$

An important problem in the two-dimensional case should be noted that since the symmetric line can have arbitrary directions, the dividing method is not unique. Thus, the range of l should be the intersection of all possible constraints. The following example demonstrates this analysis.

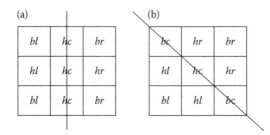

FIGURE 5.13 Two kinds of divisions of the structuring element (a) by a vertical symmetric line and (b) by a diagonal symmetric line.

Example 5.3: If we separate the structuring element of size 3×3, shown in Figure 5.13a, by a vertical symmetric line and let $k = 2$, then $S_r = S_l = 2$, $S_c = 0$, $H_r = H_l = 1$, and $H_c = 3$. According to Equation 5.64, the range for l is $11 > l > 4$. However, if we separate the structuring element by a diagonal symmetric line, shown in Figure 5.13b, then $S_r = S_l = 1$, $S_c = 2$, $H_r = H_l = 2$, and $H_c = 1$. According to Equation 5.64, the range for l is $10 > l > 5$. To satisfy both conditions, we must consider their intersection. Therefore, $10 > l > 5$. That is, $l = 6, 7, 8,$ or 9. Note that because the symmetric distribution of six parts exists for the cases of vertical and horizontal edges as well as of 45°- and 135°-diagonal edges, only vertical and 135°-diagonal edges are used in the analysis. In other cases of nonsymmetric distribution, each directional edge must be considered.

It can be easily observed that the positions of five hard-center components in the above example happen to be +-shaped; therefore, the range of parameter l in the diagonal case plays the dominant role. Alternatively, if the positions are changed to be ×-shaped, the range of l in the horizontal or vertical case will dominate.

5.4.3 Relationship between the OSSM Dilation and Erosion

According to the OSSM definitions, choosing l in an OSSM dilation is the same as choosing $T - l + 1$ in an OSSM erosion, where T represents the total number in the order list. That is, we have the following duality equation: $f \oplus [B, A, k, l] = f \ominus [B, A, k, T - l + 1]$. This relationship also applies in the two-dimensional case. Therefore, to reduce computational complexity, we can choose either an OSSM dilation or erosion depending upon which one has the smaller l value.

5.4.4 Properties of OSSM Filters and Relationship to Other Nonlinear Filters

There are some open questions. Whether do the OSSMFs possess the convergence property? Can the OSSMFs be implemented by threshold decomposition like soft morphological filters? What relationship is held between the

OSSMFs and other nonlinear filters, such as stack filters, weighted order statistic filters, and weighted median filters? In this section we will address these questions.

Definition 5.20: A filter that possesses the stacking property is called *a stack filter* (SF). It can be represented as a positive Boolean function (PBF).

Definition 5.21: A weighted order statistic filter (WOSF) is a stack filter that is based on threshold logic with nonnegative weights and a nonnegative threshold value.

Definition 5.22: A weighted median filter (WMF) is an extension of the MF which gives more weights to some values within the window. If weights of the MF are constrained to put at the center of the window, it is called *a center weighted median filter* (CWMF).

It is trivial to find that the relationship of the above filters is

$$SF \supset WOSF \supset WMF \supset CWMF \supset MF \qquad (5.65)$$

Note that the OSSMF is a subset of WOSF with the weight of the hard center being set to k and the weight of the soft boundary being 1. The soft morphological filter (SMF) is of course a subset of OSSMF. Therefore, the relationship is

$$SF \supset WOSF \supset OSSMF \supset SMF \text{ and } SF \supset WOSF \supset OSSMF \supset CWMF \quad (5.66)$$

A diagram showing the relationship between the OSSMF and the aforementioned filters is depicted in Figure 5.14. Furthermore, the properties between OSSMF and WMF are stated as follows.

Property 5.6: An OSSMF is reduced to a WMF whose weights are symmetric about the window center if the order is set to $(\sum_{i=1}^{2N+1} [w_i] + 1)/2$ for $\sum_{i=1}^{2N+1} [w_i]$ being odd. However, if the width of the hard center of OSSMF is set to 1, it is reduced to CWMF.

Property 5.7: An OSSMF is reduced to an OS filter if $k = 1$. Moreover, if l equals to the median value of the window width, it is reduced to the MF.

Property 5.8: An OSSMF is reduced to the standard SMF if $l = k$.

These properties can be easily obtained by observing the definitions. It is worth noting that since OSSMFs belong to WOSF and therefore belong to SF, all the convergence properties, deterministic properties, as well as statistic properties derived in [Wendt, Coyle, and Gallagher 1986; Yu and Liao 1994] are also valid for the OSSMFs. In soft morphology [Shih and Pu 1995], it is

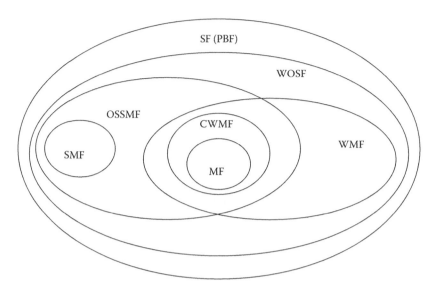

FIGURE 5.14   Relationship diagram of the OSSMF and other nonlinear filters.

important to implement the soft morphological filters by the threshold decomposition technique. Since all the filters, which can be expressed as positive Boolean functions, share the threshold decomposition property, the OSSMFs also possess the threshold decomposition property. An example is illustrated as follows.

Example 5.4: Let a 4-valued input f be [110232312] and it is undertaken an OSSM dilation by [5B3A, 2, 3]. The process by direct implementation and threshold decomposition are illustrated in Figure 5.15.

Moreover, because all the stack filters can be implemented by using the stacking property on very large-scale integration (VLSI), we can also implement the OSSMFs by these means, which also provides efficient performance.

Property 5.9: The OSSMFs are increasing. That is, for two functions f and g, when $f \leq g$, we have $OSSM(f) \leq OSSM(g)$.

Proof: To simplify the description, some notations are first introduced. Let X and Y represent two binary sets. Assume that B is a structuring element of size $2N + 1$. Using the concept introduced in Maragos and Schafer [1987b], an OSSM dilation on X with parameter B, k, and l can be described as

$$OSSM(X, B, A, k, l) = \{y \mid \mathrm{Card}(X \cap B_y) \geq l\} = OSSM(X) \qquad (5.67)$$

where B_y denotes the weighting value set that shifts B at location y. Let $X \subseteq Y$, then $z \in OSSM(X, B, A, k, l) \Leftrightarrow \mathrm{Card}(X \cap B_z) \geq l$. But, $X \cap B_z \subseteq Y \cap B_z$

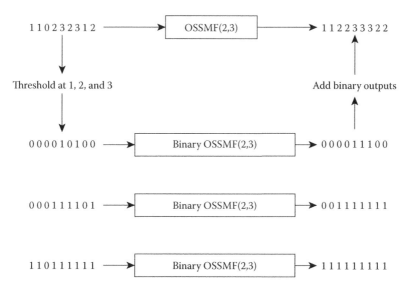

FIGURE 5.15 Threshold decomposition property of the OSSMFs.

$\Rightarrow \text{Card}(X \cap B_z) \leq \text{Card}(Y \cap B_z) \Rightarrow z \in OSSM(Y, B, A, k, l)$. Hence, $OSSM(X) \subseteq OSSM(Y)$. Now, if $f \leq g$, then $f(z) \leq g(z)$, $\forall z \in B_x$, $\forall x \in Z^m$. Thus, $OSSM(f, B, A, k, l)(x) \leq OSSM(g, B, A, k, l)(x)$, $\forall x$. □

Property 5.10: The OSSM dilation is extensive if $l \leq k$, and the origin of the structuring element is contained in the core set A. That is, for any function f, we have $f \leq OSSM(f)$.

Proof: This property can be proved by intuition as follows. Since an OSSM dilation on f at location x, $OSSM(f, B, A, k, l)(x)$, is to choose the lth largest value from f within domain B_x, we can write the descending sort list as $[f_U, k \Diamond f(x), f_L]$, where f_U and f_L represent upper [greater than $f(x)$] and lower parts, respectively. It is found that as long as $l \leq k$ is satisfied, either a greater value in f_U or $f(x)$ will be picked out. Thus, we always have $f \leq OSSM(f)$, $\forall x$. □

Property 5.11: The OSSM erosion is antiextensive if $l \leq k$, and the origin of the structuring element is contained in the core set A. That is, for any function f, we have $f \geq OSSM(f)$.

This proof is obtained immediately when the duality relationship is used.

Property 5.12: The idempotent OSSMFs exist.

Although the exact expression of the idempotent OSSMFs is unknown, the idempotent OSSMFs do exist. For example, let the structuring element be a center weighted structure with size $2N + 1$, and let $k = 2N - 1$ and $l = 2N$, then the OSSMF is idempotent.

Example 5.5: Let the structuring element be $5B1A$. This implies that $N = 2$. Let an input sequence $f = \{3, 1, 5, 7, 9, 1, 3, 5, 7, 2, 4, 6, 8\}$, $k = 3$, and $l = 4$, then the first and second iterations of the OSSM filtered output, denoted as f^1 and f^2, are $f^1 = \{3, 3, 5, 7, 7, 3, 3, 5, 5, 4, 4, 6, 8\}$ and $f^2 = \{3, 3, 5, 7, 7, 3, 3, 5, 5, 4, 4, 6, 8\}$, respectively. This result leads to an idempotent filter.

Properties 5.9 through 5.11 deal with some well-known properties that are owned by the morphological operators. To discuss the convergence property of the OSSM operator, the classification of stack filters stated in Yu and Liao [1994] is used. Assume that B is a structuring element of size $2N + 1$ and B_i represents the repetition time of each structuring element component. The convergence behaviors are analyzed as follows.

Property 5.13: The OSSM operator possesses the convergence property if it is either extensive or anti-extensive.

Proof: It is observed that since $B_{N+1} < \sum_{i=1, i \neq N+1}^{2N+1} B_i$, an extensive (anti-extensive) OSSM operator corresponds to a type-II (type-I) stack filter according to Yu and Liao [1994]. Thus, it possesses the convergence property.   □

5.4.5 Experimental Results

Figure 5.16 shows a one-dimensional signal and its corrupted versions with pepper-salt, Gaussian, and uniform distribution noises. The filters such as median, OS, SM, and OSSM filters with different parameters are used to process the signals and the results are shown in Figures 5.17 through 5.19. It should be noted that those curves with $k = 1$ correspond to the OS filter and center points of the curves correspond to the median filter. From these figures it is observed that for a signal corrupted by Gaussian or uniform distributed noise, the MF performs the restoration the best no matter what size the soft boundary and hard center are. If the size difference between hard center and soft boundary is small, the performance for each k-value curve will near each other. It tells us that we do not need to choose a higher k for computational complexity consideration since there exists little improvement for a larger k. Moreover, for each curve it can be predicted that the WMF always performs the best. If the noise is pepper-salt alike, the OSSMF performs better restoration than other WOSF.

Figure 5.20 shows a multi-pepper and its corrupted versions with pepper-salt noise with occurrence probability 0.1 and standard deviation 50, and Gaussian and uniform noises with the same standard deviation. They are processed by the aforementioned filters to test their performance and the results are shown in Figures 5.21 through 5.23. In the resultant figures it is noted that the variation of performance for each kind of filters is not as obvious as in one-dimension,

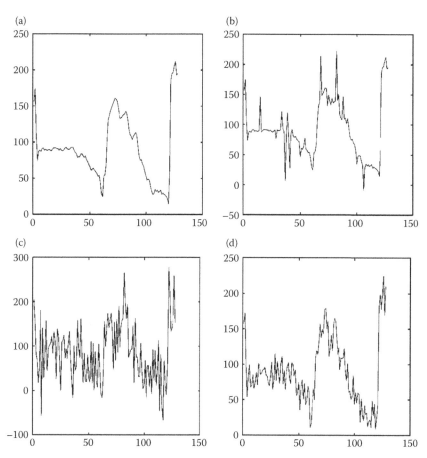

FIGURE 5.16 (a) A one-dimensional original signal and its corrupted versions by (b) pepper-salt (occurrence probability 0.2 and standard deviation 50), (c) Gaussian (mean 0 and standard deviation 50), and (d) uniform distribution noises (mean 0 and standard deviation 50).

but the observation stated in the one-dimensional case still holds. Moreover, the size of soft boundary, hard center, and the division ways do not significantly influence the performance. Thus, the OSSMFs offer more flexibility in dealing with different application requirements and with near the same performance as some popular WOS filters. Two mean square error (MSE) performance tables, Tables 5.3 and 5.4, are given to state the experimental results.

To observe the performance in edge preservation, a test method stated in You and Crebbin [1995] is adopted and is described below. An edge image of 128×128 with contrast 100 is corrupted by pepper-salt, Gaussian, and uniform distributed noises with standard deviation 20. Both the transition rate and MSE are computed after the images are filtered by the OSSMFs and other filters, and the results are shown in Figures 5.24 through 5.26 and in Table 5.5. It is observed that the OSSMFs have better performance in edge preserving than

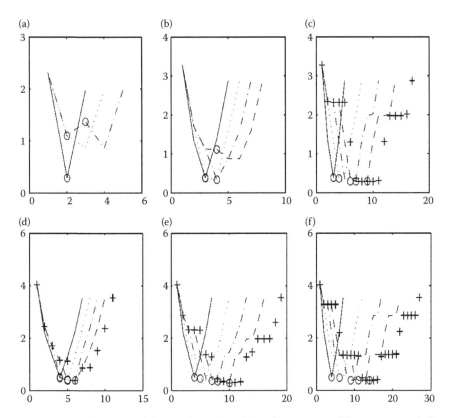

FIGURE 5.17 MSE curves of the one-dimensional signal reconstructed from pepper-salt distributed noise by different structuring elements. (a) 3B1A, (b) 5B1A, (c) 5B3A, (d) 7B1A, (e) 7B3A, and (f) 7B5A. Note that the horizontal axis denotes l value and the vertical axis denotes mean square error. Solid line, ···, —·—, – –, and + represent $k = 1$ (OSF), 2, 3, 4, and 5, respectively, and o represents MF.

other filters for the above three kinds of noises. A tradeoff between the edge preservation and the MSE should be considered.

Generally speaking, an OSSMF with $k = 2$ or 3, $l = TT + 1$, where TT represents the minimum value that satisfies the constraint, and a suitably sized structuring element (depending on applications) will have near the best performance and the fewest computational load.

5.4.6 Extensive Applications

It is noted that the impulse noise does not always occupy only one pixel but sometimes with several pixels in width. The proposed algorithm can be extended to solve this problem and delete the multiple-pixel impulse noise successfully. Another application is dedicated to changing the ramp edges into step edges for the purpose of enhancing the edges. For simplicity of

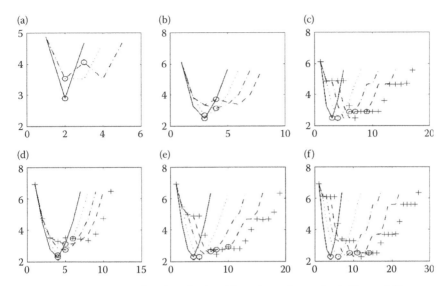

FIGURE 5.18 For Gaussian distributed noise. Also refer to the caption in Figure 5.17.

illustration, one-dimensional signals are used. The two-dimensional case can be similarly derived.

First, we deal with the multiple-pixel impulse deletion problem. The width of the impulse noise to be deleted depends on the cardinality of the structuring element. Let a structuring element have the same definition as described

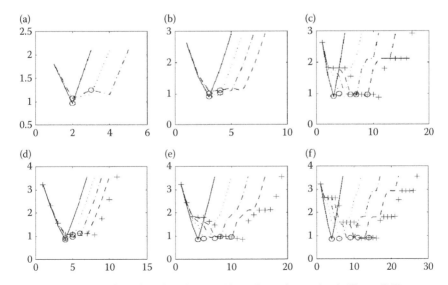

FIGURE 5.19 For uniform distributed noise. Also refer to the caption in Figure 5.17.

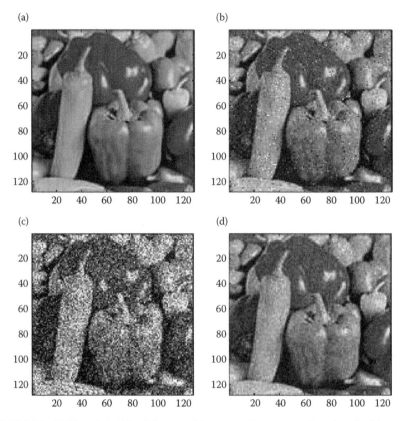

FIGURE 5.20 (a) The original "multi-pepper" image, and its corrupted versions by (b) pepper-salt (occurrence probability 0.2 and standard deviation 50), (c) Gaussian (mean 0 and standard deviation 50), and (d) uniform distribution noises (mean 0 and standard deviation 50).

in the previous section and assume that the number of noisy points in n and $n \le S + H + 1$. We can separate the noise into two parts, one located at the hard center with p pixels and the other located at the boundary with q pixels. The constraints for selecting the suitable order of l parameter are as follows.

1. To delete the positive impulse noise, it must be $l > pk + q$.
2. To delete the negative impulse noise, l should be in the range of $l < k(2H + 1 - p) + 2S - q + 1$.
3. To preserve the edges, the constraints are the same as in the previous description.

To satisfy all the constraints we have

$$\min \begin{bmatrix} k(H + 1) + S + 1, \\ k(2H + 1 - p) + 2S - q + 1 \end{bmatrix} > l > \max \begin{bmatrix} pk + q, \\ kH + S \end{bmatrix} \tag{5.68}$$

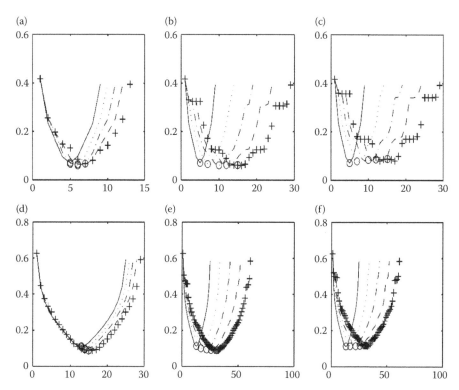

FIGURE 5.21 MSE curves of the two-dimensional signal reconstructed from pepper-salt distributed noise by different structuring elements. (a) 3×3 shape with one pixel hard center, (b) 3×3 with +-shaped hard center, (c) 3×3 with x-shaped hard center, (d) 5×5 shape with 1 pixel hard center, (e) 5×5 with +-shaped hard center, and (f) 5×5 with x-shaped hard center. Note that the horizontal axis denotes l value and the vertical axis denotes MSE. Solid line, $\cdots$, $-\cdot-$, $-\ -$, and + represent $k = 1$ (OSF), 2, 3, 4, and 5, respectively, and o represents MF.

It is worth noting that not all the problems of multiple-pixel noise deletion have a solution for parameter l. Here we give an example to depict such a situation.

Example 5.6: Let a structuring element be 5B3A and $k = 2$. The width of noise we want to remove equals 2, so we have $S = 1$, $H = 1$, $p = 2$, and $q = 0$. According to Equation 5.68, we have $5 > l > 4$. There is no solution in this case. However, if the structuring element is changed to 7B3A, the solution with range $7 > l > 4$ can be obtained. Figure 5.27 shows the results of a man-made signal with one, two, and multiple pixels impulse noise after processed by two OSSMFs with 5B3A and 7B3A.

For some signal processing applications, it is important to enhance edges of the signal for feature detection or visual perception. Due to the flexibility

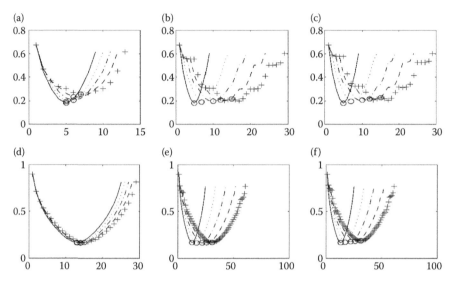

FIGURE 5.22   For Gaussian distributed noise. Also refer to the caption in Figure 5.21.

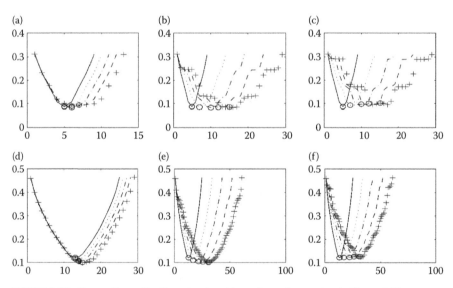

FIGURE 5.23   For uniform distributed noise. Also refer to the caption in Figure 5.21.

of OSSMFs, the ramp edges can be transformed to step edges to achieve this goal. Let a structuring element be $5B1A$ and $k = 1$ for simplicity. An example of one-dimensional ramp edge enhancement is shown in Figure 5.28 and described below.

TABLE 5.3

MSE of Different Nonlinear Filtering Results

		Uniform		Gaussian		Pepper-Salt	
3B1A	MF	0.966		2.8958		0.2899	
	$k = 2$	—	—	—	—	—	—
	$k = 3$	—	—	—	—	—	—
5B1A	MF	0.9092		2.4889		0.3849	
	$k = 2$	1.0071+	0.9357*	2.6974+	2.6974*	0.3766+	0.3468*
	$k = 3$	1.031+	1.031*	3.1368+	3.1368*	0.3374+	0.3374*
5B3A	MF	0.9092		2.4889		0.3849	
	$k = 2$	0.990+	0.8831*	2.4805+	2.4805*	0.368+	0.3111*
	$k = 3$	0.990+	0.8831*	2.4805+	2.4805*	0.368+	0.2899*
7B1A	MF	0.8545		2.2557		0.505	
	$k = 2$	0.9276+	0.9009*	2.3886+	2.6075*	0.4779+	0.4213*
	$k = 3$	1.0376+	0.964*	2.8006+	2.7233*	0.5035+	0.3884*
7B3A	MF	0.8545		2.2557		0.505	
	$k = 2$	0.8912+	0.8691*	2.2649+	2.5911*	0.4588+	0.3893*
	$k = 3$	0.9675+	0.8551*	2.4085+	2.4085*	0.445+	0.2702*
7B5A	MF	0.8545		2.2557		0.505	
	$k = 2$	0.8711+	0.8711*	2.2494+	2.2494*	0.4848+	0.4101*
	$k = 3$	0.8711+	0.8711*	2.2494+	2.2494*	0.4848+	0.3849*

Note: + represents the MSE results with minimal satisfactory *l* and * represents the minimal MSE results.

For each point $f(i)$ in the signal, its tangent Ta is first determined by $f(i) - f(i - 1)$, then l is chosen according to the tangent value. If Ta is greater than a threshold value T, l is set to 4. If Ta is smaller than $-T$, l is set to 2. Those cases with Ta ranging between $-T$ and T will set l to 3. The threshold value T controls the sensitivity of the ramp edge detection and varies with applications. The idea of RSMFs [Shih and Puttagunta 1995] is applied here. That is, the filtering output of current point depends on the previously filtered points. Figure 5.29 shows the result of a blurred Lena image and its filtered result by using the edge enhancement method. It is noted that the fuzzy edges are now enhanced with clear boundary.

5.5 RSM Filters (RSMFs)

The structuring element in morphological filters can be regarded as a template, which is translated to each pixel location in an image. These filters can be implemented in parallel because each pixel's value in the transformed

TABLE 5.4

MSE of Different Two-Dimensional Nonlinear Filtering Results

		Uniform		Gaussian		Pepper-Salt	
3 × 3 "·" shape	MF	0.0873		0.1806		0.0701	
	$k = 2$	0.1114+	0.0853*	0.2442+	0.1925*	0.0963+	0.0649*
	$k = 3$	0.1167+	0.084*	0.281+	0.2042*	0.1104+	0.0596*
3 × 3 "+" shape	MF	0.0873		0.1805		0.0701	
	$k = 2$	0.104+	0.0848*	0.2193+	0.1901*	0.0883+	0.0649*
	$k = 3$	0.1038+	0.0834*	0.2199+	0.1978*	0.0881+	0.0603*
3 × 3 "×" shape	MF	0.0873		0.1805		0.0701	
	$k = 2$	0.1089+	0.0921*	0.2234+	0.1948*	0.0946+	0.077*
	$k = 3$	0.1092+	0.0948*	0.2243+	0.2063*	0.0747+	0.0697*
5 × 5 "·" shape	MF	0.121		0.1692		0.1134	
	$k = 2$	0.1802+	0.1145*	0.2602+	0.1666*	0.1752+	0.1061*
	$k = 3$	0.1808+	0.1104*	0.2705+	0.1646*	0.1753+	0.0992*
5 × 5 "+" shape	MF	0.1256		0.1692		0.1134	
	$k = 2$	0.1224+	0.1101*	0.1834+	0.1648*	0.1148+	0.1068*
	$k = 3$	0.1169+	0.1047*	0.1824+	0.1655*	0.1079+	0.0954*
5 × 5 "×" shape	MF	0.121		0.1692		0.1134	
	$k = 2$	0.1361+	0.1211*	0.1924+	0.1734*	0.1286+	0.113*
	$k = 3$	0.1361+	0.1223*	0.1957+	0.1785*	0.1279+	0.1129*

Note:　+ represents the MSE results with minimal satisfactory l and * represents the minimal MSE results.

image is only a function of its neighboring pixels in the given image. Also, the sequence in which the pixels are processed is completely irrelevant. Thus, these parallel image operations can be applied to each pixel simultaneously if a suitable parallel architecture is available. Parallel image transformations are also referred to as *non-recursive transformations*.

In contrast to non-recursive transformations, a class of *recursive transformations* is also widely used in signal and image processing, for example, sequential block labeling, predictive coding, and adaptive dithering [Serra 1982]. The main distinction between these two classes of transformations is that in the recursive transformations, the pixel's value of the transformed image depends upon the pixel's values of both the input image and the transformed image itself. Due to this reason, some partial order has to be imposed on the underlying image domain, so that the transformed image can be computed recursively according to this imposed partial order. In other words, a pixel's value of the transformed image may not be processed until all the pixels preceding it have been processed.

Recursive filters are the filters, which use previously filtered outputs as their inputs. Let x_i and y_i denote the input and output values at location i,

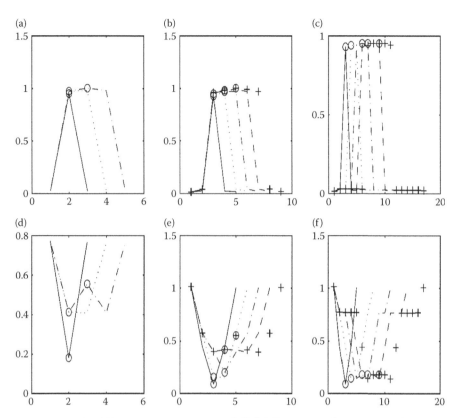

FIGURE 5.24 (a–c) Transition rate curves, and (d–f) MSE curves of a two-step edge reconstructed from pepper-salt noise (occurrence probability 0.2 and standard deviation 50) by different structuring elements. (a)(d) 3B1A, (b)(e) 5B1A, and (c)(f) 5B3A. Note that the horizontal axis denotes *l* value and the vertical axis denotes mean square error. Solid line, ···, –·–, – –, and + represent $k = 1$ (OSF), 2, 3, 4, and 5, respectively, and o represents MF.

respectively, where $i = \{0, 1, \ldots, N - 1\}$. Let the domain of the structuring element be $\langle -L, \ldots, -1, 0, 1, \ldots, R \rangle$, where L is the left margin and R is the right margin. Hence, the structuring element has the size of $L + R + 1$. Start up and end effects are accounted for by appending L samples to the beginning and R samples to the end of the signal sequence. The L appended samples are given the value of the first signal sample; similarly, the R appended samples receive the value of the last sample of the signal.

Definition 5.23: The recursive counterpart of a non-recursive filter Ψ given by

$$y_i = \Psi(x_{i-L}, \ldots, x_{i-1}, x_i, x_{i+1}, \ldots, x_{i+R}) \tag{5.69}$$

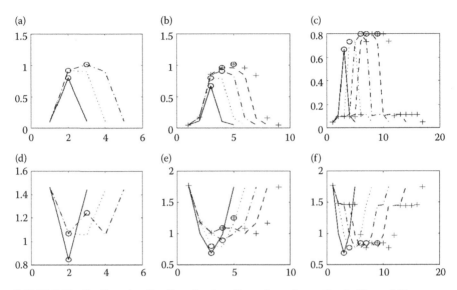

FIGURE 5.25   For Gaussian distributed noise. Also refer to the caption in Figure 5.24.

is defined as

$$y_i = \Psi(y_{i-L}, \ldots, y_{i-1}, x_i, x_{i+1}, \ldots, x_{i+R}) \tag{5.70}$$

by assuming that the values of $y_{i-L}, \ldots, y_{i-1}$ are already given.

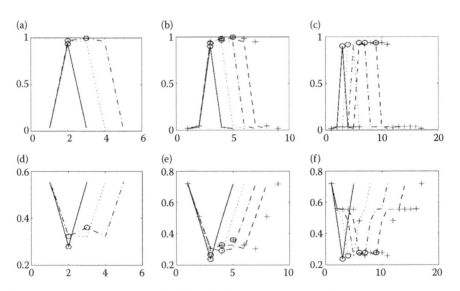

FIGURE 5.26   For uniform distributed noise. Also refer to the caption in Figure 5.24.

TABLE 5.5

Transition Rate Results of Edge Preservation Performance
by using Three Different Structuring Elements

		Uniform	Gaussian	Pepper-Salt
3*B*1*A*	MF	0.9363	0.7983	0.9523
	$k = 2$	—	—	—
	$k = 3$	—	—	—
5*B*14	MF	0.9011	0.6695	0.9307
	$k = 2$	0.9352	0.7927	0.9514
	$k = 3$	0.9676	0.9051	0.971
5*B*3*A*	MF	0.9011	0.6695	0.9307
	$k = 2$	0.92	0.7357	0.9426
	$k = 3$	0.9363	0.7983	0.9523

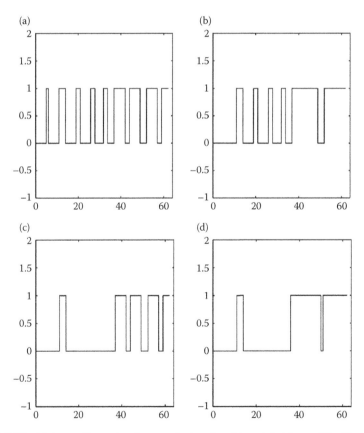

FIGURE 5.27   (a) A one-dimensional upward edge signal translated by $x = 37$ with one, two, and multiple-pixel impulse noises, and the filtered results by (b) 5*B*3*A* with $k = 2$ and $l = 4$, (c) 5*B*3*A* with $k = 2$ and $l = 5$, and (d) 7*B*3*A* with $k = 2$ and $l = 5$. Note that only the filter (d) can remove the multiple-pixel impulse noises.

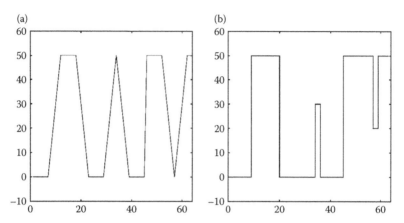

FIGURE 5.28 A one-dimensional edge enhancement example. (a) A ramp edge and (b) its filtered output by using an OSSM filter.

Example 5.7: Let $B = \langle -1,0,1 \rangle$, $A = \langle 0 \rangle$, and $k = 2$. Let the input signal $f = \{4\,7\,2\,9\,6\,8\,5\,4\,7\}$. We have

$$f \oplus_r [B, A, 2] = \{4\,7\,7\,9\,8\,8\,5\,5\,7\},$$
$$f \ominus_r [B, A, 2] = \{4\,4\,2\,6\,6\,6\,5\,4\,7\},$$

where $\oplus_r$ and $\ominus_r$ denote the recursive soft morphological dilation and erosion, respectively.

In Figure 5.30a shows a 125×125 Lena image, and Figure 5.30b shows the same image but with added Gaussian noise having zero mean and standard

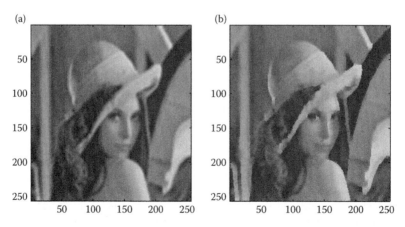

FIGURE 5.29 A two-dimensional image edge enhancement example. (a) Original image and (b) filtered image.

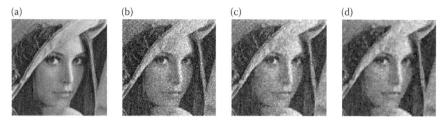

FIGURE 5.30   (a) An original image, (b) image corrupted by Gaussian noise, (c) result of a recursive soft morphological dilation, (d) followed by a recursive soft morphological erosion (i.e., closing).

deviation 20. The mean-square signal-to-noise ratio (*SNR*) is used. Figure 5.30b has $SNR = 40.83$. Figure 5.30c shows the result of applying a recursive soft morphological dilation with $B = \langle (0, -1), (-1, 0), (0, 0), (1, 0), (0, 1) \rangle$, $A = \langle (0, 0) \rangle$, and $k = 3$, where $SNR = 64.65$. Figure 5.30d shows the result of applying a recursive soft morphological erosion on Figure 5.30c (i.e., closing) with the same structuring elements and rank order, where $SNR = 101.45$.

In Figure 5.31a shows a 512×432 grayscale image, and Figure 5.31b shows the same image but with added Gaussian noise having zero mean and standard deviation 20. Figure 5.31b has $SNR = 35.54$. Figure 5.31c shows the result of applying a standard morphological closing with a structuring element of size 3×3, where $SNR = 25.15$. Figure 5.31d shows the result of applying a recursive soft morphological closing with $B = \langle (0, -1), (-1, 0), (0, 0), (1, 0), (0, 1) \rangle$, $A = \langle (0, 0) \rangle$, and $k = 3$, where $SNR = 69.41$.

5.5.1 Properties of RSMFs

We prove that RSMFs are increasing by first proving that if a SMF is increasing, and then its recursive counterpart will also increase.

Theorem 5.5: If a SMF is increasing, then the recursive soft morphological filter will also increase.

Proof: Let 2 input signals be $f(x) = \{x_0, x_1, \ldots, x_{N-1}\}$ and $g(x') = \{x'_0, x'_1, \ldots, x'_{N-1}\}$ and have the ordering relation of $f(x) \le g(x')$ for every x and x'. We need to prove that for every y_i and y'_i,

$$y_i = \Psi(y_{i-L}, \ldots, y_{i-1}, x_i, x_{i+1}, \ldots, x_{i+R})$$
$$\le \Psi(y'_{i-L}, \ldots, y'_{i-1}, x'_i, x'_{i+1}, \ldots, x'_{i+R}) = y'_i.$$

That is $y_i \le y'_i$ for every i. We prove the theorem by induction. Since the recursive and non-recursive filters of y_0 and y'_0 only depend upon the values of the input pixels $\{x_0, \ldots, x_R\}$ and $\{x'_0, \ldots, x'_R\}$, respectively, which have the ordering

FIGURE 5.31 (a) An original image, (b) image corrupted by Gaussian noise, (c) result of a standard morphological closing, (d) result of a recursive soft morphological closing.

of $\Psi(f(x)) \leq \Psi(g(x'))$, the initial condition of $i = 0$ is satisfied. Assume that the condition is true for $i = L - 1$. That is, the output values at locations $\{0, \ldots, L - 1\}$ satisfy the condition. For $i = L$, we have

$$y_L = \Psi(y_0, \ldots, y_{L-1}, x_L, x_{L+1}, \ldots, x_{L+R})$$
$$\leq \Psi(y'_0, \ldots, y'_{L-1}, x'_L, x'_{L+1}, \ldots, x'_{L+R}) = y'_L$$

simply because the output values at locations $\{0, \ldots, L - 1\}$ and the input values at locations $\{L, \ldots, L + R\}$ both satisfy the condition. This implies that the recursive counterpart of Ψ also has the increasing property.   □

Based on the above theorem and the fact that the soft morphological dilation and erosion are increasing, we conclude that the recursive soft morphological dilation and erosion are also increasing.

Theorem 5.6: Recursive soft morphological dilation and erosion are increasing.

Theorem 5.7: Recursive soft morphological dilation is extensive, and recursive soft morphological erosion is anti-extensive.

Proof: According to the definition of soft morphological dilation, the values of the multiset $\{k \lozenge f(a) \mid a \in A_x\} \cup \{f(b) \mid b \in (B \backslash A)_x\}$ are sorted in a descendent order and the kth largest is selected. If $f(x)$ is the maximum value in the set B, it is selected as the output after the repetition k times. If $f(x)$ is not the maximum value in B, the selected kth largest value must be greater than or equal to $f(x)$ after $f(x)$ is repeated k times. This implies that for every x, soft morphological dilation is extensive. We can similarly derive that the recursive soft morphological erosion is anti-extensive. □

From the definition of recursive soft morphological dilation and the properties of scalar multiplication, the result obtained by multiplying the dilated value of the input signal by a positive constant is equal to the result obtained by initially performing a recursive soft morphological dilation and then multiplying it by a positive number. The proof is straight-forward and therefore is skipped.

Property 5.14: RSMFs are scaling-invariant.

5.5.2 Idempotent RSMFs

An idempotent filter maps an arbitrary input signal into an associated set of root sequences. Each of these root signals is invariant to additional filter passes. The standard morphological opening and closing have the property of idempotency. In contrast, recursive soft morphological opening and closing are not idempotent in general. However, when the filter is designed in a specific way, recursive soft morphological opening and closing are also idempotent. We will present the idempotent RSMFs in one dimension for simplicity. Two or more dimensional filters can be similarly derived. If $B = \langle -n, (-n+1), \ldots, -1, 0, 1, \ldots, (n-1), n \rangle$ and $A = \langle (-n+1), \ldots, -1, 0, 1, \ldots, (n-1) \rangle$, where $n \geq 1$, then we denote the structuring element $[B, A, k]$ by $[n, n-1, k]$.

Property 5.15: Recursive soft morphological dilation and erosion are idempotent for the structuring element: B of length three, A the central point, and $k = 2$. That is the structuring element is $[1, 0, 2]$.

Proof: Let $f(x) = \{a\ b\ c\ d\ e\}$, where a, b, c, d, and e are arbitrary numbers. The permutation number of the five variables is $5! = 120$. Applying a recursive soft morphological dilation on f by $[1, 0, 2]$ gives the same result as we apply the dilation again on the result. The same property holds true for any size of f. Similarly, the procedures can be applied to the recursive soft morphological erosion and yield the same property. □

The proof of Property 5.16 can be similarly derived and is therefore skipped.

Property 5.16: Recursive soft morphological closing and opening by $[n, n–1, k]$ are idempotent, where $k = 1$ or 2.

Corollary 5.1: If the kth largest (smallest) value selected during the scanning of the first pixel of the input signal for a recursive soft morphological dilation (erosion) happens to be the maximum (minimum) positive impulse in the input signal, then the filter is idempotent.

5.5.3 Cascaded RSMFs

Cascaded weighted median filters were introduced by Yli-Harja, Astola, and Neuvo [1991] in which several interesting properties of weighted cascaded filters were discussed. We now present some properties of cascaded RSMFs. By cascade connection of filters F and G, we mean that the original input signal f is filtered by the filter F to produce an intermediate signal g. Then g is filtered by the filter G to produce the output signal h. Cascaded filters F and G can also be presented as a single filter H which produces the output h directly from the input f. We now discuss the properties of cascaded RSMFs.

Property 5.17: The cascaded RSMFs are not commutative.

Property 5.18: The cascaded RSMFs are associative.

Note that the output of RSMFs depends upon two factors, namely, the order index k and the length of the hard center. Although the technique to combine any two cascaded filters F and G into a single filter H is the same, no direct formula can be given to produce the combined filter H for different cascade combinations. Due to many variations, we can only present the essential idea in constructing the combined filter and illustrate it on some examples.

Given two filters $[B_1, A_1, k]$ and $[B_2, A_2, k]$ of length five and a hard center of length three, where $k \geq 2$. Let $B_1 = B_2 = \langle -2 \ -1 \ 0 \ 1 \ 2 \rangle$ and $A_1 = A_2 = \langle -1 \ 0 \ 1 \rangle$. Denote the input signal f to be $\{x_0, x_1, \ldots, x_{N-1}\}$, the output of the first filter to be $\{y_0, y_1, \ldots, y_{N-1}\}$, and the output of the second filter to be $\{z_0, z_1, \ldots, z_{N-1}\}$. Since $k \geq 2$, we have $y_0 = k$th largest of multiset $\{k \lozenge x_0, k \lozenge x_1\}$, where the multiset need not consider x_2. It means that y_0 depends upon (x_0, x_1). Again, $y_1 = k$th largest of multiset $\{k \lozenge y_0, k \lozenge x_1, k \lozenge x_2\}$. It means that y_1 depends upon (y_0, x_1, x_2). Proceeding in this way, we obtain y_i to be $y_1 = k$th largest of multiset $\{k \lozenge y_{i-1}, k \lozenge x_i, k \lozenge x_{i+2}\}$. It means that y_1 depends upon (y_{i-1}, x_i, x_{i+1}).

According to the definition of cascaded filters, the intermediate output $\{y_0, y_1, \ldots, y_i, \ldots, y_{N-1}\}$ is used as the input further processed by the filter $[B_2, A_2, k]$. Similarly, we obtain the final output z_i to be $z_i = k$th largest of multiset $\{z_{i-1}, y_i, y_{i+1}\}$. It means that z_i depends upon $(z_{i-1}, x_i, x_{i+1}, x_{i+2})$. The above derivation states that the result of the cascade combination of two aforementioned filters can be equivalently obtained by a single recursive standard morphological filter of $\langle -1 \ 0 \ 1 \ 2 \rangle$. Other examples of reducing the cascaded filters into a single recursive standard morphological filter can be similarly derived.

TABLE 5.6

Different Cascade Combination and Their Combined Filters (for $k \geq 2$)

B_1	B_2	A_1	A_2	Combined Filter
$\langle -1\,0\,1 \rangle$	$\langle -1\,0\,1 \rangle$	$\langle 0 \rangle$	$\langle -1\,0 \rangle$	$\langle -1\,0 \rangle$
$\langle -2\,-1\,0 \rangle$	$\langle -1\,0\,1 \rangle$	$\langle -1\,0 \rangle$	$\langle 0 \rangle$	$\langle -2\,-1\,0 \rangle$
$\langle -1\,0\,1 \rangle$	$\langle -2\,-1\,0\,1\,2 \rangle$	$\langle 0 \rangle$	$\langle -1\,0\,1 \rangle$	$\langle -1\,0\,1 \rangle$
$\langle -2\,-1\,0 \rangle$	$\langle -2\,-1\,0\,1\,2 \rangle$	$\langle -1\,0 \rangle$	$\langle -1\,0\,1 \rangle$	$\langle -1\,0\,1 \rangle$
$\langle -1\,0\,1 \rangle$	$\langle -2\,-1\,0\,1\,2 \rangle$	$\langle -1\,0 \rangle$	$\langle -1\,0\,1 \rangle$	$\langle -1\,0\,1 \rangle$
$\langle -2\,-1\,0\,1\,2 \rangle$	$\langle -2\,-1\,0\,1\,2 \rangle$	$\langle -1\,0\,1 \rangle$	$\langle -1\,0\,1 \rangle$	$\langle -1\,0\,1\,2 \rangle$

Note that the number of computations required is also significantly reduced when we use the combined filter obtained by the above result. For example, if we have two filters of size 5, each having a hard center of length 3 and the order index k, then the number of computations required for an input signal of length N is equal to $Nk(5k + 3)$, whereas according to Table 5.6 in the case of the combined filter of size 4, we have only $3N$.

The following property provides that when $k = 1$, the cascaded filters are reduced to be equivalent to standard morphological filters.

Property 5.19: Let the cascade combination of two filters of any lengths be denoted by $B_1 = \langle -L_1 \cdots -1\ 0\ 1 \cdots R_1 \rangle$, $B_2 = \langle -L_2 \cdots -1\ 0\ 1 \cdots R_2 \rangle$, and any A_1 and A_2, and the order index $k = 1$. Their combined recursive standard morphological filter will be $\langle -1\ 0\ 1 \cdots (R_1 + R_2) \rangle$.

5.6 ROSSM Filters

A class of ROSSM filters is introduced and their properties are described in this section. It has been shown through experimental results that the ROSSM filters, compared to the OSSMFs or other nonlinear filters, have better outcomes in signal reconstruction [Pei, Lai, and Shih 1998].

To define the recursive OSSMFs, let Φ denote an ROSSM operation and $mBnA$ be the symmetric structuring element sets with m pixels in B and n pixels in A. For example, the structuring element $[-2, -1, \underline{0}, 1, 2]$ is denoted as $5B1A$ while $[-2, \underline{-1, 0, 1,} 2]$ is denoted as $5B3A$, where the center of the underlined denotes the center of symmetry. Then, applying a ROSSM filter on input f at index i can be written as

$$y_i = \Phi(y_{i-L}, \ldots, y_{i-1}, y_i, y_{i+1}, \ldots, y_{i+R}) \qquad (5.71)$$

where $y_{i-L}, \ldots, y_{i-1}$ are already obtained by $\Phi(f_{i-L}), \ldots, \Phi(f_{i-1})$.

The two-dimensional ROSSM operation on an image F with a structuring element S is represented as

$$Y_{ij} = \Phi(Y_p, F_{ij}, F_q), \tag{5.72}$$

where $S = (S_B, S_A)$ is a two-dimensional structuring element with core set S_A and boundary set $S_B \backslash S_A$ (Note that $S_A \subseteq S_B$), and p and q are domain index sets that precede and succeed the index (i, j) in a scanning order (see Figure 5.32).

An example is given below to show the difference between the OSSM and ROSSM filters.

Example 5.8: Let the structuring element be $5B1A$. Let $k = 2$ and $l = 3$. Assume that the input sequence be $f = \{74, 14, 28, 65, 55, 8, 61, 20, 17, 65, 26, 94\}$ and the values which are outside the region of f are set to be boundary values of f. Then the output sequences obtained from the OSSM and the ROSSM dilations are

$$\text{OSSM}(f, B, A, k, l) = \{74, 65, 55, 55, 55, 55, 55, 20, 26, 65, 65, 94\}$$

and

$$\text{ROSSM}(f, B, A, k, l) = \{74, 65, 65, 65, 61, 61, 61, 61, 61, 65, 65, 94\},$$

respectively. Figures 5.33a and b show a 256×256 image and its disturbed version with additive Gaussian noise having zero mean and standard deviation 20. The reconstructed images of applying the OSSM and ROSSM filters with $S_B = [(-1, 0), (0, -1), (0, 0), (1, 0), (0, 1)]$, $S_A = [(0, 0)]$, $k = 3$, and $l = 4$ are shown in Figures 5.33c and d. The mean square signal to noise ratio, defined as

$$SNR_{ms} = \frac{\sum (y_i')^2}{\sum (y_i' - y_i)^2} \tag{5.73}$$

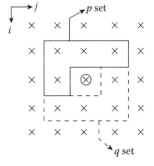

FIGURE 5.32   Two-dimensional structuring element and the preceding (succeeding) domain index sets when using the rater-scan order.

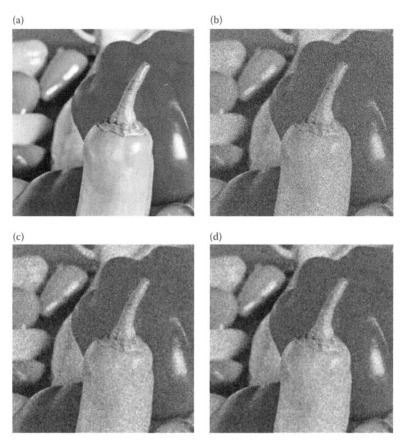

FIGURE 5.33 Visual comparison of the OSSM and ROSSM filters: (a) original image, (b) noisy image disturbed by zero mean Gaussian noise ($\sigma = 20$), (c) reconstructed image obtained from the OSSMF, (d) reconstructed image obtained from the ROSSMF.

with y and y' denoting the reconstructed and original signal respectively, is used to measure the fidelity of the reconstructed signals. The SNR_{ms} values of Figures 5.33c and d are 84.3 and 91.6, respectively. It is observed that the visual perception effect of Figure 5.33d is slightly better than Figure 5.33c.

An important feature of the OSSMF is the duality property. This relationship property also holds on the ROSSM operators. That is, choosing l in an ROSSM dilation is the same as choosing $T - l + 1$ in an ROSSM erosion, where T represents the total number in the order list. If $\tilde{\oplus}_R$ and $\tilde{\ominus}_R$ denote the ROSSM dilation and erosion, respectively, the duality relationship can be represented as $f \tilde{\oplus}_R [B, A, k, l] = f \tilde{\ominus}_R [B, A, k, T - l + 1]$. However, if the structuring element is obtained by taking the intersection (union) of two or more structuring elements, this duality property may be unavailable.

Before discussing the properties of the ROSSM filters, we should note that the first output signal value of an OSSM is the same as that of its corresponding

ROSSM operator because the left-hand side domain of the first value is the appended signal. The proofs of the following properties can be referred to in Pei, Lai, and Shih [1998].

Property 5.20: The ROSSM filters are increasing.

Property 5.21: If $l \leq k$ and the origin of set B belongs to set A, then the ROSSM dilation is extensive and the ROSSM erosion is anti-extensive.

Property 5.22: The idempotent ROSSM filters exist.

Although the exact expression of the idempotent ROSSM filters is not known, the idempotent ROSSM filters do exist. For example, let the structuring element be a center weighted structure with size $2N + 1$, $k = 2N - 1$, and $l = 2N$, then the ROSSM filter is idempotent.

Example 5.9: Let the structuring element be $5B1A$. This implies that $N = 2$. Let an input sequence $f = \{1, 3, 5, 7, 2, 4, 6, 8, 3, 1, 5, 7, 9\}$, $k = 3$, and $l = 4$, then the first and second iterations of the ROSSM filtered output, denoted as f^1 and f^2, are $f^1 = \{1, 3, 5, 5, 4, 4, 6, 6, 3, 3, 5, 7, 9\}$ and $f^2 = \{1, 3, 5, 5, 4, 4, 6, 6, 3, 3, 5, 7, 9\}$, respectively. This result leads to an idempotent filter.

Properties 5.20 through 5.22 deal with the basic properties possessed by the morphological filters. According to Property 5.22, a recursive morphological opening (closing) is obtained if the operator is idempotent and anti-extensive (extensive). In the following, some other interesting properties related to the ROSSM filters are developed.

Property 5.23: The ROSSM operator can be implemented by the threshold decomposition technique.

Example 5.10: Let the structuring element be $5B3A$, $k = 2$, and $l = 3$. The input sequence $f = \{2, 3, 1, 4, 4, 1, 2, 3\}$. The process by direct implementation and threshold decomposition to realize this ROSSM operator is illustrated in Figure 5.34.

Property 5.24: An anti-extensive (extensive) ROSSM dilation (erosion) is reduced to its corresponding OSSM dilation if the input signal f is increasing (decreasing) and $l \leq T/2$, where T denotes the total number of the sort list.

The above property is useful in offering a way to substitute a ROSSMF by an OSSMF. It can improve the efficiency because the OSSM operator can be implemented in parallel. However, if the ROSSMF is used, the result can just be obtained sequentially since each output value must take its previous output pixels into account.

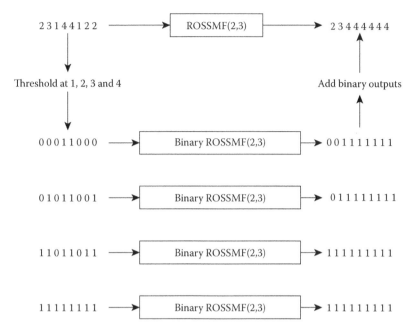

FIGURE 5.34 Threshold decomposition property of the ROSSM filters.

Property 5.25: The output signal function of the ROSSM dilation (erosion) is increasing (decreasing) if $l = 1$.

5.7 Regulated Morphological Filters

Agam and Dinstein [1999] developed regulated morphological operators and showed how the fitting property can be adapted for analyzing the map and line-drawing images. Since regulated morphology inherits many properties of the traditional standard morphology, it is feasible to apply in image processing and optimize its strictness parameters. Tian, Li, and Yan [2002] extended the regulated morphological operators by adjusting the weights in the structuring element.

Regulated morphological operations adopt a strictness parameter to control their sensitivity with respect to noise and small intrusions or protrusions on object boundary, and thereby prevent excessive dilation or erosion. The *regulated dilation* of a set A by a structuring element set B with a strictness parameter s is defined by [Agam and Dinstein 1999]:

$$A \oplus^s B = \{x \mid \#(A \cap (\hat{B})_x) \geq s, \quad s \in [1, \min(\#A, \#B)]\}, \tag{5.74}$$

where the symbol # denotes the cardinality of a set.

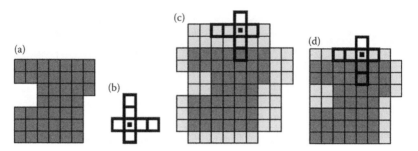

FIGURE 5.35 The difference between the standard dilation and the regulated dilation. (a) The input set, (b) the structuring element set, (c) the result of a standard dilation, (d) the result of a regulated dilation with a strictness of two. The pixels colored in light gray indicate the additional pixels generated by the dilation. (Courtesy of Agam and Dinstein 1999.)

Figure 5.35 illustrates the difference between the standard dilation and the regulated dilation, where Figure 5.35a is the input set A and Figure 5.35b is the structuring element set B (the black square indicates its origin). Figure 5.35c shows the result of a standard dilation, and Figure 5.35d shows the result of a regulated dilation with a strictness of two. The pixels colored in light gray indicate the additional pixels generated by the dilation. It is observed that the regulated dilation produces a smaller set than the standard dilation because in the regulated dilation the structuring element needs to penetrate deeper into the object to add a pixel to the dilated object.

The properties of regulated morphological dilation are introduced below. Their proofs can be referred to in Agam and Dinstein [1999].

Property 5.26: The regulated dilation is decreasing with respect to the strictness s.

$$A \oplus^{s1} B \subseteq A \oplus^{s2} B \Leftrightarrow s1 \geq s2 \tag{5.75}$$

Property 5.27: A regulated dilation generates a subset of a standard dilation.

$$A \oplus^{s} B \subseteq A \oplus B \tag{5.76}$$

Property 5.28: The regulated dilation is commutative, increasing with respect to the first and the second arguments, and translation invariant.

$$A \oplus^{s} B = B \oplus^{s} A \tag{5.77}$$

$$A \subseteq B \Rightarrow A \oplus^{s} D \subseteq B \oplus^{s} D \tag{5.78}$$

$$B \subseteq C \Rightarrow A \oplus^s B \subseteq A \oplus^s C \tag{5.79}$$

$$(A)_x \oplus^s B = (A \oplus^s B)_x \tag{5.80}$$

$$A \oplus^s (B)_x = (A \oplus^s B)_x \tag{5.81}$$

Property 5.29: The regulated dilation of a union (intersection) of sets is bigger (smaller) or equal to the union (intersection) of the regulated dilation of the individual sets.

$$(A \cup B) \oplus^s D \supseteq (A \oplus^s D) \cup (B \oplus^s D) \tag{5.82}$$

$$(A \cap B) \oplus^s D \subseteq (A \oplus^s D) \cap (B \oplus^s D) \tag{5.83}$$

The *regulated erosion* of a set A by a structuring element set B with a strictness parameter s is defined by [Agam and Dinstein 1999]

$$A \ominus^s B = \{x \mid \# (A^c \cap (B)_x) < s, \quad s \in [1, \#B]\}. \tag{5.84}$$

Figure 5.36 illustrates the difference between the standard erosion and the regulated erosion, where Figure 5.36a is the input set A and Figure 5.36b is the structuring element set B (the black square indicates its origin). Figure 5.36c shows the result of a standard erosion, and Figure 5.36d shows the result of a regulated erosion with a strictness of two. The pixels colored in light gray indicate the removed pixels affected by the erosion. It is observed that the regulated erosion produces a larger set than the standard erosion because in the regulated erosion the structuring element can get into more pixels in order to prevent their removal from the eroded object.

The properties of regulated morphological erosion are introduced below. Their proofs can be referred to in Agam and Dinstein [1999].

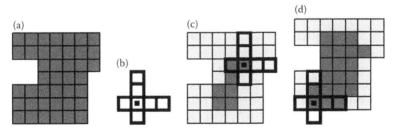

FIGURE 5.36 The difference between the standard erosion and the regulated dilation. (a) The input set, (b) the structuring element set, (c) the result of a standard erosion, (d) the result of a regulated erosion with a strictness of two. The pixels colored in light gray indicate the removed pixels affected by the erosion. (Courtesy of Agam and Dinstein 1999.)

Property 5.30: The regulated erosion is increasing with respect to the strictness s.

$$A \ominus^{s1} B \subseteq A \ominus^{s2} B \implies s1 \leq s2 \tag{5.85}$$

Property 5.31: A regulated erosion results in a superset of a standard dilation.

$$A \ominus^s B \supseteq A \ominus B \tag{5.86}$$

Property 5.32: The regulated erosion is increasing with respect to the first argument, decreasing with respect to the second argument, and translation invariant.

$$A \subseteq B \implies A \ominus^s D \subseteq B \ominus^s D \tag{5.87}$$

$$B \subseteq C \implies A \ominus^s B \supseteq A \ominus^s C \tag{5.88}$$

$$(A)_x \ominus^s B = (A \ominus^s B)_x \tag{5.89}$$

$$A \ominus^s (B)_x = (A \ominus^s B)_{-x} \tag{5.90}$$

Property 5.33: The regulated erosion of a union (intersection) of sets is bigger (smaller) or equal to the union (intersection) of the regulated dilation of the individual sets.

$$(A \cup B) \ominus^s D \supseteq (A \ominus^s D) \cup (B \ominus^s D) \tag{5.91}$$

$$(A \cap B) \ominus^s D \subseteq (A \ominus^s D) \cap (B \ominus^s D) \tag{5.92}$$

Property 5.34: The regulated dilation and erosion are dual in the same sense that exists for the standard dilation and erosion.

$$A \ominus^s B = (A^c \oplus^s \hat{B})^c \tag{5.93}$$

Figure 5.37 shows the results of a standard dilation and a regulated dilation when applying on a dashed-lines image. The original image is shown in Figure 5.37a, and the result of a standard dilation by a line-structuring element in 45° is shown in Figure 5.37b. The result of a regulated dilation by the same structuring element using a strictness parameter greater than 1 is shown in Figure 5.37c. It is observed that both the standard and regulated dilation operations manage to fill the gaps in the diagonal dashed lines. However, the regulated dilation removes the horizontal dashed lines, while the standard dilation generates noise due to the dilation of these lines.

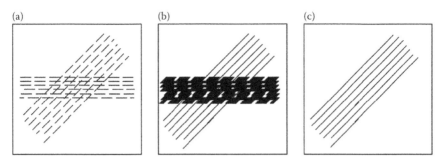

FIGURE 5.37 Results of the standard and the regulated dilations of dashed lines. (a) The original image, (b) the result of a standard dilation by a line structuring element in 45°, (c) the result of a regulated dilation using a strictness parameter greater than one. (Courtesy of Agam and Dinstein 1999.)

To maintain the same properties as standard morphological operators, the *regulated closing* of a set A by a structuring element set B with a strictness parameter s is defined by

$$A \bullet^s B = ((A \oplus^s B) \ominus B) \cup A, \tag{5.94}$$

where $A \oplus^s B = (A \oplus^s B) \cup A$ is defined as the *extensive regulated dilation*. The *regulated opening* of a set A by a structuring element set B with a strictness parameter s is defined by

$$A \circ^s B = ((A \ominus^s B) \oplus B) \cap A, \tag{5.95}$$

where $A \ominus^s B = (A \ominus^s B) \cap A$ is defined as the *anti-extensive regulated erosion*.

By adjusting the strictness parameter, we can alleviate the noise sensitivity problem and small intrusions or protrusions on the object boundary. An application of using regulated morphological filters is for corner detection. The regulated morphological corner detector is described as follows [Shih, Chuang, and Gaddipati 2005]:

Step 1: $A_1 = (A \oplus^s B) \ominus^s B$. Corner strength: $C_1 = |A - A_1|$.
Step 2: $A_2 = (A \ominus^s B) \oplus^s B$. Corner strength: $C_2 = |A - A_2|$.
Step 3: Corner detector $= C_1 \cup C_2$.

The original image is performed using a regulated dilation by a 5×5 circular structuring element with a strictness s, and then followed by a regulated erosion by the same structuring element with the same strictness. The corner strength C_1 is computed by finding the absolute value of the difference between the original and resulting images. This step is to extract concave corners. By reversing the order such that the regulated

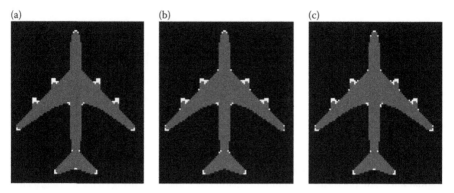

FIGURE 5.38 (a– c) Corner detection for an airplane image using the regulated morphological corner detector at strictness = 2, 3, and 4, respectively.

erosion is applied first, and then followed by the regulated dilation, the convex corners can be extracted. Finally, both types of corners are combined. Figure 5.38 shows the corner detection for an airplane image using the regulated morphological corner detector at strictness = 2, 3, and 4.

5.8 Fuzzy Morphological Filters

Fuzzy set theory has found a promising field of applications in the domain of digital image processing since fuzziness is an intrinsic property of image. The fuzzy logic has been developed to capture the uncertainties associated with human cognitive processes such as in thinking, reasoning, perception, and so on. Images are modeled as fuzzy subsets of the Euclidean plane or Cartesian grid, and the morphological operators are defined in terms of a fuzzy index function. Zadeh [1965] introduced fuzzy sets, which contain elementary operations such as intersection, union, complementation, and inclusion. The fuzzy set theoretical operations are defined as follows. Let the characteristic function of a crisp set A be denoted as $\mu_A: U \rightarrow \{0, 1\}$, is defined to be

$$\mu_A(x) = \begin{cases} 1 & \text{if } x \in A \\ 0 & \text{otherwise} \end{cases} \tag{5.96}$$

where A is any finite set. However, the object classes generally encountered in the real world are not so "precisely" or "crisply" defined. In most cases, several ambiguities arise in the determination of whether a particular element belongs to a set or not. A good example mentioned by Zadeh is a class of animals. This class clearly includes dogs, cats, tigers, and so on and excludes rocks, plants, houses, and so on. However, an ambiguity arises in the context of objects such as bacteria and starfish with respect to the class of animals.

The membership function of a fuzzy set A, denoted as $\mu_A: U \to [0, 1]$, is defined in such a way that $\mu_A(x)$ denotes the degree to which x belongs to A. Note that in the crisp set, the membership function has the value either 1 or 0, but in the fuzzy set, the member function has the value in the range of 0 to 1. The higher the value of $\mu_A(x)$, the more x belongs to A, and conversely, the smaller the value of $\mu_A(x)$, the less likelihood of x being in the set A. For further details see Zadeh [1965, 1977].

The union, intersection, difference, and complement operations on crisp as well as fuzzy sets can be defined in terms of their characteristic/membership functions as

$$\mu_{A \cup B}(x) = \max[\mu_A(x), \mu_B(x)] \tag{5.97}$$

$$\mu_{A \cap B}(x) = \min[\mu_A(x), \mu_B(x)] \tag{5.98}$$

$$\mu_{A \setminus B}(x) = \min[\mu_A(x), 1 - \mu_B(x)] \tag{5.99}$$

$$\mu_A^c(x) = 1 - \mu_A(x) \tag{5.100}$$

The subset relation can be expressed as

$$A \subseteq B \implies \mu_B = \max[\mu_A, \mu_B] \implies \mu_A = \min[\mu_A, \mu_B] \tag{5.101}$$

The support of a set A, denoted as $S(A)$, is a crisp set of those elements of U which belong to A with some certainty:

$$S(A) = \{x \mid \mu_A(x) > 0\} \tag{5.102}$$

The translation of a set A by a vector $v \in U$, denoted by $\Im(A; v)$, is defined as

$$\mu_{\Im(A;v)}(x) = \mu_A(x - v) \tag{5.103}$$

The reflection of a set A, denoted by $\hat{A}$, is defined as

$$\mu_{\hat{A}}(x) = \mu_A(-x) \tag{5.104}$$

The scalar addition of a fuzzy set A and a constant α, denoted as $A \dagger \alpha$, is defined as

$$\mu_{A \dagger \alpha}(x) = \min(1, \max[0, \mu_A(x) + \alpha]) \tag{5.105}$$

Giles [1976] proposed fuzzy operations, bold union, $X \Delta Y$ of two sets X and Y as

$$\mu_{X \Delta Y}(z) = \min[1, \ \mu_X(z), \ \mu_Y(z)] \tag{5.106}$$

and bold intersection $X \nabla Y$ as

$$\mu_{X \nabla Y}(z) = \max[0, \ \mu_X(z), \ \mu_Y(z) - 1] \tag{5.107}$$

If X and Y are crisp sets, then $X \Delta Y \equiv X \cup Y$ and $X \nabla Y \equiv X \cap Y$.
   An index function $I: 2^U \times 2^U \to \{0, 1\}$ is defined as

$$I(A,B) = \begin{cases} 1 & \text{if } A \subseteq B \\ 0 & \text{otherwise} \end{cases} \tag{5.108}$$

The above equation can be rewritten so as to express the index function directly in terms of characteristic functions as

$$\begin{aligned} I(A,B) &= \inf_{x \in A} \mu_B(x) \\ &= \min[\inf_{x \in A} \mu_B(x), \ \inf_{x \notin A} 1] \\ &= \inf_{x \in U} \mu_{A^c \Delta B}(x) \end{aligned} \tag{5.109}$$

The last relation follows because for crisp sets the bold union has the following properties:

$$x \in A \Rightarrow \mu_{A^c \Delta B}(x) = \mu_B(x)$$
$$x \notin A \Rightarrow \mu_{A^c \Delta B}(x) = 1.$$

The index function can be generalized, so that $I(A, B)$ gives the degree to which A is a subset of B. The formulation in Equation 5.109 suffices for this purpose. The properties of morphological operations will be induced by the properties of the index function.
   Consider any two fuzzy subsets $A, B \subset U$; index function $I(A, B)$ for different values of set B, and set A essentially fixed and $A \neq \phi$. The properties (axioms) for index function to satisfy are as follows:

$$I(A, B) \in [0, 1].$$

If A and B are crisp sets, then $I(A, B) \in \{0, 1\}$.

$$A \subseteq B \iff I(A, B) = 1.$$

If $B \subseteq C$, then $I(A, B) \le I(A, C)$.
If $B \subseteq C$, then $I(C, A) \le I(B, A)$.

Invariant under translation, complement, and reflection:

(a) $I(A, B) = I(\Im(A; v), \Im(B; v))$.

(b) $I(A, B) = I(A^c, B^c)$.

(c) $I(A, B) = I(\hat{A}, \hat{B})$.

If B and C are subsets of A, then so is $B \cup C$. The converse also holds:

$(B \subseteq A) \wedge (C \subseteq A) \Leftrightarrow (B \cup C) \subseteq A$; i.e., $I(B \cup C, A) = \min[I(B, A), I(C, A)]$.

$(A \subseteq B) \wedge (A \subseteq C) \Leftrightarrow A \subseteq (B \cup C)$; i.e., $I(A, B \cup C) = \min[I(A, B), I(A, C)]$.

The fuzzy morphological operations, dilation, erosion, opening, and closing are defined in terms of index function as follows [Sinha and Dougherty 1992, 1995]. The erosion of a set A by another set B, denoted as $A \ominus B$, is defined by

$$\mu_{A \ominus B}(x) = I(\Im(B; x), A). \tag{5.110}$$

The dilation of a set A by another set B, denoted as $A \oplus B$, is defined by

$$\mu_{A \oplus B}(x) = \mu_{(A^c \ominus \hat{B})^c}(x) = 1 - \mu_{A^c \ominus \hat{B}}(x)$$

$$= I^c(\Im(\hat{B}; x), A^c) = \sup_{z \in U} \max[0, \mu_{\Im(\hat{B}; x)}(z) + \mu_A(z) - 1] \tag{5.111}$$

The opening of a set A by another set B, denoted as $A \circ B$, is defined by

$$A \circ B = (A \ominus B) \oplus B. \tag{5.112}$$

The closing of a set A by another set B, denoted as $A \bullet B$, is defined by

$$A \bullet B = (A \oplus B) \ominus B. \tag{5.113}$$

Sinha and Dougherty [1992] did not just introduce the modeling of gray-scale images and the fuzzy morphological operations by simply replacing ordinary set theoretic operations by their fuzzy counterparts. Instead, they develop an intrinsically fuzzy approach to mathematical morphology. A new fitting paradigm is used by employing an index for set inclusion to measure the degree to which one image is beneath another image for erosion. Fuzzy morphological operations are defined based on this approach.

In binary morphology, erosion of image A by structuring element B is defined by

$$A \ominus B = \bigcap_{x \in -B} T(A; x) \qquad (5.114)$$

where $T(A; x)$ is translation of A by a vector. Dilation of image A by structuring element B can be defined by

$$A \oplus B = \bigcup_{x \in B} T(A; x) \qquad (5.115)$$

The definition of fuzzy morphology required an index function for set inclusion:

$$I(A, B) = \inf_{x \in A} \mu_B(x) A$$

$$= \min\left[\inf_{x \in A} \mu_B(x), \inf_{x \notin A} 1 \right] = \inf_{x \in U} \mu_{A^c \Delta B}(x) \qquad (5.116)$$

The (fuzzy) complement of the set inclusion indicator is then defined as

$$I^c(A, B) = 1 - I(A, B) = \sup_{x \in U} \mu_{A \nabla B^c}(x) \qquad (5.117)$$

Based on Equations 5.116 and 5.117, the fuzzy erosion is defined as

$$\mu_{E(A,B)}(x) = I(T(B; x), A)$$

$$= \inf_{x \in U} \min\left[1, \ 1 + \mu_A(x) - \mu_T(B; x)(x)\right] \qquad (5.118)$$

and the fuzzy dilation is defined as

$$\mu_{D(A,B)}(x) = 1 - \mu_{E(A,B)}(x)$$

$$= \sup_{x \in U} \max\left[0, \ \mu_T(-B; x)(x) + \mu_A(x) - 1\right]. \qquad (5.119)$$

The property of being dual operations between erosion and dilation is preserved.

$$\mu_{D(A,B)}(x) = \mu_{E(A^c, -B)^c}(x). \qquad (5.120)$$

The fuzzy morphological operations are illustrated in the following examples. The area of interest in a given image is represented by a rectangle, and the membership values within this rectangular region are specified in a matrix format. The membership values outside this region are assumed to be fixed, and this value is specified as a superscript of the matrix and the coordinates of the topmost-leftmost element of the matrix are used as a subscript.

Example 5.11:

$$A = \begin{bmatrix} 0.2 & 1.0 & 0.8 & 0.1 \\ 0.3 & 0.9 & 0.9 & 0.2 \\ 0.1 & 0.9 & 1.0 & 0.3 \end{bmatrix}^{0.0}_{(0,0)} \quad \text{and} \quad B = [0.8 \quad 0.9]^{0.0}_{(0,0)}$$

Since the structuring element fits entirely under the image when it translated by $(1, -1)$, $A \ominus B$ $(1, -1) = 1.0$. Similarly, $A \ominus B$ $(1, -2) = 1.0$. For vector $(1, 0)$, even though $\Im(B; 1, 0) \not\subset A$, the subset relationship almost holds:

$$A \cap \Im(B; 1, 0) = (0.8 \quad 0.8)_{(1,0)} \approx \Im(B; 1, 0).$$

Therefore, a relatively high value of $A \ominus B$ $(1, 0)$ is expected.

$$I[\Im(B; 1, 0), A] = \min\{\min[1, 1 + 1 - 0.8], \min[1, 1 + 0.8 - 0.9]\} = 0.9.$$

Thus, $A \ominus B(1, 0) = 0.9$. Proceeding along this way, the eroded image is obtained as

$$A \ominus B = \begin{bmatrix} 0.2 & 0.4 & 0.9 & 0.2 \\ 0.2 & 0.5 & 1.0 & 0.3 \\ 0.2 & 0.3 & 1.0 & 0.4 \end{bmatrix}^{0.1}_{(-1,0)}.$$

Note that the coordinates of the topmost-leftmost element are now $(-1, 0)$ instead of $(0, 0)$. By appropriately thresholding the image at a value between 0.5 and 0.9, we obtain

$$A \ominus B = \begin{bmatrix} 0.9 \\ 1.0 \\ 1.0 \end{bmatrix}^{0.0}_{(1,0)}.$$

Example 5.12:

$$A = \begin{bmatrix} 0.7 \\ 0.9 \\ 0.8 \end{bmatrix}^{0.0}_{(0,0)} \quad \text{and} \quad B = [0.8 \quad 0.9]^{0.0}_{(0,0)}$$

We obtain the reflected set: $\hat{B} = [0.9 \quad 0.8]_{(-1,0)}^{0.0}$.

$$I^c(\Im(\hat{B};x), A^c) = \begin{cases} 0.5 & \text{if } x = (0,0) \\ 0.6 & \text{if } x = (1,0), (0,-2) \\ 0.7 & \text{if } x = (0,-1), (1,-2) \\ 0.8 & \text{if } x = (1,-1) \\ 0.0 & \text{otherwise} \end{cases}$$

Hence, the fuzzy dilation is

$$A \ominus B = \begin{bmatrix} 0.5 & 0.6 \\ 0.7 & 0.8 \\ 0.6 & 0.7 \end{bmatrix}_{(0,0)}^{0.0}$$

Example 5.13: Consider the image from Example 5.11.

$$P \equiv A \ominus B = \begin{bmatrix} 0.2 & 0.4 & 0.9 & 0.2 \\ 0.2 & 0.5 & 1.0 & 0.3 \\ 0.2 & 0.3 & 1.0 & 0.4 \end{bmatrix}_{(-1,0)}^{0.1}$$

Therefore,

$$P^c \ominus \hat{B} = \begin{bmatrix} 0.8 & 0.3 & 0.2 & 0.9 \\ 0.7 & 0.2 & 0.1 & 0.8 \\ 0.9 & 0.2 & 0.1 & 0.7 \end{bmatrix}_{(0,0)}^{1.0}$$

Hence,

$$A \circ B = (P^c \ominus \hat{B})^c = \begin{bmatrix} 0.2 & 0.7 & 0.8 & 0.1 \\ 0.3 & 0.8 & 0.9 & 0.2 \\ 0.1 & 0.8 & 0.9 & 0.3 \end{bmatrix}_{(-1,0)}^{0.0}$$

By appropriately thresholding the image at a value between 0.3 and 0.7, we obtain

$$A \circ B = \begin{bmatrix} 0.7 & 0.8 \\ 0.8 & 0.9 \\ 0.8 & 0.9 \end{bmatrix}_{(-1,0)}^{0.0}$$

Gasteratos, Andreadis, and Tsalides [1998] proposed a framework, which extends the concepts of soft mathematical morphology into fuzzy sets. They studied its compatibility with binary soft mathematical morphology as well as the algebraic properties of fuzzy soft operations.

Some applications of fuzzy morphological filters to image processing are given below. Großert, Koppen, and Nickolay [1996] applied fuzzy morphology to detect infected regions of a leaf. It was shown that the fuzzy morphological operation achieves a compromise between the deletion of isolated bright pixels and the connection of scattered regions of bright pixels. Wirth and Nikitenko [2005] applied fuzzy morphology to contrast enhancement. Yang, Guo, and Jiang [2006] proposed an edge detection algorithm based on adaptive fuzzy morphological neural network. The gradient of the fuzzy morphology utilizes a set of structuring elements to detect the edge strength with a view to decrease the spurious edge and suppressed the noise.

Chatzis and Pitas [2000] proposed the generalized fuzzy mathematical morphology (GFMM) based on a novel definition of the fuzzy inclusion indicator (FII). FII is a fuzzy set used as a measure of the inclusion of a fuzzy set into another. The GFMM provides a very powerful and flexible tool for morphological operations. It can be applied on the skeletonization and shape decomposition of two-dimensional and three-dimensional objects. Strauss and Comby [2007] presented variable structuring element based fuzzy morphological operations for single viewpoint omni-directional images.

References

Agam, G. and Dinstein, I., "Regulated morphological operations," *Pattern Recognition*, vol. 32, no. 6, pp. 947–971, June 1999.

Brodatz, P., *Textures: A Photographic Album for Artists and Designers*, Dover Publications, New York, 1966.

Chatzis, V. and Pitas, I., "A generalized fuzzy mathematical morphology and its application in robust 2-D and 3-D object representation," *IEEE Trans. Image Processing*, vol. 9, no. 10, pp. 1798–1810, Oct. 2000.

Danielsson, P. E., "A new shape factor," *Comput. Graphics Image Processing*, vol. 7, no. 2, pp. 292–299, Apr. 1978.

David, H., *Order Statistics*, Wiley, New York, 1981.

Destival, I., "Mathematical morphology applied to remote sensing," *Acta Astronaut.*, vol. 13, no. 6/7, pp. 371–385, June–July 1986.

Gasteratos, A., Andreadis, I., and Tsalides, P., "A new hardware structure for implementation of soft morphological filters," Computer Analysis of Images and Patterns, Lecture Notes in Computer Science, Springer, vol. 1296, pp. 488–494, 1997.

Gasteratos, A., Andreadis, I., and Tsalides, P., "Fuzzy soft mathematical morphology," *IEE Proc. Vision, Image and Signal Processing*, vol. 145, no. 1, pp. 41–49, Feb. 1998.

Giardina, C. and Dougherty, E., *Morphological Methods in Image and Signal Processing*, Prentice-Hall, Englewood Cliffs, NJ, 1988.

Giles, R., "Lukasiewicz logic and fuzzy theory," *Intl. Journal Man-Machine Stud.*, vol. 8, no. 3, pp. 313–327, 1976.

Großert, S., Koppen, M., and Nickolay, B. A., "New approach to fuzzy morphology based on fuzzy integral and its application in image processing," *Proc. Intl. Conf. Pattern Recognition*, vol. 2, pp. 625–630, Aug. 1996.

Hamid, M. S., Harvey, N. R., and Marshall, S., "Genetic algorithm optimization of multidimensional grayscale soft morphological filters with applications in film archive restoration," *IEEE Trans. Circuits and Systems for Video Technology*, vol. 13, no. 5, pp. 406–416, May 2003.

Haralick, R., Zhuang, X., Lin, C., and Lee, J., "The digital morphological sampling theorem," *IEEE Trans. Acoust. Speech, Signal Processing*, vol. 37, no. 12, pp. 2067–2090, Dec. 1989.

Heijmans, H. J. A. M., "A new class of alternating sequential filters," in *Proc. IEEE Workshop on Nonlinear Signal and Image Processing*, vol. 2, pp. 30–33, 1995.

Heijmans, H. J. A. M., "Composing morphological filters," *IEEE Trans. Image Processing*, vol. 6, no. 5, pp. 713–723, May 1997.

Ko, S.-J., Morales, A., and Lee, K.-H., "Fast recursive algorithms for morphological operators based on the basis matrix representation," *IEEE Trans. Image Processing*, vol. 5, no. 6, pp. 1073–1077, June 1996.

Koskinen, L., Astola, J. T., and Neuvo, Y. A., "Soft morphological filters," in D. Gader and E. R. Dougherty (Eds.), *Proc. SPIE*, vol. 1568, Image Algebra and Morphological Image Processing II, pp. 262–270, July 1991.

Kuosmanen, P. and Astola, J., "Soft morphological filtering," *Journal of Mathematical Imaging and Vision*, vol. 5, no. 3, pp. 231–262, Sep. 1995.

Li, W. and Coat, V. H., "Composite morphological filters in multiresolution morphological decomposition," *Proc. IEE Workshop on Image Processing and Applications*, no. 410, pp. 752–756, July 1995.

Maragos, P. and Schafer, R., "Morphological filters—Part I: Their set-theoric analysis and relations to linear shift-invariant filters," *IEEE Trans. Acoustics, Speech, and Signal Processing*, vol. 35, no. 8, pp. 1153–1169, Aug. 1987a.

Maragos, P. and Schafer, R., "Morphological filters—Part II: Their relations to median, order-statistics, and stack filters," *IEEE Trans. Acoustics, Speech, and Signal Processing*, vol. 35, no. 8, pp. 1170–1184, Aug. 1987b.

Morales, A., Acharya, R., and Ko, S.-J., "Morphological pyramids with alternating sequential filters," *IEEE Trans. Image Processing*, vol. 4, no. 7, pp. 965–977, July 1995.

Nadadur, D. and Haralick, R. M., "Recursive binary dilation and erosion using digital line structuring elements in arbitrary orientations," *IEEE Trans. Image Processing*, vol. 9, no. 5, pp. 749–759, May 2000.

Pei, S., Lai, C., and Shih, F. Y., "An efficient class of alternating sequential filters in morphology," *Graphical Models and Image Processing*, vol. 59, no. 2, pp. 109–116, Mar. 1997.

Pei, S., Lai, C., and Shih, F. Y., "Recursive order-statistic soft morphological filters," *IEE Proceedings Vision, Image and Signal Processing*, vol. 145, no. 5, pp. 333–342, Oct. 1998.

Preteux, F., Laval-Jeantet, A. M., Roger, B., and Laval-Jeantet, M. H., "New prospects in C.T. image processing via mathematical morphology," *Eur. J. Radiol.*, vol. 5, no. 4, pp. 313–317, 1985.

Pu, C. C. and Shih, F. Y., "Threshold decomposition of gray-scale soft morphology into binary soft morphology," *Graphical Model and Image Processing*, vol. 57, no. 6, pp. 522–526, Nov. 1995.

Schonfeld, D. and Goutsias, J., "Optical morphological pattern restoration from noisy binary image," *IEEE Trans. Pattern Anal. Mach. Intelligence*, vol. 13, no. 1, pp. 14–29, Jan. 1991.

Serra, J., *Image Analysis and Mathematical Morphology*, Academic Press, New York, 1982.

Shih, F. Y., Chuang, C., and Gaddipati, V., "A modified regulated morphological corner detector," *Pattern Recognition Letters*, vol. 26, no. 7, pp. 931–937, June 2005.

Shih, F. Y., King, C. T., and Pu, C. C., "Pipeline architectures for recursive morphological operations," *IEEE Trans. Image Processing*, vol. 4, no. 1, pp. 11–18, Jan. 1995.

Shih, F. Y. and Mitchell, O. R., "Threshold decomposition of gray-scale morphology into binary morphology," *IEEE Trans. Pattern Analysis and Machine Intelligence*, vol. 11, no. 1, pp. 31–42, Jan. 1989.

Shih, F. Y. and Mitchell, O. R., "A mathematical morphology approach to Euclidean distance transformation," *IEEE Trans. Image Processing*, vol. 1, no. 2, pp. 197–204, Apr. 1992.

Shih, F. Y. and Pu, C. C., "Analysis of the properties of soft morphological filtering using threshold decomposition," *IEEE Trans. Signal Processing*, vol. 43, no. 2, pp. 539–544, Feb. 1995.

Shih, F. Y. and Puttagunta, P., "Recursive soft morphological filters," *IEEE Trans. Image Processing*, vol. 4, no. 7, pp. 1027–1032, July 1995.

Sinha, D. and Dougherty, E. R., "Fuzzy mathematical morphology," *Journal of Visual Communication and Image Representation*, vol. 3, no. 3, pp. 286–302, Sep. 1992.

Sinha, D. and Dougherty, E. R, "A general axiomatic theory of intrinsically fuzzy mathematical morphologies," *IEEE Trans. Fuzzy Systems*, vol. 3, no. 4, pp. 389–403, Nov. 1995.

Smith, S. W., *The Scientist and Engineer's Guide to Digital Signal Processing*, California Technical Pub., 1997.

Soille, P., "On morphological operators based on rank filters," *Pattern Recognition*, vol. 35, no. 2, pp. 527–535, Feb. 2002.

Sternberg, S. R., "Grayscale morphology," *Computer Vision, Graphics, and Image Processing*, vol. 35, pp. 333–355, Sep. 1986.

Stevenson, R. L. and Arce, G. R., "Morphological filters: Statistics and further syntactic properties," *IEEE Trans. Circuits and Systems*, vol. 34, no. 11, pp. 1292–1305, Nov. 1987.

Strauss, O. and Comby, F., "Variable structuring element based fuzzy morphological operations for single viewpoint omnidirectional images," *Pattern Recognition*, vol. 40, no. 12, pp. 3578–3596, Dec. 2007.

Tian, X.-H., Li, Q.-H., and Yan, S.-W., "Regulated morphological operations with weighted structuring element," *Proc. Intl. Conf. Machine Learning and Cybernetics*, pp. 768–771, Nov. 2002.

Wendt, P. D., Coyle, E. J., and Gallagher, N. C., "Stack filters," *IEEE Trans. Acoust., Speech, Signal Processing*, vol. 34, no. 4, pp. 898–911, Aug. 1986.

Wirth, M. A. and Nikitenko, D., "Applications of fuzzy morphology to contrast enhancement," *Proc. Annual Meeting of the North American Fuzzy Information Processing Society*, pp. 355–360, June 2005.

Yang, G.-Q., Guo, Y.-Y., and Jiang, L.-H., "Edge detection based on adaptive fuzzy morphological neural network," *Proc. Intl. Conf. Machine Learning and Cybernetics*, pp. 3725–3728, Aug. 2006.

Yli-Harja, O., Astola, J., and Neuvo, Y., "Analysis of the properties of median and weighted median filters using threshold logic and stack filter representation," *IEEE Trans. Signal Processing*, vol. 39, no. 2, pp. 395–409, Feb. 1991.

You, X. and Crebbin, G., "A robust adaptive estimation for filtering noise images," *IEEE Trans. Image Processing*, vol. 4, no. 5, pp. 693–699, May 1995.

Yu, P. T. and Liao, W. H., "Weighted order statistics filters—Their classification, some properties, and conversion algorithms," *IEEE Trans. Signal Processing*, vol. 42, no. 10, pp. 2678–2691, Oct. 1994.

Zadeh, L. A., "Fuzzy sets," *Information Control*, vol. 8, no. 3, pp. 338–353, June 1965.

Zadeh, L. A., "Theory of fuzzy sets," in *Encyclopedia of Computer Science and Technology* in J. Belzer, A. Holzman, and A. Kent (Eds.), Dekker, New York, 1977.

6

Distance Transformation

The interior of a closed object boundary is considered object pixels and the exterior as background pixels. Let S denote a set of object pixels in an image. The function d mapping from a binary image containing S to a matrix of nonnegative integers is called *a distance function*, if it satisfies the following three criteria:

(a) *Positive definite*: That is $d(p, q) \geq 0$, and $= 0$, if and only if $p = q$, for all $p, q \in S$.

(b) *Symmetric*: That is $d(p, q) = d(q, p)$, for all $p, q \in S$.

(c) *Triangular*: That is $d(p, r) \leq d(p, q) + d(q, r)$, for all $p, q, r \in S$.

Distance transformation (DT) converts a binary image, which consists of object (foreground) and nonobject (background) pixels, into another image where each object pixel has a value corresponding to the minimum distance from the background by a distance function [Vincent 1991]. The DT has wide applications in image analysis. One of the applications is computing a shape factor that measures shape compactness based on the ratio of total number of object pixels and the summation of all distance measures [Danielsson 1978]. Another is to obtain the medial axis (or skeleton), which is used for feature extractions [Blum 1964; Hilditch 1969; Lantuejoul 1980]. Computing the distance from an object pixel to background is a global operation, which is often prohibitively costly; therefore, a decomposition strategy using local operations is needed to speed up the computation.

Maurer, Rensheng, and Raghavan [2003] presented a linear time algorithm for computing exact Euclidean distance transforms of binary images in arbitrary dimensions. Cuisenaire [2006] presented when the structuring elements

(SEs) are balls of a metric, locally adaptable erosion and dilation can be efficiently implemented as a variant of DT algorithms. Opening and closing are obtained by a local threshold of a distance transformation, followed by the adaptable dilation. Xu and Li [2006] proposed the Euclidean distance transform of digital images in arbitrary dimensions.

The DT is a general operator forming the basis of many methods in computer vision and geometry, with great potential for practical applications. However, all the optional algorithms for the computation of the exact Euclidean DT (EDT) were proposed only since the 1990s. In Fabbri et al. [2008], state-of-the-art sequential 2D EDT algorithms were reviewed and compared. In binary images, the DT and the geometrical skeleton extraction are classic tools for shape analysis. In Coeurjolly and Montanvert [2007], the time optional algorithms are presented to solve the reverse EDT and the reversible medial axis extraction problems for d-dimensional images.

This chapter is organized as follows. Section 6.1 introduces DT by iterative operations. Section 6.2 presents DT by mathematical morphology. Section 6.3 describes the approximation of Euclidean distances. Section 6.4 discusses the decomposition of distance SEs. Section 6.5 presents the iterative erosion algorithm. Section 6.6 presents the two-scan based algorithm. Section 6.7 presents the three-dimensional Euclidean distances. Section 6.8 describes the acquiring approaches. Section 6.9 describes the deriving approaches.

6.1 DT by Iterative Operations

Three types of distance measures in digital image processing are often used: Euclidean, city-block, and chessboard. City-block and chessboard distances are easy to compute and can be recursively accumulated by considering only a small neighborhood at a time. The algorithms for city-block, chessboard, or a combination of the two (called *octagon*) have been widely developed [Rosenfeld and Pfaltz 1966; Shih and Mitchell 1987; Toriwaki and Yokoi 1981]. One distance transformation algorithm uses an iterative operation by peeling off border pixels and summing the distances layer-by-layer [Rosenfeld and Kak 1982]. This is quite efficient on a cellular array computer.

The *city-block distance* between two points $P = (x,y)$ and $Q = (u,v)$ is defined as:

$$d_4(P, Q) = |x - u| + |y - v| \tag{6.1}$$

The *chessboard distance* between P and Q is defined as:

$$d_8(P, Q) = \max(|x - u|, |y - v|) \tag{6.2}$$

The Euclidean distance between two points $P = (x,y)$ and $Q = (u,v)$ is defined as:

$$d_e(P, Q) = \sqrt{(x - u)^2 + (y - v)^2} \tag{6.3}$$

Note that the subscripts "4" and "8" indicate the 4-neighbor and 8-neighbor considerations in calculating the city-block and the chessboard distances. The subscript "e" denotes Euclidean distance.

Among them, Euclidean distance measure is most accurate, but it involves square, square-root, and floating-point computations, which are time-consuming. City-block and chessboard distance measures are not quite accurate, but can be calculated quickly. There is an octagon distance measure, which applies the city-block and the chessboard distances in an alternative calculation. Figure 6.1 shows (a) city-block, (b) chessboard, and (c) octagon distances from a center point P. From the shape of distance value "3," we observe that city-block is like a diamond, chessboard is like a square, and the combination of both is like an octagon.

The city-block and chessboard distances satisfy the following property: For all P, Q such that $d(P, Q) \geq 2$, there exists a point R, different from P and Q,

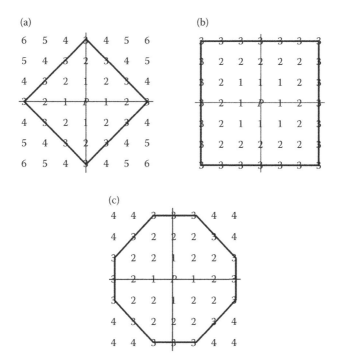

FIGURE 6.1 (a) City-block, (b) chessboard, and (c) octagon distances.

such that $d(P, Q) = d(P, R) + d(R, Q)$. Given χ_S, which is 1 at the points of S and 0 elsewhere, we define $\chi_S^{(m)}$ inductively for $m = 1, 2, \ldots$ as follows [Rosenfeld and Kak 1982]:

$$\chi_S^{(m)}(P) = \chi_S^{(0)}(P) + \min_{d(Q,P)\leq 1} \chi_S^{(m-1)}(Q) \tag{6.4}$$

where $\chi_S^{(0)} = \chi_S$. Note that the above equation requires iterative operations, which are time-consuming when an image size is large.

Example 6.1: Compute the city-block and the chessboard DTs on the following two images:

$$
\begin{array}{ccccc}
1 & 1 & 1 & 1 & 1 \\
1 & 1 & 1 & 1 & 1 \\
\text{(a)}\ \ 1 & 1 & 1 & 1 & 1 \ , \\
1 & 1 & 1 & 1 & 1 \\
1 & 1 & 1 & 1 & 1
\end{array}
\qquad
\begin{array}{ccccc}
0 & 0 & 1 & 0 & 0 \\
0 & 1 & 1 & 1 & 0 \\
\text{(b)}\ \ 1 & 1 & 1 & 1 & 1 \\
0 & 1 & 1 & 1 & 0 \\
0 & 0 & 1 & 0 & 0
\end{array}
$$

Answer: The city-block and the chessboard DTs of (a) are obtained as follows.

$$
\text{Cityblock:}\
\begin{array}{ccccc}
1 & 1 & 1 & 1 & 1 \\
1 & 2 & 2 & 2 & 1 \\
1 & 2 & 3 & 2 & 1 \\
1 & 2 & 2 & 2 & 1 \\
1 & 1 & 1 & 1 & 1
\end{array}
\ ,\qquad
\text{Chessboard:}\
\begin{array}{ccccc}
1 & 1 & 1 & 1 & 1 \\
1 & 2 & 2 & 2 & 1 \\
1 & 2 & 3 & 2 & 1 \\
1 & 2 & 2 & 2 & 1 \\
1 & 1 & 1 & 1 & 1
\end{array}
$$

Note that in this particular case both DTs output the same result. The city-block and the chessboard distance transformations of (b) are obtained as follows.

$$
\text{Cityblock:}\
\begin{array}{ccccc}
0 & 0 & 1 & 0 & 0 \\
0 & 1 & 2 & 1 & 0 \\
1 & 2 & 3 & 2 & 1 \\
0 & 1 & 2 & 1 & 0 \\
0 & 0 & 1 & 0 & 0
\end{array}
\ ,\qquad
\text{Chessboard:}\
\begin{array}{ccccc}
0 & 0 & 1 & 0 & 0 \\
0 & 1 & 1 & 1 & 0 \\
1 & 1 & 2 & 1 & 1 \\
0 & 1 & 1 & 1 & 0 \\
0 & 0 & 1 & 0 & 0
\end{array}
$$

Another algorithm only needs two scans of the image: One is in the left-to-right, top-to-bottom direction, and the other is in the right-to-left, bottom-to-top direction. This algorithm does not require a large number of iterations and can be efficiently implemented using a conventional computer. Let $N_1(P)$ be the set of (4- or 8-) neighbors that precede P in a row-by-row (left to right, top to bottom) scan of the picture, and let $N_2(P)$ be the remaining

(4- or 8-) neighbors of P. The two-step city-block and chessboard distance transformations are given as follows [Rosenfeld and Kak 1982]:

$$\chi'_S(P) = \begin{cases} 0 & \text{if } P \in \overline{S} \\ \min_{Q \in N_1} \chi'_S(Q)+1 & \text{if } P \in S \end{cases} \qquad (6.5)$$

$$\chi''_S(P) = \min_{Q \in N_2}[\chi'_S(P), \chi''_S(Q)+1] \qquad (6.6)$$

Thus we can compute χ'_S in a single left-to-right, top-to-bottom scan of the image since for each P, χ'_S has already been computed for the Qs in N_1. Similarly, we can compute χ''_S in a single reverse scan (right-to-left, bottom-to-top).

Example 6.2: Perform the two-step (a) city-block and (b) chessboard distance transformations on the following image:

```
1 1 1 1 1 1 0 0
1 1 1 1 1 1 0 0
1 1 1 1 1 1 1 0
1 1 1 0 0 1 1 1
1 1 1 0 0 1 1 1
1 1 1 1 1 1 1 0
```

Answer: (a) The first-step output image for city-block DT is:

```
1 1 1 1 1 1 0 0
1 2 2 2 2 2 0 0
1 2 3 3 3 3 1 0
1 2 3 0 0 1 2 1
1 2 3 0 0 1 2 2
1 2 3 1 1 2 3 0
```

The second-step output image for city-block DT is:

```
1 1 1 1 1 1 0 0
1 2 2 2 2 1 0 0
1 2 2 1 1 2 1 0
1 2 1 0 0 1 2 1
1 2 1 0 0 1 2 1
1 1 1 1 1 1 1 0
```

(b) The first-step output image for chessboard DT is:

```
1 1 1 1 1 1 0 0
1 2 2 2 2 1 0 0
1 2 3 3 2 1 1 0
1 2 3 0 0 1 1 1
1 2 1 0 0 1 2 1
1 2 1 1 1 1 2 0
```

The second-step output image for chessboard DT is:

```
1 1 1 1 1 1 0 0
1 2 2 2 2 1 0 0
1 2 1 1 1 1 1 0
1 2 1 0 0 1 1 1
1 2 1 0 0 1 1 1
1 1 1 1 1 1 1 0
```

Figure 6.2 shows the respective images of city-block, chessboard, and EDTs of a circle, where the distances are, respectively, rescaled to the grayscale range of [0, 255] for display purposes.

Serra [1982] applied binary morphological erosions recursively and accumulates the distance step by step. A significant disadvantage of the above city-block and chessboard algorithms is that both distance measures are very sensitive to the orientation of the object. The Euclidean distance measurement is rotation invariant. However, the square root operation is costly and the global operation is hard to decompose into small neighborhood operations because of its nonlinearity. Hence algorithms concerning the approximation of EDT have been extensively discussed [Borgefors 1984; Danielsson 1978; Vossepoel 1988]. Borgefors [1986] optimized the local distances used in 3×3, 5×5, and 7×7 neighborhoods by minimizing the maximum of the absolute value of the difference between the EDT and the proposed distance transformation. In the next section, we will introduce a mathematical morphology (MM) approach to construct the DT and apply its theorems to accomplish the decomposition of the global operation into local operations with special emphasis on the Euclidean distance computation.

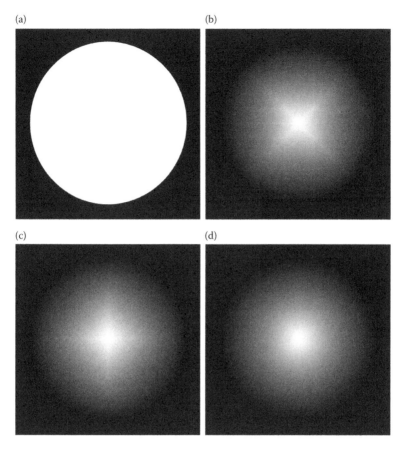

FIGURE 6.2 (a) A circle, (b) city-block distance transformation, (c) chessboard DT, and (d) EDT.

6.2 DT by Mathematical Morphology

The approaches to achieve DT do not adopt directly the definition of the minimum distance from an object pixel to all background border pixels, since their computations are extremely time-consuming. The literature of Shih and Mitchell [1992], Saito and Toriwaki [1994], Eggers [1998], Cuisenaire and Macq [1999], and Datta and Soundaralakshmi [2001] represents a sampled set of successful efforts for improving speed efficiency. In general, the algorithms of DT can be categorized into two classes: one is the *iterative* method which is efficient in a cellular array computer since all the pixels at each iteration can be processed in parallel, and the other is the *sequential* (or *recursive*) method which is suited for a conventional computer by avoiding iterations with the efficiency to be independent of object size. Using the general machines that most people have access to, sequential algorithms are often much more efficient than iterative ones.

Among different kinds of DT, the EDT is often-used because of its rotational invariance property, but it involves the time-consuming calculations such as square, square-root, and the minimum over a set of floating-point numbers. Although many techniques have been presented to obtain EDT, most of them are either inefficient or complex to implement and understand. Furthermore, they require extra cost such as special structure or storages for recording information. Cuisenaire and Macq [1999] proposed a fast EDT by propagation using multiple neighborhoods, but they need bucket sorting for calculating the EDT. Datta and Soundaralakshmi [2001] proposed a constant-time algorithm, but their algorithm is based on a special hardware structure, the reconfigurable mesh. Eggers [1998] proposed an algorithm by avoiding unnecessary calculations, but some data must be recorded in the lists. Saito and Toriwaki [1994] proposed an algorithm to compute the EDT in an n-dimensional domain, but the time-complexity is high in their algorithm.

In Shih and Mitchell [1992], Shih and Wu [1992], and Huang and Mitchell [1994], a MM approach was proposed to realize the EDT using grayscale erosions with successive small distance SEs by decomposition. Furthermore, a *squared Euclidean-distance structuring element* (SEDSE) was used to perform the *squared Euclidean distance transform* (SEDT). Shih and Wu [1992] decomposed the SEDSE into successive dilations of a set of 3×3 structuring components. Hence, the SEDT is equivalent to the successive erosions of the results at each preceding stage by each structuring component. The EDT can be finally obtained by simply a square-root operation over the entire image. In fact, the task of image analysis can directly take the result of SEDT as an input for object feature extraction and recognition.

There are six types of distance measures discussed in Borgefors [1986]: city-block, chessboard, Euclidean, octagonal, Chamfer 3-4, and Chamfer 5-7-11. The morphological DT algorithm presented here is a general method, which uses a predefined structuring element to obtain the desired type of distance measure. By placing the center of the structuring element located at the origin point P, all the points can be represented by their distances from P (so that P is represented by 0). The weights of the structuring element are selected to be negative of related distance measures because grayscale erosion is the minimum selection after subtraction. This gives a positive distance measure. Three types of distance SEs, Euclidean, city-block, and chessboard, are illustrated in Figure 6.3 in both tabular and graphical forms. Note that only those distances of 4 or smaller are shown.

A binary image f consists of two classes, object pixels (foreground) and nonobject pixels (background). Let the object pixels have the value $+\infty$ (or any number larger than the object's greatest distance) and the nonobject

(1) Euclidean distance

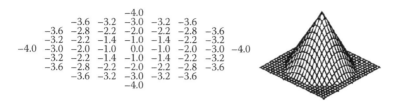

```
                        -4.0
              -3.6 -3.2 -3.0 -3.2 -3.6
         -3.6 -2.8 -2.2 -2.0 -2.2 -2.8 -3.6
         -3.2 -2.2 -1.4 -1.0 -1.4 -2.2 -3.2
    -4.0 -3.0 -2.0 -1.0  0.0 -1.0 -2.0 -3.0 -4.0
         -3.2 -2.2 -1.4 -1.0 -1.4 -2.2 -3.2
         -3.6 -2.8 -2.2 -2.0 -2.2 -2.8 -3.6
              -3.6 -3.2 -3.0 -3.2 -3.6
                        -4.0
```

(2) City-block distance

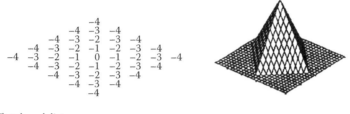

```
                  -4
              -4 -3 -4
           -4 -3 -2 -3 -4
        -4 -3 -2 -1 -2 -3 -4
     -4 -3 -2 -1  0 -1 -2 -3 -4
        -4 -3 -2 -1 -2 -3 -4
           -4 -3 -2 -3 -4
              -4 -3 -4
                  -4
```

(3) Chessboard distance

```
-4 -4 -4 -4 -4 -4 -4 -4 -4
-4 -3 -3 -3 -3 -3 -3 -3 -4
-4 -3 -2 -2 -2 -2 -2 -3 -4
-4 -3 -2 -1 -1 -1 -2 -3 -4
-4 -3 -2 -1  0 -1 -2 -3 -4
-4 -3 -2 -1 -1 -1 -2 -3 -4
-4 -3 -2 -2 -2 -2 -2 -3 -4
-4 -3 -3 -3 -3 -3 -3 -3 -4
-4 -4 -4 -4 -4 -4 -4 -4 -4
```

FIGURE 6.3 Three distance SEs. Only those distances of 4 or smaller are shown.

pixels have the value 0. Let k be the selected distance structuring element. The DT algorithm is as follows:

DT algorithm: The binary image is eroded by the distance structuring element; that is, $g = f \ominus_g k$.

The reason for assigning a large number to all the object pixels is to avoid interfering with the minimum selection of grayscale erosion that is designed for replacing the object value by a smaller distance value. Note that the size of the distance structuring element should be at least as large as the largest expected object size in the image. Otherwise, the central portion of the object pixels will not be reached by the minimum selection. The big size of the neighborhood computation is global and prohibitively costly. Most morphological hardware implementations are limited to a fixed size of the structuring element. The decomposition technique of a big structuring element into the combination of small ones can be referred to in Shih and Mitchell [1991, 1992], and Shih and Wu [2004a, b, c, 2005].

The DT has wide applications in image analysis. One application is to compute a shape factor, which is a measure of the compactness of the shape based on the ratio of the number of total object pixels and the summation of all distance measures. Another application is to obtain the medial axis (or skeleton), which is used for feature extraction.

Algorithm verification: According to the erosion definition and the fact that the pixels of the binary image f are only represented by two values $+\infty$ and 0, we have

$$f \ominus k(x) = \min\{\infty - k(z), 0 - k(z')\}, \quad \text{where } z, z' \in K. \tag{6.7}$$

Because the differences of all $k(z)$ and $k(z')$ are finite, $\infty - k(z)$ is always greater than $0 - k(z')$. Hence,

$$f \ominus k(x) = \min\{-k(z')\}, \tag{6.8}$$

where for any $x \in F$, all of z' satisfy that the pixel of $x + z'$ is located in the background (since a "0" in f corresponds to the background). Because the weights of the structuring element are replaced by the negative of the related distance measures from the center of the structuring element, the binary image f eroded by the distance structuring element k is exactly equal to the minimum distance from the object point to the outer boundary.

6.3 Approximation of Euclidean Distances

The DT is essentially a global computation. All distance measures are positive definite, symmetrical, and satisfy the triangle inequality. Hence global distances in the image can be approximated by propagating the local distance; that is, distances between neighboring pixels. The city-block distance is based on 4-neighborhood computation, and the chessboard distance is based on 8-neighborhood computation. Both distance measures are linearly additive and can be easily decomposed into the accumulation of neighboring distances. The Euclidean distance is a nonlinear measure. However, we can still make a reasonable approximation.

Borgefors [1986] optimized the local distances used in 3×3, 5×5, and 7×7 neighborhoods by minimizing the maximum of the absolute value of the difference between the EDT and the DT proposed. Vossepoel [1988] determined the coefficients of a distance transform by minimizing the above maximum absolute difference and minimizing the root-mean-square difference between the small neighborhoods DT and the EDT.

In the following discussion we present three different approximation methods. The first one is to combine the two outputs of relative x- and

y-coordinates, where the minimum city-block and chessboard distances accumulate individually from each object point, and to map into the Euclidean distance by a look-up table. During the city-block and chessboard DT, we also accumulate the relative x- and y-coordinates with respect to the current point and the iteration will stop when there is no distance change. After the iteration stops, we obtain the relative x- and y-coordinates of the closest point on the boundary which has the shortest path in the image. Let (dx_1, dy_1) and (dx_2, dy_2) denote the relative coordinates of the closest points on the boundary for city-block and chessboard, respectively. The quasi-Euclidean distance is computed by the following formula to construct the mapping look-up table with 5-tuples $(dx_1, dy_1, dx_2, dy_2, d_e)$:

$$d_e = \min\left(\sqrt{dx_1^2 + dy_1^2}, \sqrt{dx_2^2 + dy_2^2}\right). \tag{6.9}$$

It can be easily proved from the definitions of distance measure that the relationship among city-block (d_4), chessboard (d_8), and Euclidean (d_e) is $d_4 \geq d_e \geq d_8$. In most of the situations the Euclidean distance can be exactly obtained by Equation 6.3 except when the closest point on the boundary for the Euclidean distance is not located at the same position as that for the city-block or for the chessboard distances (e.g., see Figure 6.4).

The second approximation to the Euclidean distance is to combine the city-block and chessboard DTs together and to map them into the Euclidean distance. The city-block and chessboard distance SEs are expressed as follows:

$$k_4 = \begin{bmatrix} -2 & -1 & -2 \\ -1 & 0 & -1 \\ -2 & -1 & -2 \end{bmatrix}, \quad k_8 = \begin{bmatrix} -1 & -1 & -1 \\ -1 & 0 & -1 \\ -1 & -1 & -1 \end{bmatrix}. \tag{6.10}$$

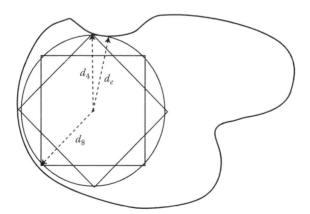

FIGURE 6.4 Special case when the closest boundary point for Euclidean distance (d_e) does not occur at the same position as that for city-block (d_4) or for chessboard (d_8).

We can calculate the Euclidean distance by the following formula:

$$d_e = \sqrt{(d_4 - d_8)^2 + (d_8)^2}. \tag{6.11}$$

Because the city-block distance is the summation of two absolute values in x- and y-coordinates and the chessboard distance is the maximum of them, the subtraction $d_4 - d_8$ is equal to the minimum of two absolute values. Equation 6.11 is exactly equivalent to the definition of Euclidean distance. In determining a minimum distance, this equality does not hold when the city-block and chessboard distances are measured from the different closest boundary points.

The third approximation is to improve the second method by separating the neighborhoods into four quadrants. We use four city-block and another four chessboard distance SEs. The masks applied are given as

$$(k_4)_1 = \begin{bmatrix} -1 & -2 \\ \underline{0} & -1 \end{bmatrix}, \ (k_4)_2 = \begin{bmatrix} -2 & -1 \\ -1 & \underline{0} \end{bmatrix}, \ (k_4)_3 = \begin{bmatrix} -1 & \underline{0} \\ -2 & -1 \end{bmatrix}, \ (k_4)_4 = \begin{bmatrix} \underline{0} & -1 \\ -1 & -2 \end{bmatrix} \tag{6.12}$$

$$(k_8)_1 = \begin{bmatrix} -1 & -1 \\ \underline{0} & -1 \end{bmatrix}, \ (k_8)_2 = \begin{bmatrix} -1 & -1 \\ -1 & \underline{0} \end{bmatrix}, \ (k_8)_3 = \begin{bmatrix} -1 & \underline{0} \\ -1 & -1 \end{bmatrix}, \ (k_8)_4 = \begin{bmatrix} \underline{0} & -1 \\ -1 & -1 \end{bmatrix}. \tag{6.13}$$

Note that the location of the origin is underlined. The origins of the first pair, $(k_4)_1$ and $(k_8)_1$, are located at the bottom-left corner. The origins of the second, third, and fourth pairs are located at the bottom-right, upper-right, and upper-left corners, respectively. The Euclidean distance can be obtained by computing four pairs of city-block and chessboard distances according to Equation 6.8 and then selecting the minimum value of the four. This method has the advantage of confining the location of the closest boundary points to be within the same quadrant. However, the exceptional case of Figure 6.4 is still a problem.

The first of the three approximations which extracts the relative x- and y-coordinates of the closest points on the boundary to any interior point using city-block and chessboard computation and then calculating the Euclidean distance, gives the most precise Euclidean distance measure but requires extra computation. The second one is the simplest for implementation but has the least precision. Its accuracy is improved by the third approximation but requires additional computation or hardware.

Different approximations to Euclidean distance measure, such as octagonal using the alternative city-block and chessboard [Rosenfeld and Kak 1982], Chamfer 3-4 using a 3×3 kernel, and Chamfer 5-7-11 using a 5×5 kernel [Borgefors 1984], can also be adapted to the morphological approach.

The SEs of the Chamfer 3-4 and 5-7-11 distance measures are constructed as follows:

$$\text{Chamfer 3-4: } \begin{bmatrix} -4 & -3 & -4 \\ -3 & 0 & -3 \\ -4 & -3 & -4 \end{bmatrix}$$

$$\text{Chamfer 5-7-11: } \begin{bmatrix} -14 & -11 & -10 & -11 & -14 \\ -11 & -7 & -5 & -7 & -11 \\ -10 & -5 & 0 & -5 & -10 \\ -11 & -7 & -5 & -7 & -11 \\ -14 & -11 & -10 & -11 & -14 \end{bmatrix}$$

6.4 Decomposition of Distance SEs

Implementation difficulties arise when an image is operated using a structuring element of a large size as in the DT algorithm. To solve this problem, the decomposition of these large-sized distance SEs into recursive operations of small-sized structuring components is critical for the efficiency of the algorithm. All the distance measures have this property: their values increase outwardly from the center point and are symmetrical with respect to the central point.

6.4.1 Decomposition of City-Block and Chessboard Distance SEs

According to the properties of grayscale operations, when an image f is eroded by a large sized structuring element k, which can be decomposed into dilations of several small structuring components k_i, we may obtain the same result by sequential erosions with these small structuring components [Serra 1982; Shih and Mitchell 1989]. This can be expressed as follows:

$$k = k_1 \oplus k_2 \oplus \cdots \oplus k_n \tag{6.14}$$

$$f \ominus k = [\cdots [[f \ominus k_1] \ominus k_2] \cdots] \ominus k_n. \tag{6.15}$$

City-block and chessboard distance SEs have a linear separable slope [Shih and Mitchell 1991] in any direction from the center. The k_is used for decomposition are identical to the k_4 and k_8 expressed in Equation 6.10. That is, if the dimension of the city-block or chessboard distance structuring element is $(2n + 3) \times (2n + 3)$, then it can be decomposed into n successive morphological dilations using the 3×3 components. Examples of the city block and chessboard distance decompositions are shown below.

Example 6.3: Let $k(x,y) = -(|x|+|y|)$, which is a city-block distance structuring element. For simplicity, only the structuring element of size 9×9 is shown:

$$k_{(9\times9)} = \begin{bmatrix} -8 & -7 & -6 & -5 & -4 & -5 & -6 & -7 & -8 \\ -7 & -6 & -5 & -4 & -3 & -4 & -5 & -6 & -7 \\ -6 & -5 & -4 & -3 & -2 & -3 & -4 & -5 & -6 \\ -5 & -4 & -3 & -2 & -1 & -2 & -3 & -4 & -5 \\ -4 & -3 & -2 & -1 & 0 & -1 & -2 & -3 & -4 \\ -5 & -4 & -3 & -2 & -1 & -2 & -3 & -4 & -5 \\ -6 & -5 & -4 & -3 & -2 & -3 & -4 & -5 & -6 \\ -7 & -6 & -5 & -4 & -3 & -4 & -5 & -6 & -7 \\ -8 & -7 & -6 & -4 & -4 & -5 & -6 & -7 & -8 \end{bmatrix}$$

We select a 3×3 structuring element k as

$$k_{(3\times3)} = \begin{bmatrix} -2 & -1 & -2 \\ -1 & 0 & -1 \\ -2 & -1 & -2 \end{bmatrix}$$

Hence,

$$f \ominus k_{9\times9} = (((f \ominus k_{3\times3}) \ominus k_{3\times3}) \ominus k_{3\times3}) \ominus k_{3\times3}.$$

Example 6.4: Let $k(x,y) = -\max(|x|,|y|)$ which is a chessboard distance structuring element. For simplicity, only the structuring element of size 9×9 is shown.

$$k_{(9\times9)} = \begin{bmatrix} -4 & -4 & -4 & -4 & -4 & -4 & -4 & -4 & -4 \\ -4 & -3 & -3 & -3 & -3 & -3 & -3 & -3 & -4 \\ -4 & -3 & -2 & -2 & -2 & -2 & -2 & -3 & -4 \\ -4 & -3 & -2 & -1 & -1 & -1 & -2 & -3 & -4 \\ -4 & -3 & -2 & -1 & 0 & -1 & -2 & -3 & -4 \\ -4 & -3 & -2 & -1 & -1 & -1 & -2 & -3 & -4 \\ -4 & -3 & -2 & -2 & -2 & -2 & -2 & -3 & -4 \\ -4 & -3 & -3 & -3 & -3 & -3 & -3 & -3 & -4 \\ -4 & -4 & -4 & -4 & -4 & -4 & -4 & -4 & -4 \end{bmatrix}$$

We select a 3×3 structuring element k as

$$k_{(3\times3)} = \begin{bmatrix} -1 & -1 & -1 \\ -1 & 0 & -1 \\ -1 & -1 & -1 \end{bmatrix}$$

Hence,

$$f \ominus k_{9\times9} = (((f \ominus k_{3\times3}) \ominus k_{3\times3}) \ominus k_{3\times3}) \ominus k_{3\times3}.$$

6.4.2 Decomposition of the Euclidean Distance Structuring Element

The decomposition of the Euclidean distance structuring element (EDSE) cannot be treated as dilations of small structuring components, since it poses an additional problem in the off-axis and off-diagonal directions. However, if a EDSE k can be segmented into the pointwise maximum selection of multiple linearly-sloped structuring components k_i (see Equation 6.16), then the gray-scale erosion of an image with a Euclidean structuring element is equivalent to the minimum of the outputs (see Equation 6.17) when the image is individually eroded by these structuring components. This can be expressed as

$$k = \max(k_1, k_2, \ldots, k_n) \tag{6.16}$$

$$f \ominus k = \min(f \ominus k_1, f \ominus k_2, \ldots, f \ominus k_n). \tag{6.17}$$

Since each structuring component k_i has a linear slope, it can be further decomposed by Equation 6.14 into the dilation of its structuring subcomponents k_{ij}. The procedure to construct these structuring components and sub-components is described below.

6.4.2.1 Construction Procedure

We describe the construction procedure by illustrating a 9×9 Euclidean distance structuring element. Any $N \times N$ EDSE can be decomposed into the operations of 3×3 ones by following a similar procedure. Detailed generalization and mathematical proof of the decomposition technique can be found in Shih and Mitchell [1991]. A 9×9 EDSE is represented as

$$k_{e(9\times9)} = \begin{bmatrix} -d_4 & -d_3 & -d_2 & -d_1 & -d_0 & -d_1 & -d_2 & -d_3 & -d_4 \\ -d_3 & -c_3 & -c_2 & -c_1 & -c_0 & -c_1 & -c_2 & -c_3 & -d_3 \\ -d_2 & -c_2 & -b_2 & -b_1 & -b_0 & -b_1 & -b_2 & -c_2 & -d_2 \\ -d_1 & -c_1 & -b_1 & -a_1 & -a_0 & -a_1 & -b_1 & -c_1 & -d_1 \\ -d_0 & -c_0 & -b_0 & -a_0 & 0 & -a_0 & -b_0 & -c_0 & -d_0 \\ -d_1 & -c_1 & -b_1 & -a_1 & -a_0 & -a_1 & -b_1 & -c_1 & -d_1 \\ -d_2 & -c_2 & -b_2 & -b_1 & -b_0 & -b_1 & -b_2 & -c_2 & -d_2 \\ -d_3 & -c_3 & -c_2 & -c_1 & -c_0 & -c_1 & -c_2 & -c_3 & -d_3 \\ -d_4 & -d_3 & -d_2 & -d_1 & -d_0 & -d_1 & -d_2 & -d_3 & -d_4 \end{bmatrix},$$

where $a_0 = 1$, $a_1 = \sqrt{2}$, $b_0 = 2$, $b_1 = \sqrt{5}$, $b_2 = 2\sqrt{2}$, $c_0 = 3$, $c_1 = \sqrt{10}$, $c_2 = \sqrt{13}$, $c_3 = 3\sqrt{2}$, $d_0 = 4$, $d_1 = \sqrt{17}$, $d_2 = 2\sqrt{5}$, $d_3 = 5$, $d_4 = 4\sqrt{2}$.

1. Select the central 3×3 window to be k_{11}:

$$k_{11} = \begin{bmatrix} -a_1 & -a_0 & -a_1 \\ -a_0 & 0 & -a_0 \\ -a_1 & -a_0 & -a_1 \end{bmatrix}$$

The k_{11} corresponds to the structuring component k_1 in Equations 6.16 and 6.17.

2. Because b_0, b_1, and b_2 do not represent equal increments, we need to account for the increasing difference offset. The k_{21} consists of $-b_1$ and $-b_2$. The k_{22} consisting of $(b_1 - b_0)$ and 0 is used as the offset. These matrices are shown as follows:

$$k_{21} = \begin{bmatrix} -b_2 & -b_1 & -b_2 \\ -b_1 & \times & -b_1 \\ -b_2 & -b_1 & -b_2 \end{bmatrix}$$

$$k_{22} = \begin{bmatrix} 0 & b_1 - b_0 & 0 \\ b_1 - b_0 & \times & b_1 - b_0 \\ 0 & b_1 - b_0 & 0 \end{bmatrix},$$

where $\times$ means a don't care term. The dilation of k_{21} and k_{22} will construct the 5×5 k_2 corresponding to Equations 6.16 and 6.17 because the extended values outside the window are regarded as $-\infty$. The k_2, which is shown below, has the same values on the boundary as those of k_e and has smaller values in the interior than does k_e. The maximum of k_1 and k_2 will generate exactly the values as the 5×5 Euclidean distance structuring element.

$$k_2 = \begin{bmatrix} -b_2 & -b_1 & -b_0 & -b_1 & -b_2 \\ -b_1 & -b_0 & -b_1 & -b_0 & -b_1 \\ -b_0 & -b_1 & -b_0 & -b_1 & -b_0 \\ -b_1 & -b_0 & -b_1 & -b_0 & -b_1 \\ -b_2 & -b_1 & -b_0 & -b_1 & -b_2 \end{bmatrix}$$

3. Repeat the selection procedure in step 2 until the total size of the EDSE is reached. For example, the k_{31}, k_{32}, and k_{33} are chosen as follows:

$$k_{31} = \begin{bmatrix} -c_3 & -c_2 & -c_3 \\ -c_2 & \times & -c_2 \\ -c_3 & -c_2 & -c_3 \end{bmatrix}$$

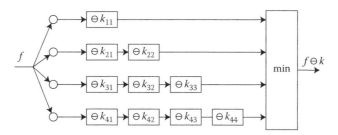

FIGURE 6.5 The decomposition diagram of a 9×9 Euclidean distance structuring element.

$$k_{32} = \begin{bmatrix} 0 & c_2 - c_1 & 0 \\ c_2 - c_1 & \times & c_2 - c_1 \\ 0 & c_2 - c_1 & 0 \end{bmatrix}$$

$$k_{33} = \begin{bmatrix} 0 & c_1 - c_0 & 0 \\ c_1 - c_0 & \times & c_1 - c_0 \\ 0 & c_1 - c_0 & 0 \end{bmatrix}$$

Hence, we have

$$k_{e(9 \times 9)} = \max(k_{1(3 \times 3)}, k_{2(5 \times 5)}, k_{3(7 \times 7)})$$
$$= \max(k_{1(3 \times 3)}, k_{21(3 \times 3)} \oplus k_{22(3 \times 3)}, k_{31(3 \times 3)} \oplus k_{32(3 \times 3)} \oplus k_{33(3 \times 3)})$$

An application of the EDSE may be found in Borgefors [1984].

A decomposition diagram for a 9×9 EDSE is shown in Figure 6.5 in which a binary image f is simultaneously fed to four levels and sequentially processed with the grayscale erosion. Finally all the outputs are combined by a minimum operator.

6.4.2.2 Computational Complexity

In general, suppose that the image size is $(2n + 1) \times (2n + 1)$, then the Euclidean distance computation needs $n(n + 1)/2$ grayscale erosions. For the 9×9 example of Figure 6.5, $n = 4$. The decomposition can be implemented quite efficiently on a parallel pipelined computer. The input goes to all n levels simultaneously and is sequentially pipelined to the next operator at each level. After appropriate delays for synchronization, all outputs are fed to the minimum operator. Alternatively, we can compare the minimum of first- and second-level outputs, and again compare the result with the third-level output, and so forth. Hence, the time complexity of this decomposition algorithm is of order n. The real-valued DTs are considered during the process and the result is approximated to the closest integer in the final output image.

6.5 Iterative Erosion Algorithm

The EDSE is expressed by $k(x, y) = -\sqrt{x^2 + y^2}$. The SEDSE is expressed by $k^2(x, y) = -(x^2 + y^2)$. The following is an example of SEDSE with size of 9×9:

$$k^2_{(9\times9)} = \begin{bmatrix}
-32 & -25 & -20 & -17 & -16 & -17 & -20 & -25 & -32 \\
-25 & -18 & -13 & -10 & -9 & -10 & -13 & -18 & -25 \\
-20 & -13 & -8 & -5 & -4 & -5 & -8 & -13 & -20 \\
-17 & -10 & -5 & -2 & -1 & -2 & -5 & -10 & -17 \\
-16 & -9 & -4 & -1 & 0 & -1 & -4 & -9 & -16 \\
-17 & -10 & -5 & -2 & -1 & -2 & -5 & -10 & -17 \\
-20 & -13 & -8 & -5 & -4 & -5 & -8 & -13 & -20 \\
-25 & -18 & -13 & -10 & -9 & -10 & -13 & -18 & -25 \\
-32 & -25 & -20 & -17 & -16 & -17 & -20 & -25 & -32
\end{bmatrix} \quad (6.18)$$

Since the SEDSE changes uniformly both in the vertical and horizontal directions, the aforementioned structuring element can be decomposed into

$$k^2_{(9\times9)} = \begin{bmatrix} -2 & -1 & -2 \\ -1 & 0 & -1 \\ -2 & -1 & -2 \end{bmatrix} \oplus \begin{bmatrix} -6 & -3 & -6 \\ -3 & 0 & -3 \\ -6 & -3 & -6 \end{bmatrix} \oplus \begin{bmatrix} -10 & -5 & -10 \\ -5 & 0 & -5 \\ -10 & -5 & -10 \end{bmatrix} \oplus \begin{bmatrix} -14 & -7 & -14 \\ -7 & 0 & -7 \\ -14 & -7 & -14 \end{bmatrix}$$

$$(6.19)$$

Let the structuring component be denoted by $g(l)$ and expressed as

$$g(l) = \begin{bmatrix}
-(4l-2) & -(2l-1) & -(4l-2) \\
-(2l-1) & 0 & -(2l-1) \\
-(4l-2) & -(2l-1) & -(4l-2)
\end{bmatrix} \quad (6.20)$$

where l indicates the iteration numbers. Hence, the EDT is the iterative erosions by a set of small structuring components and then a square-root operation. Given a binary image f, which is represented by $+\infty$ (or a large number) and 0. The following is the *iterative erosion algorithm* (IEA) [Shih and Wu 1992] and its flowchart is shown in Figure 6.6.

1. Set the value of foreground to a large number, and the value of background to 0.
2. Initialize $l = 1$.
3. $d = f \ominus g(l)$.

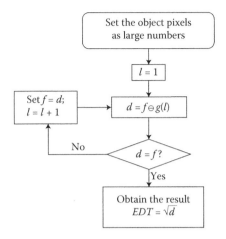

FIGURE 6.6 The flowchart of the IEA.

4. If $d \neq f$, let $f = d$ and l++.

5. Repeat steps 3 and 4 until $d = f$.

6. Take a square root of d (the distance image). That is $EDT = \sqrt{d}$.

Example 6.5: Let a binary image f of size 12×11 be:

$$f = \begin{bmatrix} 0 & 0 & 0 & 0 & 0 & 0 & 0 & 0 & 0 & 0 & 0 & 0 \\ 0 & 0 & 0 & 0 & 0 & 0 & 0 & 0 & 0 & 0 & 0 & 0 \\ 0 & 0 & 255 & 255 & 255 & 255 & 255 & 255 & 255 & 255 & 0 & 0 \\ 0 & 0 & 255 & 255 & 255 & 255 & 255 & 255 & 255 & 255 & 0 & 0 \\ 0 & 0 & 255 & 255 & 255 & 255 & 255 & 255 & 255 & 255 & 0 & 0 \\ 0 & 0 & 255 & 255 & 255 & 255 & 255 & 255 & 0 & 0 & 0 & 0 \\ 0 & 0 & 255 & 255 & 255 & 255 & 255 & 255 & 0 & 0 & 0 & 0 \\ 0 & 0 & 255 & 255 & 255 & 255 & 0 & 0 & 0 & 0 & 0 & 0 \\ 0 & 0 & 255 & 255 & 255 & 255 & 0 & 0 & 0 & 0 & 0 & 0 \\ 0 & 0 & 0 & 0 & 0 & 0 & 0 & 0 & 0 & 0 & 0 & 0 \\ 0 & 0 & 0 & 0 & 0 & 0 & 0 & 0 & 0 & 0 & 0 & 0 \end{bmatrix}, \quad (6.21)$$

where 255 denotes the object pixel and 0 the background pixel. In the first step of applying the erosion by

$$g(1) = \begin{bmatrix} -2 & -1 & -2 \\ -1 & 0 & -1 \\ -2 & -1 & -2 \end{bmatrix},$$

we have

$$d = \begin{bmatrix}
0 & 0 & 0 & 0 & 0 & 0 & 0 & 0 & 0 & 0 & 0 & 0 \\
0 & 0 & 0 & 0 & 0 & 0 & 0 & 0 & 0 & 0 & 0 & 0 \\
0 & 0 & 1 & 1 & 1 & 1 & 1 & 1 & 1 & 1 & 0 & 0 \\
0 & 0 & 1 & 255 & 255 & 255 & 255 & 255 & 255 & 1 & 0 & 0 \\
0 & 0 & 1 & 255 & 255 & 255 & 255 & 1 & 1 & 1 & 0 & 0 \\
0 & 0 & 1 & 255 & 255 & 255 & 255 & 1 & 0 & 0 & 0 & 0 \\
0 & 0 & 1 & 255 & 255 & 2 & 1 & 1 & 0 & 0 & 0 & 0 \\
0 & 0 & 1 & 255 & 255 & 1 & 0 & 0 & 0 & 0 & 0 & 0 \\
0 & 0 & 1 & 1 & 1 & 1 & 0 & 0 & 0 & 0 & 0 & 0 \\
0 & 0 & 0 & 0 & 0 & 0 & 0 & 0 & 0 & 0 & 0 & 0 \\
0 & 0 & 0 & 0 & 0 & 0 & 0 & 0 & 0 & 0 & 0 & 0
\end{bmatrix}$$

After applying the erosion by

$$g(2) = \begin{bmatrix}
-6 & -3 & -6 \\
-3 & 0 & -3 \\
-6 & -3 & -6
\end{bmatrix},$$

we have

$$d = \begin{bmatrix}
0 & 0 & 0 & 0 & 0 & 0 & 0 & 0 & 0 & 0 & 0 & 0 \\
0 & 0 & 0 & 0 & 0 & 0 & 0 & 0 & 0 & 0 & 0 & 0 \\
0 & 0 & 1 & 1 & 1 & 1 & 1 & 1 & 1 & 1 & 0 & 0 \\
0 & 0 & 1 & 4 & 4 & 4 & 4 & 4 & 4 & 1 & 0 & 0 \\
0 & 0 & 1 & 4 & 255 & 255 & 5 & 2 & 1 & 1 & 0 & 0 \\
0 & 0 & 1 & 4 & 8 & 5 & 4 & 1 & 0 & 0 & 0 & 0 \\
0 & 0 & 1 & 4 & 5 & 2 & 1 & 1 & 0 & 0 & 0 & 0 \\
0 & 0 & 1 & 4 & 4 & 1 & 0 & 0 & 0 & 0 & 0 & 0 \\
0 & 0 & 1 & 1 & 1 & 1 & 0 & 0 & 0 & 0 & 0 & 0 \\
0 & 0 & 0 & 0 & 0 & 0 & 0 & 0 & 0 & 0 & 0 & 0 \\
0 & 0 & 0 & 0 & 0 & 0 & 0 & 0 & 0 & 0 & 0 & 0
\end{bmatrix}$$

After applying the erosion by

$$g(3) = \begin{bmatrix}
-10 & -5 & -10 \\
-5 & 0 & -5 \\
-10 & -5 & -10
\end{bmatrix},$$

we obtain the result below. Continuously applying the erosion by $g(4)$, the result does not change. Therefore, this result is exactly the square of its Euclidean distance.

$$d = \begin{bmatrix} 0 & 0 & 0 & 0 & 0 & 0 & 0 & 0 & 0 & 0 & 0 & 0 \\ 0 & 0 & 0 & 0 & 0 & 0 & 0 & 0 & 0 & 0 & 0 & 0 \\ 0 & 0 & 1 & 1 & 1 & 1 & 1 & 1 & 1 & 1 & 0 & 0 \\ 0 & 0 & 1 & 4 & 4 & 4 & 4 & 4 & 4 & 1 & 0 & 0 \\ 0 & 0 & 1 & 4 & 9 & 9 & 5 & 2 & 1 & 1 & 0 & 0 \\ 0 & 0 & 1 & 4 & 8 & 5 & 4 & 1 & 0 & 0 & 0 & 0 \\ 0 & 0 & 1 & 4 & 5 & 2 & 1 & 1 & 0 & 0 & 0 & 0 \\ 0 & 0 & 1 & 4 & 4 & 1 & 0 & 0 & 0 & 0 & 0 & 0 \\ 0 & 0 & 1 & 1 & 1 & 1 & 0 & 0 & 0 & 0 & 0 & 0 \\ 0 & 0 & 0 & 0 & 0 & 0 & 0 & 0 & 0 & 0 & 0 & 0 \\ 0 & 0 & 0 & 0 & 0 & 0 & 0 & 0 & 0 & 0 & 0 & 0 \end{bmatrix}$$

6.5.1 Redundant Calculations in the IEA

The IEA performs the erosion on the entire image. Figure 6.7 illustrates its redundant calculations. Figure 6.7a is an image of size 13×13. Note that 0 and 1 indicate the background and object pixels, respectively. Set the object value to be a large number (e.g., 100) as shown in Figure 6.7b1. Then, the morphological structuring component, $g(l)$, can be obtained from Equation 6.20 by setting $l = 1, 2, 3, 4, 5,$ and 6 as shown in Figures 6.7b2, c2, d2, e2, f2, and g2. We obtain Figure 6.7c1, using an erosion of Figures 6.7b1, b2. Similarly, Figures 6.7d1, e1, f1, g1, and h can be obtained, respectively, by erosions of Figures 6.7c1, c2, d1, d2, e1, e2, f1, f2, and g1, g2.

Note that it is not necessary to calculate all the object pixels in each iteration. For example, comparing Figures 6.7b1 and c1, the pixels within the bold rectangle in Figure 6.7c1, are not changed. Therefore, the calculations inside the bold rectangle are useless. On the other hand, comparing Figures 6.7c1 and d1, the changes are only on the outside border of the bold rectangle in Figure 6.7d1. That is, the values are changed from 100 to 4. Therefore, the calculations inside the bold rectangle and the pixels with value 1 are redundant.

6.5.2 An Improved Iterative Erosion Algorithm

To reduce the aforementioned redundant calculations, a *chessboard distance transformation* is applied first. Then, a morphological erosion is applied by the structuring component $g(l)$ in Equation 6.20. In each iteration, we simply process a portion of object pixels that have the chessboard distance l or a

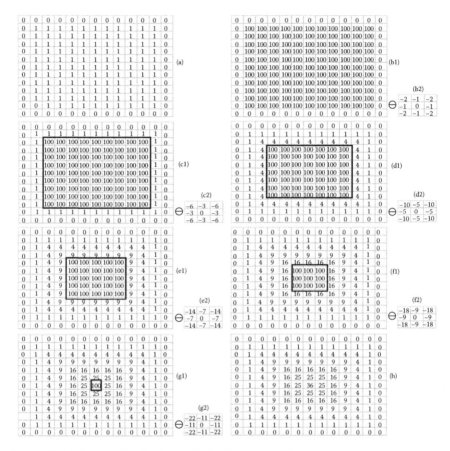

FIGURE 6.7   The example of redundant calculations of EDT.

few smaller. We describe the details in pseudo codes below. Its flowchart is shown in Figure 6.8. Let D_{ij} represent the output pixel's chessboard distance at row i and column j.

Improved Iterative Erosion Algorithm

Let $start\_d_8 = 1$, and let the chessboard DT of the original image be D_{ij}.

```
for l = 1, max(D_ij), 1 do
begin
  for m = start_d8, l, 1 do
  begin
    flag = 0;
    for i = 1, length, 1 do
      for j = 1, width, 1 do
        if (D_ij = m) then
        begin
          new_image_ij = (old_image ⊖ g(m))_ij;
```

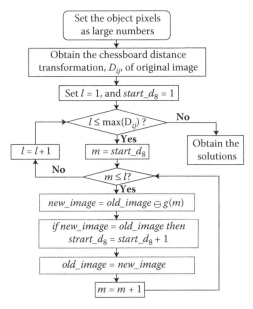

FIGURE 6.8   The flowchart of the IIEA.

```
        if (new_image_ij ≠ old_image) then flag = 1;
     end
   if (flag = 0) then start_d_8 = start_d_8 + 1;
end
for i = 1, length, 1 do
 for j = 1, width, 1 do
 old_image_ij = new_image_ij;
end
```

Proof: The IEA based on the MM properties has been proved to be able to extract the exact EDT in Shih and Wu [1992]. We now prove the correctness of the improved algorithm using mathematical induction. Steps 1 and 2 are the two-scan algorithm that achieves the *chessboard DT*. The fourth step is a square-root operation. Therefore in this proof, we emphasize the operation in step 3.

(a) **Base Case:** When $l = 1$

That is $D_{ij} = 1$ and $g(1) = \begin{bmatrix} -2 & -1 & -2 \\ -1 & 0 & -1 \\ -2 & -1 & -2 \end{bmatrix}$. If an object pixel is 8-adjacent,

but not 4-adjacent to the background, its output will be two. If an object pixel is 4-adjacent to the background, its output will be 1. The remaining object pixels (i.e., their chessboard distances > 1) will remain unchanged (i.e., $+\infty$). Thus the result of an erosion applied to all background and object pixels is equivalent to that of an erosion performed simply on the object pixels which have the chessboard distance 1.

(b) **Induction Hypothesis:** When $l = k$ and $k < \max(D_{ij})$.

The erosion is applied simply on the object pixels, which have D_{ij} from m to k, where $m \leq k$. An assumption is made that the object pixels with $D_{ij} < m$ have already reached their minimum *squared Euclidean distances*.

(c) **Induction Case:** When $l = k + 1$.

The variable *start_d$_8$* will be greater than or equal to m. Let $m' = start\_$ d_8. From inductive hypothesis we know that the object pixels with $D_{ij} < m'$ have already reached the minimum squared Euclidean distances. All the object pixels can be categorized into three classes according to their chessboard distances:

1. $D_{ij} < m'$
2. $D_{ij} > k + 1$
3. $m' \leq D_{ij} \leq k + 1$

When an erosion is applied on the pixels in the first class, the output remains unchanged (i.e., they are the squared Euclidean distances) according to the inductive hypothesis. When an erosion is performed on the pixels in the second class, the output remains unchanged (i.e., $+\infty$). Thus the result of an erosion applied to all background and object pixels in IEA is equivalent to that of an erosion performed simply on the object pixels, which have the chessboard distances as in the third class. Finally we conclude that our improved algorithm can extract the exact EDT.   □

6.5.3 An Example of Improved Iterative Erosion Algorithm

Figure 6.9 shows an example of the Improved Iterative Erosion Algorithm (IIEA). Figure 6.9a is a binary image where value 1 indicates object pixels and blank denotes background pixels. Figure 6.9b is the chessboard DT of Figure 6.9a. We obtain Figure 6.9c by changing the object pixels into 100. Figure 6.9d is the result after an erosion of Figure 6.9c with the structuring element $g(1)$. Note that the erosion is performed merely on the object pixels having the chessboard distance 1. Figure 6.9e is the result after another erosion of Figure 6.9d, with $g(2)$. Note that the erosion is operated on the object pixels having the chessboard distances 1 and 2. Similarly, Figures 6.9f and g are the results when the erosion is performed on the object pixels having the chessboard distances (2, 3) and (3, 4), respectively. There are totally 17×17 pixels in Figure 6.9. If we use the iterative erosion algorithm, the total pixels involved in the erosion are 1156. However, the total pixels

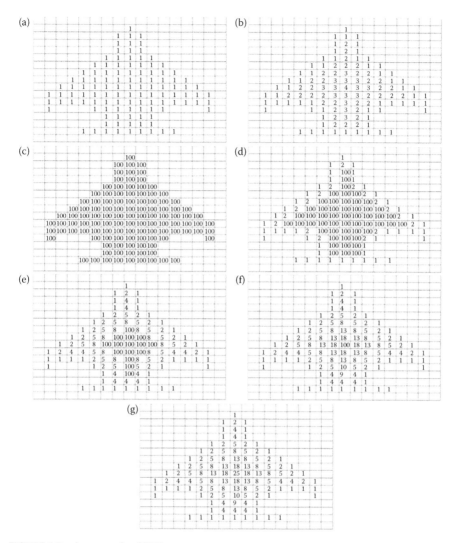

FIGURE 6.9 An example of IIEA.

calculated are 225 when IIEA is used. The computational complexity is significantly reduced.

To obtain the exact EDT, a very important strategy used here is that we must continually calculate the pixels with smaller chessboard distances in each iteration unless they are not changed anymore; that is, they have reached the minimum EDT. If we skip the calculation of the pixels with smaller chessboard distances, the resulting EDT is incorrect. Figure 6.10 illustrates its difference from the correct EDT. Figure 6.10a shows a binary image, and

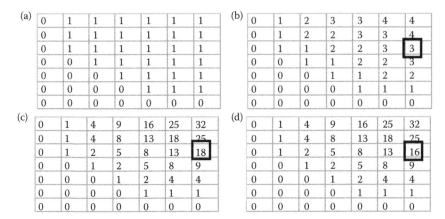

FIGURE 6.10 The example for achieving exact EDT by considering more pixels in IIEA.

Figure 6.10b shows its chessboard DT. If the structuring element is operated on the pixels with the chessboard distance l in the lth iteration, the result is shown in Figure 6.10c. To obtain the exact EDT as shown in Figure 6.10d, the pixels whose chessboard distance is equal to 3 in the fourth iteration must be calculated. Note that an error occurs as shown in a square box.

6.6 Two Scan–Based Algorithm

In this section, we present EDT by only two scans, just like the two-scan algorithm used for city-block and chessboard DTs.

6.6.1 Double Two-Scan Algorithm

The recursive morphological operation intends to feed back the output of the current pixel to overwrite its input and the new input is used for computation in the next scanning pixels. Therefore, each output obtained depends on the updated values, which reside preceding the current pixel. The resulting output image in recursive morphology inherently varies with respect to image scanning sequences. Let $N(p)$ be the set of neighbors preceding pixel p plus itself in a scanning sequence within the window of a structuring element. The *recursive erosion* of f by k, denoted by $f \circledcirc k$, is defined as

$$(f \circledcirc k)(x, y)= \min\{[(f \circledcirc k)(x + m, y + n)] - k(m, n)\}, \tag{6.22}$$

for all $(m, n) \in K$ and $(x + m, y + n) \in (N \cap F)$.

The SEs used in the first two steps of Double Two-Scan Algorithm (DTSA) are to achieve the chessboard distance transformation. Let k_1 and k_2 be given as follows:

$$k_1 = \begin{bmatrix} -1 & -1 & -1 \\ -1 & 0 & \times \\ \times & \times & \times \end{bmatrix}, \quad k_2 = \begin{bmatrix} \times & \times & \times \\ \times & 0 & -1 \\ -1 & -1 & -1 \end{bmatrix}. \tag{6.23}$$

Note that the resulting image using recursive morphology relies on the scanning order. Here, the scanning sequence of using k_1 is left-to-right top-to-bottom, and of using k_2 is right-to-left bottom-to-top.

The SEs $g_1(l)$ and $g_2(l)$ used in the third and fourth steps are functions of parameter l. That is

$$g_1(l) = \begin{bmatrix} -(4l-2) & -(2l-1) & -(4l-2) \\ -(2l-1) & 0 & \times \\ \times & \times & \times \end{bmatrix}, \tag{6.24}$$

$$g_2(l) = \begin{bmatrix} \times & \times & \times \\ \times & 0 & -(2l-1) \\ -(4l-2) & -(2l-1) & -(4l-2) \end{bmatrix}. \tag{6.25}$$

The pseudo code of the DTSA is given below and its flowchart is shown in Figure 6.11.

DTSA:
```
1. D = f ⊖ k₁
2. E = f ⊖ k₂
3. for i = 1, length, 1 do
     for j = 1, width, 1 do
       l = D_ij,
       ij = [f ⊖ g₁(l)]_ij
4. for i = length, 1, -1 do
     for j = width, 1, -1 do
       l = E_ij,
       Q_ij = [R ⊖ g₂(l)]_ij
5. for i = 1, length, 1 do
     for j = 1, width, 1 do
       Q_ij = √Q_ij
```

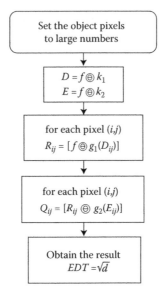

FIGURE 6.11   The flowchart of DTSA.

Note that DTSA cannot obtain correct results in some special cases. Figure 6.12 indicates the errors in DTSA.

Example 6.6: Let the binary image f be the same as that given in Equation 6.21. After the first step, we obtain

$$
D = \begin{bmatrix}
0 & 0 & 0 & 0 & 0 & 0 & 0 & 0 & 0 & 0 & 0 & 0 \\
0 & 0 & 0 & 0 & 0 & 0 & 0 & 0 & 0 & 0 & 0 & 0 \\
0 & 0 & 1 & 1 & 1 & 1 & 1 & 1 & 1 & 1 & 0 & 0 \\
0 & 0 & 1 & 2 & 2 & 2 & 2 & 2 & 2 & 1 & 0 & 0 \\
0 & 0 & 1 & 2 & 3 & 3 & 3 & 3 & 2 & 1 & 0 & 0 \\
0 & 0 & 1 & 2 & 3 & 4 & 4 & 3 & 0 & 0 & 0 & 0 \\
0 & 0 & 1 & 2 & 3 & 4 & 4 & 1 & 0 & 0 & 0 & 0 \\
0 & 0 & 1 & 2 & 3 & 4 & 0 & 0 & 0 & 0 & 0 & 0 \\
0 & 0 & 1 & 2 & 3 & 1 & 0 & 0 & 0 & 0 & 0 & 0 \\
0 & 0 & 0 & 0 & 0 & 0 & 0 & 0 & 0 & 0 & 0 & 0 \\
0 & 0 & 0 & 0 & 0 & 0 & 0 & 0 & 0 & 0 & 0 & 0
\end{bmatrix}
$$

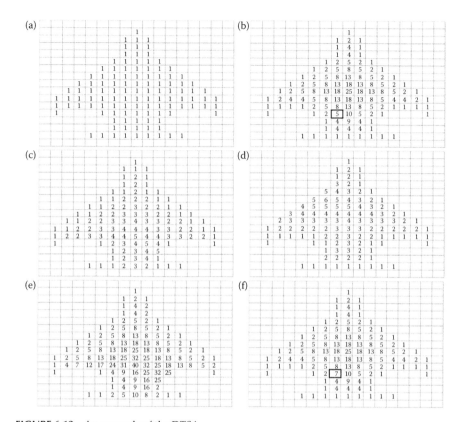

FIGURE 6.12 An example of the DTSA.

After the second step, the result is

$$
E=
\begin{bmatrix}
0 & 0 & 0 & 0 & 0 & 0 & 0 & 0 & 0 & 0 & 0 & 0 \\
0 & 0 & 0 & 0 & 0 & 0 & 0 & 0 & 0 & 0 & 0 & 0 \\
0 & 0 & 1 & 2 & 3 & 3 & 3 & 3 & 2 & 1 & 0 & 0 \\
0 & 0 & 1 & 2 & 3 & 3 & 2 & 2 & 2 & 1 & 0 & 0 \\
0 & 0 & 1 & 2 & 3 & 3 & 2 & 1 & 1 & 1 & 0 & 0 \\
0 & 0 & 1 & 2 & 2 & 2 & 2 & 1 & 0 & 0 & 0 & 0 \\
0 & 0 & 1 & 2 & 2 & 1 & 1 & 1 & 0 & 0 & 0 & 0 \\
0 & 0 & 1 & 2 & 2 & 1 & 0 & 0 & 0 & 0 & 0 & 0 \\
0 & 0 & 1 & 1 & 1 & 1 & 0 & 0 & 0 & 0 & 0 & 0 \\
0 & 0 & 0 & 0 & 0 & 0 & 0 & 0 & 0 & 0 & 0 & 0 \\
0 & 0 & 0 & 0 & 0 & 0 & 0 & 0 & 0 & 0 & 0 & 0
\end{bmatrix}
$$

After the third step, we obtain the result

$$
R = \begin{bmatrix}
0 & 0 & 0 & 0 & 0 & 0 & 0 & 0 & 0 & 0 & 0 & 0 \\
0 & 0 & 0 & 0 & 0 & 0 & 0 & 0 & 0 & 0 & 0 & 0 \\
0 & 0 & 1 & 1 & 1 & 1 & 1 & 1 & 1 & 1 & 0 & 0 \\
0 & 0 & 1 & 4 & 4 & 4 & 4 & 4 & 4 & 2 & 0 & 0 \\
0 & 0 & 1 & 4 & 9 & 9 & 9 & 9 & 7 & 2 & 0 & 0 \\
0 & 0 & 1 & 4 & 9 & 16 & 16 & 14 & 0 & 0 & 0 & 0 \\
0 & 0 & 1 & 4 & 9 & 16 & 23 & 2 & 0 & 0 & 0 & 0 \\
0 & 0 & 1 & 4 & 9 & 16 & 0 & 0 & 0 & 0 & 0 & 0 \\
0 & 0 & 1 & 4 & 9 & 2 & 0 & 0 & 0 & 0 & 0 & 0 \\
0 & 0 & 0 & 0 & 0 & 0 & 0 & 0 & 0 & 0 & 0 & 0 \\
0 & 0 & 0 & 0 & 0 & 0 & 0 & 0 & 0 & 0 & 0 & 0
\end{bmatrix}
$$

After the fourth step, the result is

$$
R = \begin{bmatrix}
0 & 0 & 0 & 0 & 0 & 0 & 0 & 0 & 0 & 0 & 0 & 0 \\
0 & 0 & 0 & 0 & 0 & 0 & 0 & 0 & 0 & 0 & 0 & 0 \\
0 & 0 & 1 & 1 & 1 & 1 & 1 & 1 & 1 & 1 & 0 & 0 \\
0 & 0 & 1 & 4 & 4 & 4 & 4 & 4 & 4 & 1 & 0 & 0 \\
0 & 0 & 1 & 4 & 9 & 9 & 5 & 2 & 1 & 1 & 0 & 0 \\
0 & 0 & 1 & 4 & 8 & 5 & 4 & 1 & 0 & 0 & 0 & 0 \\
0 & 0 & 1 & 4 & 5 & 2 & 1 & 1 & 0 & 0 & 0 & 0 \\
0 & 0 & 1 & 4 & 4 & 1 & 0 & 0 & 0 & 0 & 0 & 0 \\
0 & 0 & 1 & 1 & 1 & 1 & 0 & 0 & 0 & 0 & 0 & 0 \\
0 & 0 & 0 & 0 & 0 & 0 & 0 & 0 & 0 & 0 & 0 & 0 \\
0 & 0 & 0 & 0 & 0 & 0 & 0 & 0 & 0 & 0 & 0 & 0
\end{bmatrix}
$$

At last, by taking the square root for all object pixels, the result is the EDT.

Figures 6.12a and b show the original image and its SEDT, respectively. Figures 6.12c and d are the forward ($D = f \odot k_1$) and backward ($E = f \odot k_2$) *chessboard distance transformations* obtained by steps 1 and 2 of DTSA, respectively. Figures 6.12e and f are the forward and backward results obtained by steps 3 and 4, respectively. Note that Figure 6.12f presents an error in a square box.

6.6.2 Basic Ideas of Two-Scan Algorithms

Generally, the approaches for achieving EDT are in the way of acquiring the minimum Euclidean distance by checking neighborhood. That is, the "acquiring" is to acquire the distance information from its neighbors to obtain its current Euclidean distance. For example, IIEA calculates EDT using a morphological erosion of the original image by a structuring component $g(l)$. Figure 6.13 illustrates the acquiring procedure. Figure 6.13a is the original image in which the object pixels are changed into 100. Figures 6.13b and c are the structuring components, $g(l)$, when $l = 1$ and 2. Figure 6.13d shows the condition of achieving EDT by using $g(1)$ in the top-left object pixel. Note that the calculations are in the direction of left-to-right, top-to-bottom. Figure 6.13e shows the condition when calculating EDT by $g(2)$ in the center of the object. From Figures 6.13d and e, the EDT of each pixel is achieved based on its neighborhood.

To achieve EDT efficiently without extra cost, the deriving approach is adopted. The deriving is to propagate its current distance information to its neighbors for updating their distance values. That is, each pixel decides the *Euclidean distance* of its children (i.e., pixels in its neighborhood). Figure 6.14 shows the examples of the deriving approach. Figures 6.14a and c indicate the relation between a pixel and its children. In Figure 6.14a, the scanning path is in the direction of left-to-right, top-to-bottom. On the contrary, the scanning path in Figure 6.14c is in the direction of right-to-left,

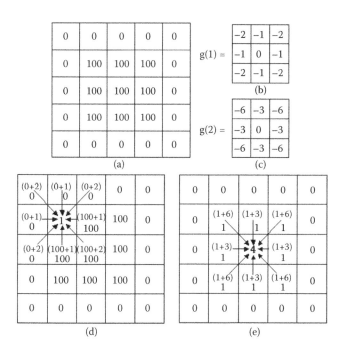

FIGURE 6.13 Examples of the acquiring procedure.

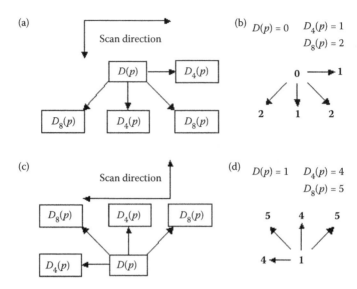

FIGURE 6.14 Examples of the deriving approach.

bottom-to-top. Let $D(p)$ be the value of SED of a pixel p. Let $D_4(p)$ and $D_8(p)$ be the SED values of 4- and 8-neighbors of p, respectively, and can be calculated by $D(p)$. The method for calculating $D_4(p)$ and $D_8(p)$ will be described later. Note that the concerned children of p are only 4 pixels in either scanning direction. Figures 6.14b and d are the examples of the ED value of children of p when $D(p) = 0$ and 1, respectively.

Figure 6.15 shows the examples when calculating the EDT in Figure 6.13a, based on the deriving approach. Figure 6.15a shows the deriving scheme when dealing with the top-left pixel, say (1,1). Only the object pixels are changed if there exists a smaller ED value. Therefore, the value in (2,2) is changed from 100 to 2. Continually dealing with the next pixel (2,1) as shown in Figure 6.14b, the value in (2,2) is changed once more since there exists a smaller value 1. Similarly, the value in (3,2) is changed from 100 to 2. In Figure 6.15c, the value in (3,2) is not changed because the original value 1 in (3,2) is smaller than the deriving value 4 from pixel (2,2). In the same way, the value in (2,3) is not changed. In Figure 6.15d, there are only two pixels changed in (3,3) and (4,3) because of the smaller deriving value from pixel (3,2). Note that the deriving values are shown in parenthesis.

6.6.3 Fundamental Lemmas

In this section, we present three lemmas.

Lemma 6.1: Pixels in the SEDT can be divided into a summation of two squared integers.

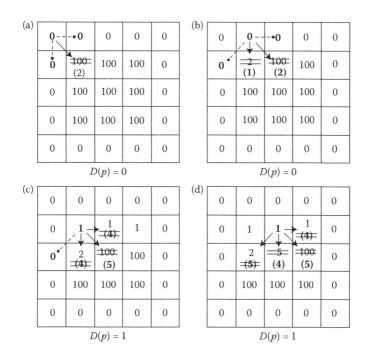

FIGURE 6.15 The examples when calculating an image based on the deriving approach.

Proof: Let two points be $p = (x, y)$ and $q = (u, v)$ in a digital image, where x, y, u, and v are integers. According to the definition of SEDT

$$d_s(p, q) = d_e^2(p, q) = (x - u)^2 + (y - v)^2,$$

it is obvious that $d_s(p, q)$ is the summation of $(x - u)^2$ and $(y - v)^2$. Therefore, Lemma 6.1 is proved by the definition of SEDT. □

Lemma 6.2: If $d_s(p, q)$ is divided into I_{max} and I_{min} where $d_s(p, q) = (I_{max})^2 + (I_{min})^2$ and $I_{max} \geq I_{min}$, $D_4(p)$ and $D_8(p)$ can be obtained by the following equations:

$$D_4(p) = \begin{cases} (I_{max} + 1)^2 + (I_{min})^2, & \text{or} \\ (I_{max})^2 + (I_{min} + 1)^2 \end{cases} \tag{6.26}$$

$$D_8(p) = (I_{max} + 1)^2 + (I_{min} + 1)^2 \tag{6.27}$$

Proof: If $d_s(p, q) = (I_{max})^2 + (I_{min})^2$, it means that the vertical and horizontal pixels between p and q are I_{max} and I_{min} (or I_{min} and I_{max}) as shown in Figure 6.16a. In Figure 6.16b, the SEDT of the 4-adjacent pixels of p can be obtained by increasing one pixel in the vertical or horizontal direction. Similarly, the SEDT of the 8-adjacent pixels of p can be obtained by increasing one pixel in

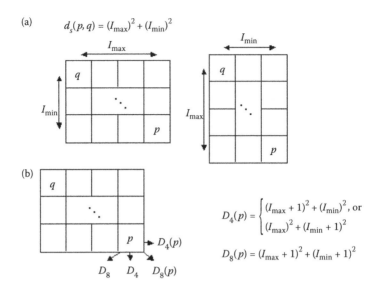

FIGURE 6.16   The $D_4\,(p)$ and $D_8\,(p)$ of pixel p.

the vertical direction and one pixel in the horizontal direction. Therefore, Equations 6.26 and 6.27 are proved.   □

   Lemma 6.3: If $d_s(p, q)$ can be divided into more than one summation of two squared integers, that is,

$$d_s(p,q)=\begin{cases}(I_{max\_1})^2+(I_{min\_1})^2\\(I_{max\_2})^2+(I_{min\_2})^2\\\vdots\\(I_{max\_n})^2+(I_{min\_n})^2\end{cases}, \tag{6.28}$$

we can check the neighborhood of pixel p to obtain the exact one.

   Proof: In Figure 6.17, the value of SEDT of pixel p can be obtained by its parents (H_1, H_2, and H_3) according to Lemma 6.2. Although $d_s(p, q)$ can be divided into more than one summation of two squared integers, the exact solution can be obtained by checking each pair of squared integers.

   Suppose there are n possibilities when dividing $d_s(p, q)$. If p is derived from H_1, $d_s(H_1, q)$ should be equal to $(I_{max} - 1)^2 + (I_{min})^2$ or $(I_{max})^2 + (I_{min} - 1)^2$. Similarly, If p is derived from H_3, $d_s(H_3, q)$ should be equal to $(I_{max} - 1)^2 + (I_{min})^2$ or $(I_{max})^2 + (I_{min} - 1)^2$. If p is derived from H_2, $d_s(H_2, q)$ should be equal to $(I_{max} - 1)^2 + (I_{min} - 1)^2$.   □

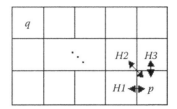

FIGURE 6.17   The relation between p and its parents.

6.6.4 TS₁—A Two Scan–Based EDT Algorithm for General Images

If only eight neighbors are used for EDT, there exist some errors from $24.47°$ as described in Danielsson [1980]. By reflecting with respect to x-, y-, $45°$-, and $135°$-axes, seven more error cases appear. On the other hand, the acquiring approach needs additional space to record all the neighborhood information. To achieve the correct EDT in the two-scan based approach for general images (i.e., without obstacles), the following two skills are adopted. First, the deriving approach is used instead of the acquiring approach. Second, the neighborhood window is dynamically adjusted based on the Euclidean distance value.

Cuisenaire and Macq [1999] pointed out the relationship between the value of SEDT and the neighborhood window. That is, the larger SEDT value a pixel has, the bigger neighborhood window it uses. The cases indicating the relationship of the SEDT and its corresponding neighborhood windows are shown in Table 6.1. For example, if the value of SED of a pixel is greater than or equal to 116, a 5×5 neighborhood window is adopted. Figure 6.18 shows the neighborhood window of p, where the symbols $\times$ and $\vee$ indicate the concerned neighbors in the forward and backward raster scans, respectively. Note that because of no error occurrence, the neighbors in the directions $0°$, $45°$, $90°$, and $135°$ are not concerned when the sizes of the neighborhood window are greater than 3×3.

TABLE 6.1

The Relationship between the Value of SEDT and the Neighborhood Window

Neighborhood	I_{max}	I_{min}	SEDT
5*5	10	4	116
7*7	22	6	520
9*9	44	9	2017
11*11	67	11	4610
⋮	⋮	⋮	⋮

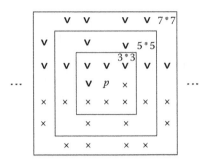

FIGURE 6.18 The example of the neighborhood window of pixel p.

Let the object pixels be a large number (larger than the square of a half of the image size) and the background pixels be 0. The two-scan based EDT algorithm, TS_1, is described below, and its flowchart is shown in Figure 6.19. Let $N_F(i, j)$ and $N_B(i, j)$ be the concerned neighbors of pixel (i, j) in forward and backward raster-scans, respectively.

TS_1 Algorithm:

A. (Forward) Raster-Scan:

```
1. for (i = 0;  i ≤ length;  i++)
2.    for (j = 0;  j ≤ length;  j++)
3.    { Divide f(i,j) into I_max and I_min, where
         f(i,j) = (I_max)² + (I_min)²;
4.       if there exists more than one division, check the
         parents of f(i,j) for exact I_max and I_min
5.       Decide the size of neighborhood window based on the
         value of f(i,j)
6.       for all q ∈ N_F(i,j), decide the SED value based on
         f(i,j)
      }
```

B. Backward (Reverse) Raster-Scan:

```
7. for (i = length;  i ≥ 0;  i--)
8. for (j = length;  j ≥ 0;  j--)
9.  {Divide f(i,j) into I_max and I_min, where
        f(i,j) = (I_max)² + (I_min)²;
10.      if there exists more than one division, check the
         parents of f(i,j) for exact I_max and I_min
11.  Decide the size of neighborhood window based on f(i,j)
12.      for all q ∈ N_B(i,j), decide the SED value based on
         pixel f(i,j)
13.      E(f(i,j)) = √f(i,j);
      }
```

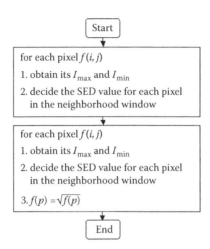

FIGURE 6.19 The flowchart of the TS_1 algorithm.

The rules for deciding the SED value of the neighbor of a pixel (i, j) are described below. Let pixel q with coordinates $(i + x, j + y)$ be one of the neighbors of pixel f with coordinates (i, j), where x and y are the distances. Let D_{max} and D_{min} be the values of $max(abs(x), abs(y))$ and $min(abs(x), abs(y))$, respectively. Let $f(i, j)$ and $f(i + x, j + y)$ be the SED value of pixels f and q, respectively. The equation for determining the SED value of q is

$$f(i+x, j+y) = \min(f(i+x, j+y),\ (I_{max} + D_{max})^2 + (I_{min} + D_{min})^2) \quad (6.29)$$

The examples for obtaining the SED value are given below.

$$f(i+1, j) = \min(f(i+1, j),\ (I_{max} + 1)^2 + (I_{min})^2) \quad (6.30)$$

$$f(i-1, j+1) = \min(f(i-1, j+1),\ (I_{max} + 1)^2 + (I_{min} + 1)^2) \quad (6.31)$$

$$f(i-2, j+1) = \min(f(i-2, j+1),\ (I_{max} + 2)^2 + (I_{min} + 1)^2) \quad (6.32)$$

To speed up the two-scan algorithm, we use a lookup table to replace the calculation of splitting a SED value into the summation of two squared integers. Table 6.2 shows a portion of the lookup table. Let N_4 and N_8 be the horizontal/vertical and diagonal neighbors of p, respectively. For example, if the SED value of p is 25, based on the lookup table, we can understand that 25 can be divided into two possible combinations; that is, $9 + 16$ and $0 + 25$. Then we check the SED values of its N_4 and N_8 neighbors to determine the correct combination. Note that $\times$ denotes that for some cases, N_4 and N_8 need not be checked.

TABLE 6.2

The Lookup Table for Obtaining the SED Values of Neighbors of a Pixel p

p (Value)	I_{min}	I_{max}	N_4	N_8	D_{min}	D_{max}	New SED
0	0	0	×	×	0	1	1
0	0	0	×	×	1	1	2
1	0	1	×	×	0	1	4
1	0	1	×	×	1	1	5
2	1	1	×	×	0	1	5
⋮	⋮	⋮	⋮	⋮	⋮	⋮	⋮
25	9	16	18	13	0	1	35
25	9	16	18	13	1	1	41
25	0	25	16	×	0	1	36
25	0	25	16	×	1	1	37
⋮	⋮	⋮	⋮	⋮	⋮	⋮	⋮

6.6.5 TS$_\infty$—A Two Scan–Based EDT Algorithm for Images with Obstacles

Generally, the images used in the shortest path planning contain some obstacles. Assume that an object of arbitrary shape moves from a starting point to an ending point in a finite region with arbitrarily shaped obstacles in it. To relate this model to mathematical morphology, the finite region consists of a free space set with values equal to ones, starting and ending points with values equal to zeros, an obstacle set with values equal to negative ones, and the moving object is modeled as a structuring element [Pei, Lai, and Shih 1998]. Thus, the shortest path finding problem is equivalent to applying a morphological erosion to the free space, followed by a DT on the region with the grown obstacles excluded, and then tracing back the resultant distance map from the target point to its neighbors with the minimum distance until the starting point is reached.

When there are obstacles existence in an image, many algorithms cannot obtain the correct EDT. Figure 6.20 shows an example when an image has obstacles, some algorithms cannot obtain the exact EDT. For simplicity, the SEDT is adopted to illustrate the example. Figures 6.20a and b are the images with and without obstacles, respectively. Both Eggers's [1998] and Shih and Mitchell's [1992] algorithms cannot obtain the correct EDT. Figures 6.20c2, d2, and e2 are the rules based on Eggers's algorithm [1998, p. 109]. Based on the rules, we obtain Figures 6.20c1, d1, and e1 as the results of achieving SEDT of Figure 6.20a in the first, second, and third iterations. Figures 6.20c4, d4, and e4 are the morphological structuring components. Similarly, we obtain Figures 6.20c6, d6, and e6. We obtain Figures 6.20c3, d3, e3, c5, d5, and e5 using morphological erosions. By observing Figures 6.20e1 and e3, we understand that the value in the square box, say 15, is obtained by the

Original image	1 1 1 -1 1 / 1 0 1 -1 1 / 1 1 1 -1 1 / 1 1 1 -1 1 / 1 1 1 1 1 (a)		1 1 1 1 1 / 1 0 1 1 1 / 1 1 1 1 1 / 1 1 1 1 1 / 1 1 1 1 1 (b)	
Iteration	Eggers's algorithm	Shih and Mitchell's	Shih and Mitchell's	Two-scan based
(I) iter=1	= +1 (c2), = +2 (c1) 2 1 2 -1 1000 1 0 1 -1 1000 2 1 2 -1 1000 1000 1000 1000 -1 1000 1000 1000 1000 1000 1000	(c4) -2 -1 -2 / -1 0 -1 / -2 -1 -2 (c3) 2 1 2 -1 1000 1 0 1 -1 1000 2 1 2 -1 1000 1000 1000 1000 -1 1000 1000 1000 1000 1000 1000	(c6) -2 -1 -2 / -1 0 -1 / -2 -1 -2 (c5) 2 1 2 1000 1000 1 0 1 1000 1000 2 1 2 1000 1000 1000 1000 1000 1000 1000 1000 1000 1000 1000 1000	
(II) iter=2	= +3 (d2), = +6 (d1) 2 1 2 -1 1000 1 0 1 -1 1000 2 1 2 -1 1000 5 4 5 -1 1000 1000 1000 1000 1000 1000	(d4) -6 -3 -6 / -3 0 -3 / -6 -3 -6 (d3) 2 1 2 -1 1000 1 0 1 -1 1000 2 1 2 -1 1000 5 4 5 -1 1000 1000 1000 1000 1000 1000	(d6) -6 -3 -6 / -3 0 -3 / -6 -3 -6 (d5) 2 1 2 5 1000 1 0 1 4 1000 2 1 2 5 1000 5 4 5 8 1000 1000 1000 1000 1000 1000	
(III) iter=3	= +5 (e2), = +10 (e1) 2 1 2 -1 1000 1 0 1 -1 1000 2 1 2 -1 1000 5 4 (5) -1 1000 10 9 10 [15] 1000	(e4) -10 -5 -10 / -5 0 -5 / -10 -5 -10 (e3) 2 1 2 -1 1000 1 0 1 -1 1000 2 1 2 -1 1000 5 4 (5) -1 1000 10 9 10 [15] 1000	(e6) -10 -5 -10 / -5 0 -5 / -10 -5 -10 (e5) 2 1 2 5 10 1 0 1 4 9 2 1 2 (8) 10 5 4 5 8 13 10 9 10 [13] 18	(f) 2 1 2 -1 58 1 0 1 -1 45 2 1 2 -1 34 5 4 5 -1 25 10 9 10 [13] 20

FIGURE 6.20 The example that some algorithms cannot obtain the exact EDT.

summation of its neighbor with value 5 and the distance 10. Both of them are not correct SED. The correct SED in this position should be 13, as shown in Figure 6.20e5. Figure 6.20f, shows the correct result obtained by the two-scan based algorithm.

The two-scan based EDT algorithm, TS_∞, for dealing with images with obstacles is described below. Let P be the SEDT image, which stores the previous image.

TS_∞ Algorithm:

1. Set $P = f$.
2. Obtain SEDT of f by the TS_1 algorithm except step 13.
3. Repeat steps 1 and 2 until $P = f$.
4. Take a square root of P (the distance image). That is $EDT = \sqrt{P}$.

6.6.6 Computational Complexity

For an image of size $n \times n$, the complexity for the simple approximate algorithm developed by Danielsson [1980] is $O(n^2)$. The exact EDT algorithm proposed by Eggers [1998] is highly dependent on the image content and can vary between $O(n^2)$ and $O(n^3)$. The complexity of the Improved Iteration Erosion is $O(dn^2)$, where d is the maximum number of chessboard distances. For the 2-scan based EDT algorithm, the complexity is $O(n^2)$. Furthermore, it can successfully achieve the correct EDT for the images with obstacles.

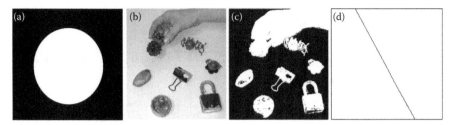

FIGURE 6.21   The testing images.

Three images shown in Figure 6.21 are experimented in a PC with Pentium III/866 MHz and 256 megabytes RAM. Figure 6.21a shows a disk of variable diameter created for test 1. The size of Figure 6.21a is 1024×1024. Figure 6.21b shows a grayscale image with size of 327×300. Figure 6.21c shows the binary image after thresholding for test 2. Figure 6.21d shows the straight line across the image with size of 1024×1024 for test 3.

There are six different algorithms experimented in test 1, and the results are shown in Figure 6.22. Following are the symbols used and their

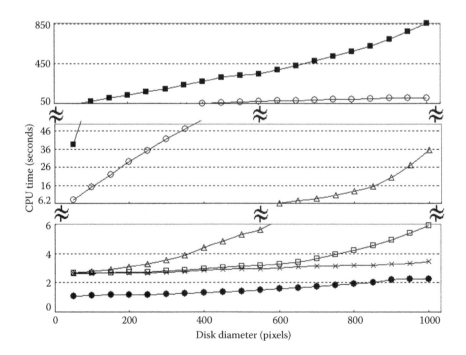

FIGURE 6.22   Test 1: diameters vary from 50 to 1000 pixels.

TABLE 6.3

The Experimental Results for Tests 2 and 3

Algorithm	Test 2		Test 3	
	Time (Seconds)	Errors (Pixels)	Time (Seconds)	Errors (Pixels)
Danielsson	0.45	3	2.1	0
TS_1 (approximate)	0.52	1	2.7	0
TS_1	0.54	0	5.8	0
Eggers	0.58	0	32.7	0
IIEA	2.87	0	62.5	0
IEA	12.06	0	842.3	0

corresponding algorithm: • (Danielsson), × (TS_1 with a fixed 3×3 window), □ (TS_1), ∆ (Eggers), ○ (IIEA), and ■ (traditional morphological erosion). It is obvious that the traditional morphological erosion is time-consuming. Since the improved IEA is implemented in a conventional computer (not a cellular array computer), we cannot see its performance. Although Eggers's algorithm successfully improved the performance, the execution time is high when the diameter of the disk becomes large. TS_1 reduces the execution time dramatically. We also implement TS_1 with a fixed 3×3 neighborhood window as an approximate algorithm. Although the execution time is larger than Danielsson's algorithm, but the correct rate is much higher than Danielsson's algorithm.

The experimental results for tests 2 and 3 are shown in Table 6.3. For a complex image with obstacles as shown in Figure 6.23, we execute eight times of TS_1 for obtaining the exact EDT.

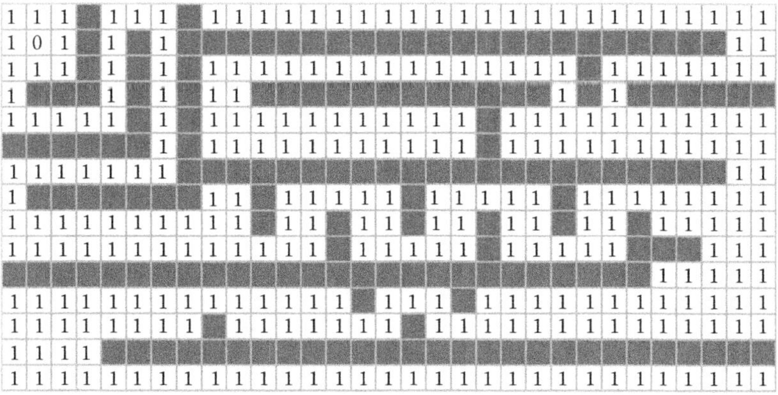

FIGURE 6.23 The image with obstacles.

6.7 Three-Dimensional Euclidean Distance

6.7.1 Three-Dimensional Image Representation

A three-dimensional digitized image can be represented as a three-dimensional matrix: $T^3[x][y][z]$ (or $T^3[x, y, z]$) of dimensions $X \times Y \times Z$, where x, y, and z, respectively, denote row, column, and height coordinates, as shown in Figure 6.24. Each voxel has the physical size of $D_x \times D_y \times D_z$ in physical units (e.g., mm or μm). For a binary image, the values of voxels are composed of only two kinds (0 and 1). For a grayscale image, the voxels are composed of 256 values (0, 1, 2, ... , 255). Note that the representation can be extended to the images of an arbitrarily dimensional domain as shown in Figure 6.25, where an n-D digitized domain can be represented as an n-D matrix $T^n[t_1][t_2] \cdots [t_n]$ of dimensions $t_1 \times t_2 \times \cdots \times t_n$.

6.7.2 Distance Definition Functions in the Three-Dimensional Domain

Let $p = (p_1, p_2, p_3)$ and $q = (q_1, q_2, q_3)$ denote two points in a three-dimensional digital image, where p_1, p_2, p_3, q_1, q_2, and q_3 are integers. Three frequently-used distance functions are defined as follows:

(a) *City-block distance*: $d_{ci}(p, q) = |p_1 - q_1| + |p_2 - q_2| + |p_3 - q_3|$.
(b) *Chessboard distance*: $d_{ch}(p, q) = \max(|p_1 - q_1|, |p_2 - q_2|, |p_3 - q_3|)$.
(c) *Euclidean distance*: $d_e(p, q) = \sqrt{(p_1 - q_1)^2 + (p_2 - q_2)^2 + (p_3 - q_3)^2}$.

Note that the voxels with *city-block distance* 1 counting from p correspond to 6-neighbors of p, and those with *chessboard distance* 1 correspond to

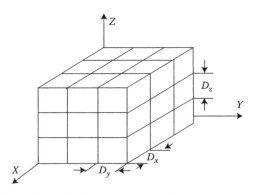

FIGURE 6.24 The three-dimensional image representation.

$$t_1 = \{ \quad -\infty \quad \cdots \quad 0 \quad \cdots \quad +\infty \quad \}$$

$$t_2 = \{ \quad -\infty \quad \cdots \quad 0 \quad \cdots \quad +\infty \quad \}$$

$$t_3 = \{ \quad -\infty \quad \cdots \quad 0 \quad \cdots \quad +\infty \quad \}$$

$$\vdots \quad \vdots \quad \vdots \quad \vdots \quad \vdots \quad \vdots \quad \vdots \quad \vdots$$

$$t_n = \{ \quad -\infty \quad \cdots \quad 0 \quad \cdots \quad +\infty \quad \}$$

FIGURE 6.25 The representation of n-dimensional space.

26-neighbors. These d_{ci} and d_{ch} are integer-valued; however, d_e is not. The SED is defined as

$$d_s(p, q) = [d_e(p, q)]^2 \tag{6.33}$$

Note that the distance functions above can be similarly extended to an n-D domain.

6.7.3 A Three-Dimensional Neighborhood in the EDT

In three-dimensional space, we denote the 26 neighbors of pixel p by $N^3(p) = \{h_1^3, h_2^3, \ldots, h_{26}^3\}$, as illustrated in Figure 6.26. They can be categorized into three groups: $N^{3\_1}(p)$, $N^{3\_2}(p)$, and $N^{3\_3}(p)$, in which the corresponding

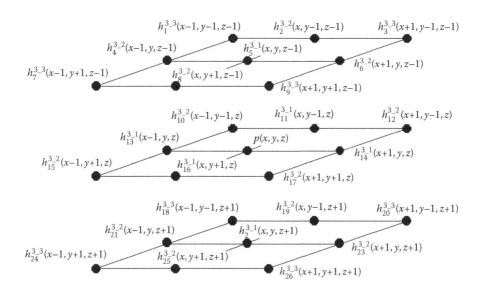

FIGURE 6.26 The 26 neighbors of pixel p.

$h^{3\_d}_{index}$ adopts the notations *index* and *d*, representing the ordering number in the neighborhood and the number of moves between p and its neighbor, respectively. Generally, the number of neighbors of p is $3^n - 1$ in the n-D domain, and they can be categorized into n groups.

6.8 Acquiring Approaches

Basically, the approaches for achieving DT can be categorized into acquiring and deriving approaches. The acquiring approach is to acquire the minimum distance of each pixel from its neighborhood as illustrated in Figure 6.27a. The deriving approach uses the strategy that each pixel broadcasts relative distances to its children (i.e., the pixels in its neighborhood) as illustrated in Figure 6.27b. The acquiring approaches for city-block, chessboard, and Euclidean distances are presented in this section. The deriving approach for Euclidean distance is described in the next section.

6.8.1 Acquiring Approaches for City-Block and Chessboard DT

Three-dimensional city-block and chessboard distances are very easy to compute since they can be recursively accumulated by considering only 6- and 26-neighbors at one time, respectively. The iterative algorithm for city-block and chessboard distances is described below. Given an X_S, which is 1 at the pixels of foreground S and 0 elsewhere. X_S^m is defined recursively for $m = 1, 2, \ldots$ as

$$X_S^m(p) = X_S^0 + \min_{d(q,p)\leq 1} X_S^{m-1}(q), \tag{6.34}$$

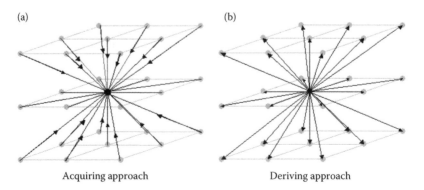

(a)   (b)

Acquiring approach   Deriving approach

FIGURE 6.27   The two approaches for achieving the EDT.

where $X_S^0 = X_S$, as the initial binary image. Thus, X_S^m can be computed by performing a local operation on the pair of arrays X_S^0 and X_S^{m-1} at each point.

The iterative algorithm is very efficient on a cellular array computer since each iteration can be performed at all points in parallel and the number of iterations is at most the radius of the image. However, each iteration requires the processing of the entire image in a conventional computer that presents inefficiency. A two-scan algorithm is therefore developed below.

Let $N_1(p)$ be the set of 6 or 26 neighbors that precede p in the left-to-right, top-to-bottom, near-to-far scan of the image, and let $N_2(p)$ be the remaining neighbors of p. The algorithm is computed by

$$X_S'(p) = \begin{cases} 0 & \text{if } P \in \bar{S} \\ \min_{q \in N_1(p)} \chi_S'(q) + 1 & \text{if } P \in S \end{cases} \tag{6.35}$$

$$X_S''(p) = \min_{q \in N_2(p)} [\chi_S'(p), \chi_S''(q) + 1] \tag{6.36}$$

$X_S'(p)$ is computed in a left-to-right, top-to-bottom, near-to-far scan since for each p, X_S' has already been computed for the q in N_1. Similarly, X_S'' is computed in a reverse scan.

6.8.2 Acquiring Approaches for EDT

Shih and Wu [1992] developed a morphology-based algorithm by decomposing the big two-dimensional SEDSE into successive dilations of a set of 3×3 structuring components. Hence, the SEDT is equivalent to the successive erosions of the results at each preceding stage by the iteration-related structuring components. Note that the input binary image is represented by $+\infty$ (or a large number) for foreground and 0 for background. In DT, the three-dimensional images and SEs are not represented in a temporal stack. By extending the structuring components into three dimensions, $g^3(l)$ is obtained as

$$g^3(l) = \begin{bmatrix} g_a^3(l) \\ g_b^3(l) \\ g_c^3(l) \end{bmatrix} \tag{6.37}$$

where $g_a^3(l) = g_c^3(l) = \begin{bmatrix} -(6l-3) & -(4l-2) & -(6l-3) \\ -(4l-2) & -(2l-1) & -(4l-2) \\ -(6l-3) & -(4l-2) & -(6l-3) \end{bmatrix}$ and

$$g_b^3(l) = \begin{bmatrix} -(4l-2) & -(2l-1) & -(4l-2) \\ -(2l-1) & 0 & -(2l-1) \\ -(4l-2) & -(2l-1) & -(4l-2) \end{bmatrix}.$$

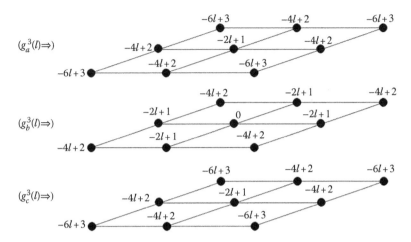

FIGURE 6.28 The extended structuring components in the three-dimensional space.

Figure 6.28 illustrates the three-dimensional structuring components. The SEDT is the iterative erosions by a set of small structuring components and a square-root operation. The following is the three-dimensional EDT algorithm:

1. Let the value of foreground be a large number, and let the value of background be 0.
2. Initialize $l = 1$.

27	22	19	18	19	22	27
22	17	14	13	14	17	22
19	14	11	10	11	14	19
18	13	10	9	10	13	18
19	14	11	10	11	14	19
22	17	14	13	14	17	22
27	22	19	18	19	22	27

Layer 1

22	17	14	13	14	17	22
17	12	9	8	9	12	17
14	9	6	5	6	9	14
13	8	5	4	5	8	13
14	9	6	5	6	9	14
17	12	9	8	9	12	17
22	17	14	13	14	17	22

Layer 2

19	14	11	10	11	14	19
14	9	6	5	6	9	14
11	6	3	2	3	6	11
10	5	2	1	2	5	10
11	6	3	2	3	6	11
14	9	6	5	6	9	14
19	14	11	10	11	14	19

Layer 3

18	13	10	9	10	13	18
13	8	5	4	5	8	13
10	5	2	1	2	5	10
9	4	1	0	1	4	9
10	5	2	1	2	5	10
13	8	5	4	5	8	13
18	13	10	9	10	13	18

Layer 4

27	22	19	18	19	22	27
22	17	14	13	14	17	22
19	14	11	10	11	14	19
18	13	10	9	10	13	18
19	14	11	10	11	14	19
22	17	14	13	14	17	22
27	22	19	18	19	22	27

Layer 7

22	17	14	13	14	17	22
17	12	9	8	9	12	17
14	9	6	5	6	9	14
13	8	5	4	5	8	13
14	9	6	5	6	9	14
17	12	9	8	9	12	17
22	17	14	13	14	17	22

Layer 6

19	14	11	10	11	14	19
14	9	6	5	6	9	14
11	6	3	2	3	6	11
10	5	2	1	2	5	10
11	6	3	2	3	6	11
14	9	6	5	6	9	14
19	14	11	10	11	14	19

Layer 5

FIGURE 6.29 An SEDT example.

3. $d = f \ominus g(l)$.

4. If $d \neq f$, let $f = d$ and $l++$; repeat step 3 until $d = f$.

5. $EDT = \sqrt{d}$.

Figure 6.29 shows the resulting image of SEDT using the three-dimensional iterative algorithm with step 5 skipped when the input contains only one background pixel in the center of layer 4. Note that due to the symmetric property, layer 3 = layer 5, layer 2 = layer 6, and layer 1 = layer 7.

6.9 Deriving Approaches

Let o denote a background pixel. The method of calculating $d_s(N^{3\text{-}1}(p), o)$, $d_s(N^{3\text{-}2}(p), o)$, and $d_s(N^{3\text{-}3}(p), o)$ is described below. Note that the respective children of pixel p are in forward and backward scans, each containing 13 pixels as illustrated in Figures 6.30 and 6.31. For simplicity in notations, $d_s(p, q)$ is substituted by $d_s(p)$, and $d_s(h^{3\text{-}d}_{index}, q)$ is replaced by $d_s(h^3_{index})$ if q is a background pixel.

$$d_s(p) = x^2 + y^2 + z^2$$

$$d_s(h^3_{14}) = (x+1)^2 + y^2 + z^2$$

$$d_s(h^3_{15}) = (x+1)^2 + (y+1)^2 + z^2$$

$$d_s(h^3_{16}) = x^2 + (y+1)^2 + z^2$$

$$d_s(h^3_{17}) = (x+1)^2 + (y+1)^2 + z^2$$

$$d_s(h^3_{18}) = (x+1)^2 + (y+1)^2 + (z+1)^2$$

$$d_s(h^3_{19}) = x^2 + (y+1)^2 + (z+1)^2$$

$$d_s(h^3_{20}) = (x+1)^2 + (y+1)^2 + (z+1)^2$$

$$d_s(h^3_{21}) = (x+1)^2 + y^2 + (z+1)^2$$

$$d_s(h^3_{22}) = x^2 + y^2 + (z+1)^2$$

$$d_s(h^3_{23}) = (x+1)^2 + y^2 + (z+1)^2$$

$$d_s(h^3_{24}) = (x+1)^2 + (y+1)^2 + (z+1)^2$$

$$d_s(h^3_{25}) = x^2 + (y+1)^2 + (z+1)^2$$

$$d_s(h^3_{26}) = (x+1)^2 + (y+1)^2 + (z+1)^2$$

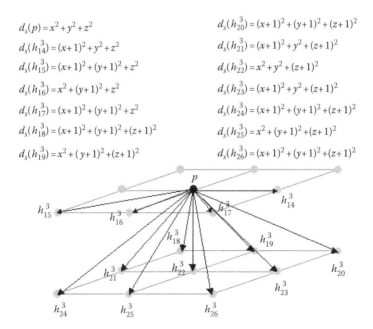

FIGURE 6.30 The 13 children in the forward scan.

$$d_s(p) = x^2 + y^2 + z^2$$

$$d_s(h_1^3) = (x+1)^2 + (y+1)^2 + (z+1)^2$$

$$d_s(h_2^3) = x^2 + (y+1)^2 + (z+1)^2$$

$$d_s(h_3^3) = (x+1)^2 + (y+1)^2 + (z+1)^2$$

$$d_s(h_4^3) = (x+1)^2 + y^2 + (z+1)^2$$

$$d_s(h_5^3) = x^2 + y^2 + (z+1)^2$$

$$d_s(h_6^3) = (x+1)^2 + y^2 + (z+1)^2$$

$$d_s(h_7^3) = (x+1)^2 + (y+1)^2 + (z+1)^2$$

$$d_s(h_8^3) = x^2 + (y+1)^2 + (z+1)^2$$

$$d_s(h_9^3) = (x+1)^2 + y^2 + z^2$$

$$d_s(h_{10}^3) = (x+1)^2 + (y+1)^2 + z^2$$

$$d_s(h_{11}^3) = x^2 + (y+1)^2 + z^2$$

$$d_s(h_{12}^3) = (x+1)^2 + (y+1)^2 + z^2$$

$$d_s(h_{13}^3) = (x+1)^2 + y^2 + z^2$$

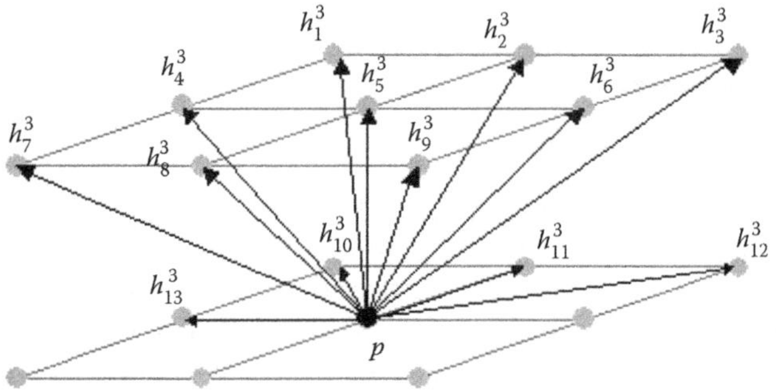

FIGURE 6.31   The 13 children in the backward scan.

6.9.1 Fundamental Lemmas

Let q denote a background pixel. The following three lemmas describe the relationship between an object pixel p and its neighbors.

Lemma 6.4: The SEDT of each pixel can be divided into the summation of three squared integers.

Note that Lemma 6.4 can be easily proved from the definition of SED. Therefore, let $d_s(p, q) = (I_{max})^2 + (I_{mid})^2 + (I_{min})^2$, where I_{max}, I_{mid}, and I_{min} are integers and $I_{max} \geq I_{mid} \geq I_{min}$.

Lemma 6.5: The SEDT of the further children of p can be obtained by the following three equations depending on its group.

$$d_s(N^{3-1}(p), q) = \begin{cases} (I_{max} + 1)^2 + (I_{mid})^2 + (I_{min})^2, \text{ or} \\ (I_{max})^2 + (I_{mid} + 1)^2 + (I_{min})^2, \text{ or} \\ (I_{max})^2 + (I_{mid})^2 + (I_{min} + 1)^2 \end{cases} \qquad (6.38)$$

$$d_s(N^{3-2}(p), q) = \begin{cases} (I_{max} + 1)^2 + (I_{mid} + 1)^2 + (I_{min})^2, \text{ or} \\ (I_{max} + 1)^2 + (I_{mid})^2 + (I_{min} + 1)^2, \text{ or} \\ (I_{max})^2 + (I_{mid} + 1)^2 + (I_{min} + 1)^2 \end{cases} \qquad (6.39)$$

$$d_s(N^{3\_3}(p), q) = (I_{max} + 1)^2 + (I_{mid} + 1)^2 + (I_{min} + 1)^2 \qquad (6.40)$$

Proof: From $d_s(p, q) = (I_{max})^2 + (I_{mid})^2 + (I_{min})^2$, we know that the relative distance between p and q in each of the three dimensions is, respectively, I_{max}, I_{mid}, and I_{min}, as shown in Figure 6.32. Note that the relative distance between $N^{3\_1}(p)$ and p is one pixel in one dimension. Therefore, the SEDT of $N^{3\_1}(p)$ can be obtained by Equation 6.38. Similarly, the relative distances between $(N^{3\_2}(p), p)$ and $(N^{3\_3}(p), p)$ are one pixel in each of the two and three dimensions, respectively. Therefore, Equations 6.39 and 6.40 are used to obtain the SEDT of $N^{3\_2}(p)$ and $N^{3\_3}(p)$, respectively. $\qquad \square$

Lemma 6.6: If $d_s(p, q)$ can be decomposed into more than one set of the summation of three squared integers; that is,

$$d_s(p,q) = \begin{cases} (I_{max\_1})^2 + (I_{mid\_1})^2 + (I_{min\_1})^2 \\ (I_{max\_2})^2 + (I_{mid\_2})^2 + (I_{min\_2})^2 \\ \vdots \\ (I_{max\_n})^2 + (I_{mid\_n})^2 + (I_{min\_n})^2 \end{cases},$$

we can check the parents of p to determine the correctly decomposed set.

Proof: According to Lemma 6.5, the value of SEDT of pixel p can be obtained from its parents. Suppose that there are n possibilities when $d_s(p, q)$ is decomposed into the summation of three squared integers. If p is derived from H_1, $d_s(H_1, q)$ should be equal to $(I_{max} - 1)^2 + (I_{mid})^2 + (I_{min})^2$, $(I_{max})^2 + (I_{mid} - 1)^2 + (I_{min})^2$, or $(I_{max})^2 + (I_{mid})^2 + (I_{min} - 1)^2$. Note that H_1 is the parent of p and has

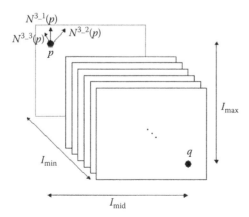

FIGURE 6.32 The $d_s(N^{3\_1}(p), q)$, $d_s(N^{3\_2}(p), q)$, and $d_s(N^{3\_3}(p), q)$.

only one-dimensional difference to p. Similarly, if p is derived from H_2 which has two-dimensional difference to p, then $d_s(H_2, q)$ should be equal to $(I_{max} - 1)^2 + (I_{mid} - 1)^2 + (I_{min})^2$, $(I_{max})^2 + (I_{mid} - 1)^2 + (I_{min} - 1)^2$ or $(I_{max} - 1)^2 + (I_{mid})^2 + (I_{min} - 1)^2$. If p is derived from H_3 which has the three-dimensional difference to p, then $d_s(H_3, q)$ should be equal to $(I_{max} - 1)^2 + (I_{mid} - 1)^2 + (I_{min} - 1)^2$. Therefore, by checking the parents of p, we can find out the exact decomposition when dividing $d_s(p, q)$.   □

6.9.2 Two Scan–Based Algorithm for Three-Dimensional EDT

In fact, both concepts in the acquiring and deriving approaches are similar. In the acquiring approach, it is obvious that the minimum EDT of a pixel can be obtained by one of its neighbors with the minimum EDT value. In the deriving approach, a pixel can derive the minimum EDT value to its neighbors based on the aforementioned lemmas. To correctly achieve the EDT in two scans to overcome the time-consuming iterative operations, the deriving approach is adopted to substitute for the acquiring approach. Let the object pixels be a large number (i.e., larger than the square of a half of the image size) and the background pixels be 0. The algorithm is described below, and its flowchart is shown in Figure 6.33.

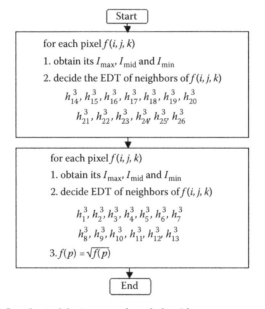

FIGURE 6.33   The flowchart of the two-scan based algorithm.

A. *Forward Scan*: left-to-right, top-to-bottom, and near-to-far

```
for (i = 0; i ≤ length; i++)
      for (j = 0; j ≤ length; j++)
            for (k = 0; k ≤ length;  k++)
{
Decompose f(i, j, k) into Iₘₐₓ, Iₘᵢ𝒹, and Iₘᵢₙ, where f(i, j) =
(Iₘₐₓ )² + (Iₘᵢ𝒹)² + (Iₘᵢₙ)²;
If there exists more than one decomposition, check the
parents of f(i,j,k) to choose the exact Iₘₐₓ, Iₘᵢ𝒹, and Iₘᵢₙ;
For the following three types of neighborhoods, perform:
```

$f(N^{3\_1}(p)) = \min(f(N^{3\_1}(p)), (I_{max}+1)^2 + (I_{mid})^2 + (I_{min})^2);$

$f(N^{3\_2}(p)) = \min(f(N^{3\_2}(p)), (I_{max}+1)^2 + (I_{mid}+1)^2 + (I_{min})^2);$

$f(N^{3\_3}(p)) = \min(f(N^{3\_3}(p)), (I_{max}+1)^2 + (I_{mid}+1)^2 + (I_{min}+1)^2)$

```
}
```

B. *Backward Scan*: far-to-near, bottom-to-top, and right-to-left

```
for (i = length; i ≥ 0; i--)
      for (j = length; j ≥ 0; j--)
            for (k = length; k ≥ 0; k--)
{
Decompose f(i, j, k) into Iₘₐₓ, Iₘᵢ𝒹, and Iₘᵢₙ, where
f(i, j) = (Iₘₐₓ)² + (Iₘᵢ𝒹)² + (Iₘᵢₙ)²;
If there exists more than one decomposition, check the
parents of f(i, j, k) to choose the exact Iₘₐₓ, Iₘᵢ𝒹, and Iₘᵢₙ;
For the following three types of neighborhoods, perform:
```

$f(N^{3\_1}(p)) = \min(f(N^{3\_1}(p)), (I_{max}+1)^2 + (I_{mid})^2 + (I_{min})^2);$

$f(N^{3\_2}(p)) = \min(f(N^{3\_2}(p)), (I_{max}+1)^2 + (I_{mid}+1)^2 + (I_{min})^2);$

$f(N^{3\_3}(p)) = \min(f(N^{3\_3}(p)), (I_{max}+1)^2 + (I_{mid}+1)^2 + (I_{min}+1)^2)$

$E(f(i, j, k)) = \sqrt{f(i, j, k)};$

```
}
```

Figures 6.34 and 6.35 show the forward and backward scanning results of the two-scan based algorithm on an image of $7 \times 7 \times 7$. Note that there are only three background pixels in the image, which are located at the centers of layers 2, 4, and 6. Also note that for some image with obstacle, we need repeating the above algorithm for several times to achieve the exact EDT. Let P be the SEDT image, which stores the previous image. The algorithm is

1. Set $P = f$.
2. Obtain the SEDT of f by the above two-scan algorithm with the last step being excluded.
3. Repeat steps 1 and 2 until $P = f$.
4. Take a square root of P (the distance image). That is $EDT = \sqrt{P}$.

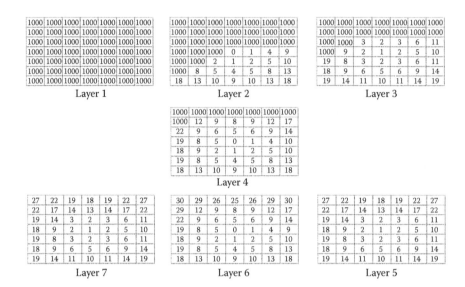

FIGURE 6.34 The forward scan of the two-scan based algorithm.

Figure 6.36 shows an example of an image of size 300×350. Figures 6.36a and b are the original image and its EDT result, respectively. Note that the pixels within the hammer are set to the object pixels. In Figure 6.36b, the brighter pixels indicate the larger EDT values. Note that we rescale the EDT values to the range of 0 to 255 for the display clarity.

Layer 1

19	14	11	10	11	14	19
14	9	6	5	6	9	14
11	6	3	2	3	6	11
10	5	2	1	2	5	10
11	6	3	2	3	6	11
14	9	6	5	6	9	14
19	14	11	10	11	14	19

Layer 2

18	13	10	9	10	13	18
13	8	5	4	5	8	13
10	5	2	1	2	5	10
9	4	1	0	1	4	9
10	5	2	1	2	5	10
13	8	5	4	5	8	13
18	13	10	9	10	13	18

Layer 3

19	14	11	10	11	14	19
14	9	6	5	6	9	14
11	6	3	2	3	6	11
10	5	2	1	2	5	10
11	6	3	2	3	6	11
14	9	6	5	6	9	14
19	14	11	10	11	14	19

Layer 4

18	13	10	9	10	13	18
13	8	5	4	5	8	13
10	5	2	1	2	5	10
9	4	1	0	1	4	9
10	5	2	1	2	5	10
13	8	5	4	5	8	13
18	13	10	9	10	13	18

Layer 7

19	14	11	10	11	14	19
14	9	6	5	6	9	14
11	6	3	2	3	6	11
10	5	2	1	2	5	10
11	6	3	2	3	6	11
14	9	6	5	6	9	14
19	14	11	10	11	14	19

Layer 6

18	13	10	9	10	13	18
13	8	5	4	5	8	13
10	5	2	1	2	5	10
9	4	1	0	1	4	9
10	5	2	1	2	5	10
13	8	5	4	5	8	13
18	13	10	9	10	13	18

Layer 5

19	14	11	10	11	14	19
14	9	6	5	6	9	14
11	6	3	2	3	6	11
10	5	2	1	2	5	10
11	6	3	2	3	6	11
14	9	6	5	6	9	14
19	14	11	10	11	14	19

FIGURE 6.35 The backward scan of the two-scan based algorithm.

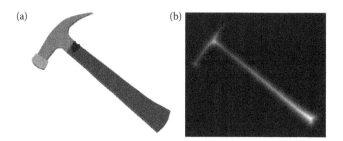

(a)   (b)

FIGURE 6.36   The EDT example of an image.

6.9.3 Complexity of the Two-Scan-Based Algorithm

Given an image of size $n \times n \times n$, the computational complexity of the approximation algorithm by Danielsson [1980] is $O(n^3)$. The exact EDT algorithm proposed by Eggers [1998] is highly dependent on the image content, and its complexity is varied between $O(n^3)$ and $O(n^4)$. The complexity of the proposed two-scan based algorithm is $O(n^3)$.

References

Blum, H., "A transformation for extracting new descriptors of shape," *Proc. of Symposium on Models for the Perception of Speech and Visual Forms*, Boston, MA, MIT Press, Nov. 1964.

Borgefors, G., "Distance transformations in arbitrary dimensions," *Computer Vision, Graphics, and Image Processing*, vol. 27, no. 3, pp. 321–345, Sep. 1984.

Borgefors, G., "Distance transformations in digital images," *Computer Vision, Graphics, and Image Processing*, vol. 34, no. 3, pp. 344–371, June 1986.

Coeurjolly, D. and Montanvert, A., "Optimal separable algorithms to compute the reverse Euclidean distance transformation and discrete medial axis in arbitrary dimension," *IEEE Trans. Pattern Analysis and Machine Intelligence*, vol. 29, no. 3, pp. 437–448, March 2007.

Cuisenaire, O., "Locally adaptable mathematical morphology using distance transformations," *Pattern Recognition*, vol. 39, no. 3, pp. 405–416, March 2006.

Cuisenaire, O. and Macq, B., "Fast Euclidean distance transformation by propagation using multiple neighborhoods," *Computer Vision and Image Understanding*, vol. 76, no. 2, pp. 163–172, Nov. 1999.

Danielsson, P. E., "A new shape factor," *Computer Graphics and Image Processing*, vol. 7, no. 2, pp. 292–299, Apr. 1978.

Danielsson, P. E., "Euclidean distance mapping," *Computer Graphics Image Processing*, vol. 14, no. 3, pp. 227–248, Nov. 1980.

Datta, A. and Soundaralakshmi, S., "Constant-time algorithm for the Euclidean distance transform on reconfigurable meshes," *Journal of Parallel and Distributed Computing*, vol. 61, no. 10, pp. 1439–1455, Oct. 2001.

Eggers, H., "Two fast Euclidean distance transformations in Z^2 based on sufficient propagation," *Computer Vision and Image Understanding*, vol. 69, no. 1, pp. 106–116, Jan. 1998.

Fabbri, R., Costa, L., Torelli, J., and Bruno, O., "2D Euclidean distance transform algorithms: A comparative survey," *ACM Computing Surveys*, vol. 40, no. 1, pp. 2:1–2:44, Feb. 2008.

Hilditch, J., "Linear skeletons from square cupboards," in Meltzer and D. Michie (Eds), *Machine Intelligence*, vol. 4, American Elsevier Publishiers Co., pp. 403–420, 1969.

Huang, C. T. and Mitchell, O. R., "A Euclidean distance transform using grayscale morphology decomposition," *IEEE Trans. Pattern Analysis and Machine Intelligence*, vol. 16, no. 4, pp. 443–448, Apr. 1994.

Lantuejoul, C., "Skeletonization in quantitative metallography," in R. M. Haralick and J. C. Simon (Eds), *Issues in Digital Image Processing*, Sijthoff & Noordhoff Publishers, Maryland, pp. 107–135, 1980.

Maurer, C. R., Rensheng, Q., and Raghavan, V., "A linear time algorithm for computing exact Euclidean distance transforms of binary images in arbitrary dimensions," *IEEE Trans. Pattern Analysis and Machine Intelligence*, vol. 25, no. 2, pp. 265–270, Feb. 2003.

Pei, S.-C., Lai, C.-L., and Shih, F. Y., "A morphological approach to shortest path planning for rotating objects," *Pattern Recognition*, vol. 31, no. 8, pp. 1127–1138, Aug. 1998.

Rosenfeld, A. and Pfaltz, J. L., "Sequential operations in digital picture processing," *J. ACM*, vol. 13, no. 4, pp. 471–494, Oct. 1966.

Rosenfeld, A. and Kak, A. C., *Digital Picture Processing*, Academic Press, New York, 1982.

Saito, T. and Toriwaki, J.-I., "New algorithms for Euclidean distance transformation of an n-dimensional digitized picture with applications," *Pattern Recognition*, vol. 27, no. 11, pp. 1551–1565, Nov. 1994.

Serra, J., *Image Analysis and Mathematical Morphology*, Academic Press, New York, 1982.

Shih, F. Y. and Liu, J., "Size-invariant four-scan Euclidean distance transformation," *Pattern Recognition*, vol. 31, no. 11, pp. 1761–1766, Nov. 1998.

Shih, F. Y. and Mitchell, O. R., "Skeletonization and distance transformation by grayscale morphology," *Proc. of SPIE Symposium on Automated Inspection and High Speed Vision Architectures*, Cambridge, MA, pp. 80–86, Nov. 1987.

Shih, F. Y. and Mitchell, O. R., "Threshold decomposition of grayscale morphology into binary morphology," *IEEE Trans. Pattern Analysis and Machine Intelligence*, vol. 11, no. 1, pp. 31–42, Jan. 1989.

Shih, F. Y. and Mitchell, O. R., "Decomposition of grayscale morphological structuring elements," *Pattern Recognition*, vol. 24, no. 3, pp. 195–203, March 1991.

Shih, F. Y. and Mitchell, O. R., "A mathematical morphology approach to Euclidean distance transformation," *IEEE Trans. Image Processing*, vol. 1, no. 2, pp. 197–204, April 1992.

Shih, F. Y. and Wu, H., "Optimization on Euclidean distance transformation using grayscale morphology," *Journal of Visual Communication and Image Representation*, vol. 3, no. 2, pp. 104–114, June 1992.

Shih, F. Y. and Wu, Y., "Three-dimensional Euclidean distance transformation and its application to shortest path planning," *Pattern Recognition*, vol. 37, no. 1, pp. 79–92, Jan. 2004a.

Shih, F. Y. and Wu, Y., "Fast Euclidean distance transformation in two scans using a 3×3 neighborhood," *Computer Vision and Image Understanding*, vol. 93, no. 2, pp. 195–205, Feb. 2004b.

Shih, F. Y. and Wu, Y., "The efficient algorithms for achieving Euclidean distance transformation," *IEEE Trans. Image Processing*, vol. 13, no. 8, pp. 1078–1091, Aug. 2004c.

Shih, F. Y. and Wu, Y., "Decomposition of arbitrarily grayscale morphological structuring elements," *Pattern Recognition*, vol. 38, no. 12, pp. 2323–2332, Dec. 2005.

Toriwaki, J. and Yokoi, S., "Distance transformations and skeletons of digitized pictures with applications," in L. N. Kanal and A. Rosenfeld (Eds), *Progress in Pattern Recognition*, North-Holland Publishers, pp. 187–264, 1981.

Vincent, L., "Exact Euclidean distance function by chain propagations," *Proc. IEEE Computer Vision and Pattern Recognition*, Hawaii, pp. 520–525, 1991.

Vossepoel, A. M., "A note on distance transformations in digital images," *Computer Vision, Graphics, and Image Processing*, vol. 43, no. 1, pp. 88–97, July 1988.

Xu, D. and Li, H., "Euclidean distance transform of digital images in arbitrary dimensions," in Y. Zhuang et al. (Eds), *Lecture Notes in Computer Science*, Springer, LNCS 4261, pp. 72–79, 2006

7

Feature Extraction

Feature extraction is an essential pre-processing step to pattern recognition and machine learning problems. It transforms an image of containing redundant data into a reduced representation set of features. Many morphological approaches for feature extraction or image segmentation have been proposed. A morphological tool for image segmentation uses the watershed transformation [Vincent and Soille 1991; Gauch 1999]. Applying watershed transformation on the gradient image generates an initial segmentation, but resulting in over-segmentation, so that region merging needs to be performed as the post-processing. A morphological clustering method is illustrated in Yun, Park, and Lee [1998]. A morphology based supervised segmentation is introduced in Gu and Lee [1998]. They created the searching area where the real object resides by applying the morphological operators to the initial boundary. Video segmentation using mathematical morphology can be referred to Wang [1998].

In this chapter we discuss feature extraction based on MFs. It is organized as follows. Section 7.1 introduces edge linking by mathematical morphology. Section 7.2 presents corner detection by regulated morphology. Section 7.3 describes the shape database with hierarchical features. Section 7.4 discusses the corner and circle detection. Section 7.5 presents size histogram.

7.1 Edge Linking by MM

In edge or line linking, two assumptions are usually made: (1) True edge and line points in a scene follow some continuity patterns, whereas the noisy pixels do not follow any such continuity; (2) the strengths of the true edge or line pixels are greater than those of the noisy pixels. The "strength" of pixels is defined differently for different applications.

Nevatia [1976] presented an algorithm to link edge points by fitting straight lines. Linking direction is based on the direction of edge elements within a defined angular interval. However, portions of small curved edge segments may be neglected. Nalwa and Pauchon [1987] presented an approach based upon local information. The defined *edgels* (i.e., short, linear edge elements, each characterized by a direction and a position) as well as curved segments are linked based solely on proximity and relative orientation. No global factor has been taken into consideration, and there is little concern about noise reduction. Liu, Lin, and Liang [1988] proposed an edge-linking algorithm to fill gaps between edge segments. The filling operation is performed in an iterative manner rather than a single step. They first connect so-called *tip ends* (i.e., the set of edge pixels which are very likely to be the knots of some contour containing the one being considered) of two segments with a line segment and then try to modify the resulting segment by straight-line fitting. In each iteration, they define the dynamic threshold, and noises are removed gradually. However, it is difficult to get accurate tip ends. Another method [Russ 1992] locates all end points of broken edges and uses a relaxation method to link them up such that line direction is maintained. Lines are not allowed to cross, and the points that are closer are matched first. However, this method may fail if unmatched end points or noises are present.

7.1.1 Adaptive MM

Traditional morphological operators process an image using a structuring element of fixed shape and size over the entire image pixels. In adaptive morphology, rotation, and scaling factors are incorporated. The structuring elements are adjusted according to local properties of an image. To introduce adaptive MM, the following terminologies need to be first introduced.

Assume that the sets in consideration are always connected and bounded. Let the boundary ∂B of a set B be the set of points, all of whose neighborhoods intersect both B and its complement B^c. If a set B is connected and has no holes, it is called *simply-connected*. If it is connected but has holes, it is called *multiply-connected*. The concept of ∂B is referred to as the continuous boundary of a structuring element in Euclidean plane and only the simply-connected structuring element is considered. Therefore, the transformations performed on the structuring element B are replaced by the transformations on its boundary ∂B in Euclidean plane and followed by a positive filling operator. The positive filling $[\partial B]_+$ of a set B is defined as the set of points that are inside ∂B.

Shih and Cheng [2004] presented the adaptive morphology approach to edge linking. Verly and Delanoy [1993] proposed the adaptive MM for range imagery. Cheng and Venetsanopoulos [1992] presented the adaptive morphology for image processing. Bouaynaya and Schonfeld [2008] presented theoretical foundations of spatially-variant MM.

Definition 7.1: Let A and B be subsets of E^N. The *adaptive morphological dilation* of A by a structuring element B is defined by

$$A \hat{\oplus} B = \{c \in E^N | c = a + \hat{b}, \text{ for some } a \in A \text{ and } \hat{b} \in [R(a)S(a)\partial B]_+\}, \quad (7.1)$$

where S and R are the scaling and rotation matrices, respectively.

Definition 7.2: Let A and B be subsets of E^N. The *adaptive morphological erosion* of A by a structuring element B is defined by

$$A \hat{\ominus} B = \{c \in E^N | c + \hat{b} \in A, \quad \text{for some } a \in A \text{ and every } \hat{b} \in [R(a)S(a)\partial B]_+\}. \quad (7.2)$$

The broken edge segments can be linked gradually using the adaptive morphological dilation as shown in Figure 7.1, where: (a) is the input signal with gaps and; (b) shows an adaptive dilation at the endpoints. The reason for choosing the elliptical structuring element is that by using appropriate major and minor axes, all kinds of curves can be linked smoothly. Figure 7.1c shows the result of an adaptive dilation operation.

7.1.2 Adaptive Morphological Edge-Linking Algorithm

A complete edge-linking algorithm consists of removing noisy edge segments, checking endpoints, applying adaptive dilation, thinning, and pruning. If large gaps occur, this algorithm needs to be applied several times until no more gap exists or a predefined number of iterations has been reached.

Step 1: *Removing noisy edge segments.* If the length of one edge segment is shorter than a threshold value, then this segment is removed. A reasonable threshold value of three is used.

Step 2: *Detecting all the endpoints.* An endpoint is defined as any pixel having only one 8-connected neighbor. All the edge points are checked to extract the entire endpoint set.

Step 3: *Applying adaptive dilation at each endpoint.* From each endpoint, a range of pixels along the edge are chosen. From the properties of this set of pixels, the rotation angle and size of the elliptical structuring element are obtained. In the program, the range of pixels used for each iteration increases in size. It is referred to as an adjustable parameters denoting the size of the range of pixels from the endpoint. Two pixels are taken from this range of pixels, the left-most side $p_1(x_1, y_1)$ and the right-most side $p_2(x_2, y_2)$. The equation: slope = $(y_2 - y_1)/(x_2 - x_1)$ is used to compute the slope.

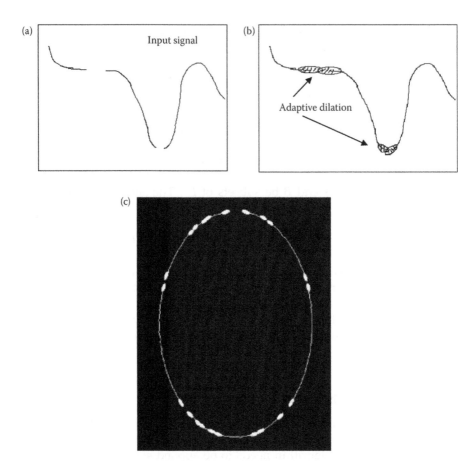

FIGURE 7.1 (a) Input signal with gaps, (b) adaptive dilation using elliptical structuring elements, and (c) the result of an adaptive dilation operation.

An ellipse can be represented as

$$\frac{x^2}{a^2} + \frac{y^2}{b^2} = 1, \tag{7.3}$$

where a and b denote the major and minor axes, respectively. If the center of the ellipse is shifted from the origin to (x_0, y_0), the equation becomes

$$\frac{(x - x_0)^2}{a^2} + \frac{(y - y_0)^2}{b^2} = 1. \tag{7.4}$$

If the center is at (x_0, y_0) and the rotation angle is θ $(-\pi/2 \le \theta \le \pi/2)$, the equation becomes:

$$\frac{[(x - x_0)\cos\theta - (y - y_0)\sin\theta]^2}{a^2} + \frac{[(x - x_0)\sin\theta - (y - y_0)\cos\theta]^2}{b^2} = 1. \quad (7.5)$$

Because the image is discrete, the rounded values are used in defining the elliptical structuring element. At each endpoint, an adaptive dilation is performed using the elliptical structuring element. Therefore, the broken edge segment will be extended along the slope direction by the shape of ellipse.

The rotation angle of the elliptical structuring element can be obtained by the equation: $\theta = \tan^{-1}$ (*slope*). The size of the elliptical structuring element is adjusted according to the number of the pixels in the set, the rotation angle θ, and the distance between p_1 and p_2. In the program, the elliptical structuring element with fixed $b = 3$ and a changing from 5 to 7 are used. These values are based on experimental results. For a big gap, the gap can be linked gradually in several iterations using $a = 7$. For a small gap, using $a = 5$ is better than using 7. If $a = 3$ or 4 is used for a small gap, the structuring element will be almost a circle, and the thinning algorithm does not work well.

Step 4: *Thinning.* After the adaptive dilation is applied at each endpoint, the edge segments are extended in the direction of the local slope. Because the elliptical structuring element is used, the edge segments grow a little fat. Morphological thinning is used to obtain the edge of one pixel wide.

Step 5: *Branch pruning.* The adaptively dilated edge segments after thinning may contain noisy short branches. These short branches must be pruned away. The resulting skeletons after thinning have width of one pixel. A root point is defined as a pixel having at least three pixels in 8-connected neighbors. From each endpoint, it is traced back along the existing edge. If the length of this branch is shorter than a threshold value after it reaches a root point, the branch is pruned.

Step 6: *Decision.* The program terminates when no endpoint exists or when a predefined number of iterations has been reached. In the program, eight iterations are used as the limit.

7.1.3 Experimental Results

As shown in Figure 7.1c, the image contains the black background and the white edge pixels. For the purpose of clear display, we invert the color

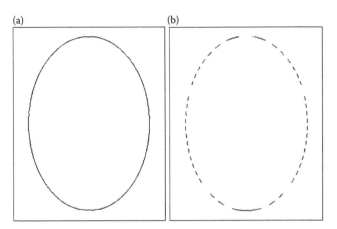

FIGURE 7.2   (a) An elliptical edge and (b) its randomly discontinuous edge.

of the image, so that the background is now in white and the edge pixels
are in black. Figure 7.2a shows an elliptical edge, and 7.2b shows its randomly discontinuous edge. The edge-linking algorithm is experimented on
Figure 7.2b.

1. *Using circular structuring elements.* Figure 7.3 shows the results of
   using circular structuring elements with $r = 3$ and $r = 5$, respectively,
   in five iterations. As comparing with the ellipse in Figure 7.2a, if the
   gap is larger than the radius of the structuring element, it would be
   difficult to link smoothly. However, if a very big circular structuring
   element is used, the edge will look hollow and protuberant. Besides,
   it can obscure the details of the edge.

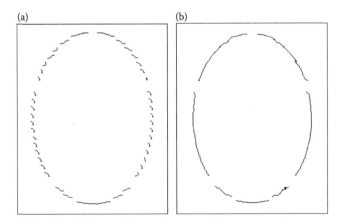

FIGURE 7.3   Using circular structuring elements in five iterations with (a) $r = 3$ and (b) $r = 7$.

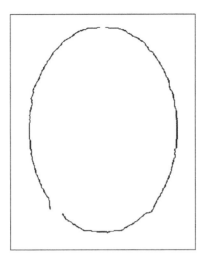

FIGURE 7.4 An elliptical structuring with $a = 5$, $b = 3$, and $s = 7$.

2. *Using a fixed sized elliptical structuring element and a fixed range of pixels to measure the slope.* In this experiment, an elliptical structuring element with a fixed minor axis $b = 3$ and a fixed major axis $a = 5, 7, 9$ are used. A fixed range of pixels is used to measure the local slope for each endpoint. That is, the same range of pixels counting from the endpoint in each iteration is used. The parameter s denotes the range of pixels from endpoint.

   Figure 7.4 shows the result of using an elliptical structuring element with $a = 5$, $b = 3$, $s = 7$. The result has some big gaps because $a = 5$ is too small. Figure 7.5 shows the result of using $a = 7, b = 3, s = 9$.

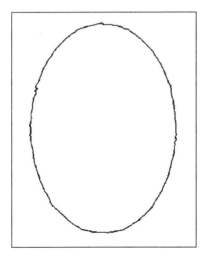

FIGURE 7.5 An elliptical structuring element with $a = 7, b = 3$, and $s = 9$.

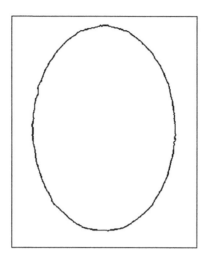

FIGURE 7.6 An elliptical structuring element with $a = 7$, $b = 3$, and $s = 11$.

Figure 7.6 shows the result of using $a = 7$, $b = 3$, $s = 11$. There is not much difference between Figures 7.5 and 7.6. Compared with the original ellipse in Figure 7.2a, Figures 7.5 and 7.6 have a few shortcomings, but they are much better than Figure 7.4. Figure 7.7 shows the result of using $a = 9$, $b = 3$, $s = 11$. Therefore, using $a = 7$ or $a = 9$, a reasonably good result is obtained except for a few shifted edge pixels.

3. *Using a fixed sized elliptical structuring element for every endpoint, but using adjustable sized range of pixels to measure local slope in each iteration.* Figure 7.8 shows the result of using $a = 5$, $b = 3$, and an

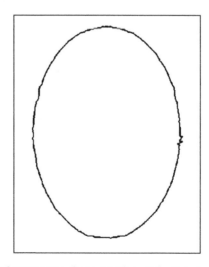

FIGURE 7.7 An elliptical structuring element with $a = 9$, $b = 3$, and $s = 11$.

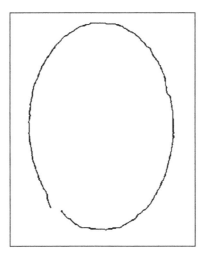

FIGURE 7.8 Using an elliptical structuring element with $a = 5$, $b = 3$, and an adjustable s.

adjustable s, and Figure 7.9 shows the result of using $a = 7$, $b = 3$, and an adjustable s. Compared with Figures 7.6 and 7.7, Figures 7.8 and 7.9 are better in terms of elliptical smoothness. This is true because after each iteration, the range of pixels used to measure local slope is increased and more information from the original edge is taken into account.

4. *Using adjustable sized elliptical structuring elements for every endpoint and using adjustable sized range of pixels to measure local slope in each iteration (the adaptive morphological edge-linking algorithm).* In this

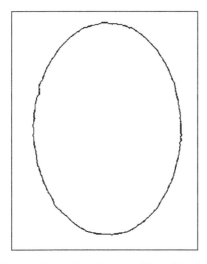

FIGURE 7.9 Using an elliptical structuring element with $a = 7$, $b = 3$, and an adjustable s.

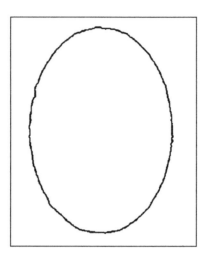

FIGURE 7.10   Using the adaptive morphological edge-linking algorithm.

experiment, the adjustable sized elliptical structuring elements with
a changing from 5 to 7, $b = 3$, and an adjustable s are used.

Figure 7.10 shows the result of using the adaptive morphological
edge-linking algorithm. Compared with Figures 7.6 and 7.9, Figure
7.10 seems about the same. However, the adaptive method has the
advantage of adjusting a and s automatically, while the parameters
in Figures 7.6 and 7.9 are fixed; that is, they work well in a certain
case, but may not work well in other cases.

Figure 7.11a shows the elliptical shape with added uniform noise.
Figure 7.11b shows the shape after removing the noise. Figure 7.11c
shows the result of using the adaptive morphological edge-linking
algorithm. Compared with Figure 7.10, Figure 7.11b has several shortcomings. This is because after removing the added noise, the shape
is changed at some places.

7.2 Corner Detection by Regulated Morphology

Corners are frequently employed in pattern recognition. They are the pixels
whose slope changes abruptly; that is, the absolute curvature is high. Given
an image, a typical approach to detect corners involves segmenting the
object from background, representing the object boundary by chain codes,
and searching for significant turns in the boundary. Local measurement
such as non-maxima suppression is used to obtain a binary corner map.

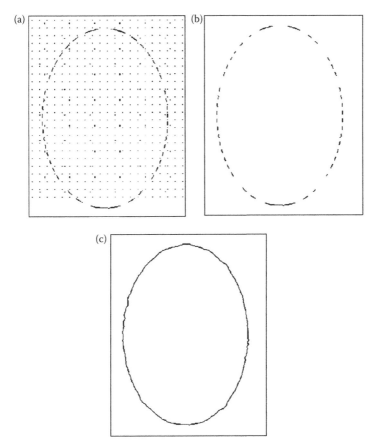

FIGURE 7.11 (a) The elliptical shape with added uniform noise, (b) the shape after removing noise, and (c) using the adaptive morphological edge-linking algorithm.

The algorithms of corner detection can be categorized into boundary-based and gray-level approaches. The boundary-based approach detects corners based on the boundary information of objects. Tsai, Hou, and Su [1999] proposed a method using eigenvalues of the covariance matrix of data points on a curve segment. The gray-level approach directly works on gray-level images using the corner template matching or the gradients at edge pixels. Lee and Bien [1996] developed a real-time gray-level corner detector using fuzzy logic. Singh and Shneier [1990] proposed a fusion method by combining template-based and gradient-based techniques. Zheng, Wang, and Teoh [1999] proposed a gradient-direction corner detector. Kitchen and Rosenfeld [1982] demonstrated that gray-level schemes perform better than boundary-based techniques.

In the corner detection based on morphological operations, Laganiere [1998] proposed a method using the *asymmetrical closing* by four different structuring elements including plus, lozenge, cross, and square. Lin, Chu, and Hsueh [1998] proposed a morphological corner detector to find convex and concave corner points using integer computation. However, their disadvantages include: (1) It is difficult to choose a suitable structuring element; (2) as the size of the structuring element increases, the computational cost increases. Zhang and Zhao [1995] developed a morphological corner detector, but it can only detect convex corners and the result is sensitive to the structuring element size. Sobania, Paul, and Evans [2005] presented morphological corner detector from segmented areas using MM employing paired triangular structuring elements. Shih, Chuang, and Gaddipati [2005] presented a modified regulated morphological corner detector. Dinesh and Guru [2006] presented a morphological approach for detecting corner points using morphological skeleton. Convex corner points are obtained by intersecting the morphological boundary and the corresponding skeleton, whereas the concave corner points are obtained by intersecting the boundary and the skeleton of the complement image.

7.2.1 A Modified Laganiere's Operator

Laganiere [1998] proposed corner detection using the *asymmetrical closing*, which is defined as a dilation of an image by a structuring element followed by an erosion by another structuring element. The idea is to make the dilation and erosion complementary and correspond to variant types of corners. Two structuring elements, cross + and lozenge ◊, are used. Let the asymmetrical closing of an image A by structuring elements + and ◊ be denoted by

$$A^c_{+,\diamond} = (A \oplus +) \ominus \diamond. \tag{7.6}$$

The corner strength is computed by

$$C_+(A) = |A - A^c_{+,\diamond}|. \tag{7.7}$$

For different types of corners, another corner strength (which is a 45°-rotated version) is computed as follows:

$$C_\times(A) = |A - A^c_{\times,\square}|. \tag{7.8}$$

By combining the four structuring elements in Figure 7.12, the corner detector is represented as:

$$C_{+,\times}(A) = |A^c_{+,\diamond} - A^c_{\times,\square}|. \tag{7.9}$$

The disadvantages of the above corner detector are that it may miss obtuse-angle corners and remove sharp-angle corner pixels. Therefore, a modified

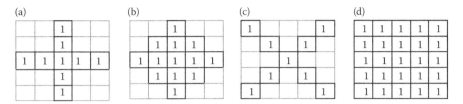

FIGURE 7.12 The four structuring elements (a) +, (b) ◇, (c) ×, and (d) □.

corner detector uses: $C_{+\times}(A) = |A - A^c_{+\diamond}| \cup |A - A^c_{\times\square}|$ [Shih, Chuang, and Gaddipati 2005]. Two examples of comparing Laganiere's detector and the modified detector are shown in Figures 7.13 and 7.14. It is observed that the modified detector can locate the corner pixels more accurately.

7.2.2 Modified Regulated Morphology for Corner Detection

Agam and Dinstein [1999] developed regulated morphological operators and showed how the fitting property can be adapted for analyzing the map and

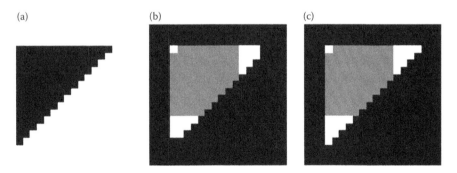

FIGURE 7.13 (a) A triangle image, (b) Laganiere's detector, and (c) the modified detector.

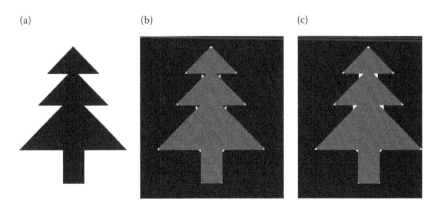

FIGURE 7.14 (a) A tree image, (b) Laganiere's detector, and (c) the modified detector.

line-drawing images. Since regulated morphology inherits many properties of ordinary morphology, it is feasible to apply in image processing and optimize its strictness parameters using some criteria. Tian, Li, and Yan [2002] extended the regulated morphological operators by adjusting the weights in the structuring element.

Note that the morphological hit-and-miss transform [Zhang and Zhao 1995] can also be used to extract corners. For example, four different SEs can be designed as consisting of four types of right-angle corners and perform the hit-and-miss transform with each structuring element. The four results are combined to obtain all the corners. However, the corners extracted are only the locations of all right-angle convex corners in four directions.

The *regulated dilation* of a set A by a structuring element set B with a strictness parameter s is defined by

$$A \oplus^s B = \{x \,|\, \#\,(A \cap (\hat{B})_x) \geq s, \quad s \in [1, \min(\#A, \#B)]\}, \tag{7.10}$$

where the symbol # denotes the cardinality of a set. The *regulated erosion* of a set A by a structuring element set B with a strictness parameter s is defined by

$$A \ominus^s B = \{x \,|\, \#\,(A^c \cap (B)_x) < s, \quad s \in [1, \#B]\}. \tag{7.11}$$

To maintain some properties the same as ordinary morphological operators, the *regulated closing* of a set A by a structuring element set B with a strictness parameter s is defined by

$$A \bullet^s B = ((A \oplus^s B) \ominus B) \cup A, \tag{7.12}$$

where $A \oplus^s B = (A \oplus^s B) \cup A$ is defined as the *extensive regulated dilation*. The *regulated opening* of a set A by a structuring element set B with a strictness parameter s is defined by

$$A \circ^s B = ((A \ominus^s B) \oplus B) \cap A, \tag{7.13}$$

where $A \ominus^s B = (A \ominus^s B) \cap A$ is defined as the *anti-extensive regulated erosion*.

By adjusting the strictness parameter, the noise sensitivity problem and small intrusions or protrusions on the object boundary can be alleviated. For corner detection, regulated openings or closings are not used because the resulting image will stay the same as the original image after the strictness parameter exceeds a certain value. The modified regulated morphological corner detector is described as follows:

Step 1: $A_1 = (A \oplus^s B) \ominus^s B$. Corner strength: $C_1 = |A - A_1|$.
Step 2: $A_2 = (A \ominus^s B) \oplus^s B$. Corner strength: $C_2 = |A - A_2|$.
Step 3: Corner detector $= C_1 \cup C_2$.

The original image is performed by a regulated dilation by a 5×5 circlular structuring element with a strictness s, and then followed by a regulated erosion by the same structuring element with the same strictness. The corner strength C_1 is computed by finding the absolute value of the difference between the original and resulting images. This step is to extract concave corners. By reversing the order such that the regulated erosion is applied first, and then followed by the regulated dilation, the convex corners can be extracted. Finally, both types of corners are combined.

7.2.3 Experimental Results

The comparisons of results by Laganiere's method and by the modified regulated morphology are shown in Figures 7.15 through 7.17. From Figures 7.15 and 7.16, the modified method can detect corners more accurately. From Figure 7.17, the modified method can detect corners more completely; for example, Laganiere's method can only detect 8 corners, but the modified method can detect all 12 corners at a strictness of 3. Using a Pentium IV

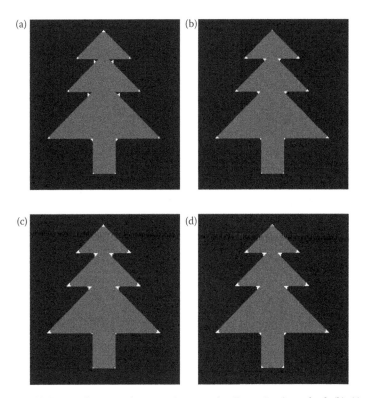

FIGURE 7.15 (a) Corner detection for a tree image using Laganiere's method, (b), (c), and (d) corner detection using the modified method at strictness = 2, 3, and 4, respectively.

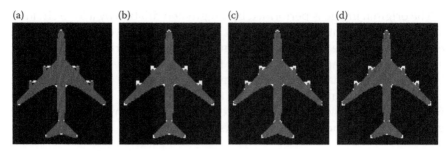

FIGURE 7.16 (a) Corner detection for an airplane image using Laganiere's method, (b), (c), and (d) corner detection using the modified method at strictness = 2, 3, and 4, respectively.

2.0 GHz PC, it takes 2.28 seconds to process Figure 7.17 by Laganiere's method; however, it spends 2.01 seconds by the proposed method.

A square is rotated by 30°, 45°, and 60° as illustrated in Figure 7.18. Experimental results show that the modified method can detect corners correctly. Table 7.1 summarizes the results of corner detection using Laganiere's and the modified methods with different values of strictness. From Table 7.1 and Figures 7.15 through 7.17, strictness of three is a better choice because it produces more accurate corners than strictness of 2 and generates less false-positive corners than strictness of 4.

In experiments, a 5×5 circuluar structuring element is adopted. A 3×3 structuring element is applied on Figure 7.17, and the results are illustrated in Figure 7.19. It is observed that the 3×3 structuring element is worse than the 5×5 structuring element because some corners are missed.

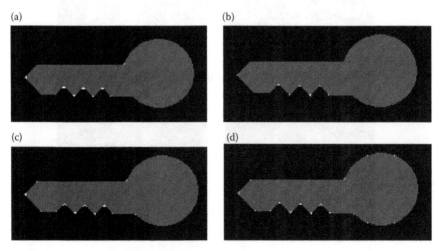

FIGURE 7.17 (a) Corner detection for a key image using Laganiere's method, (b), (c), and (d) corner detection using the modified method at strictness = 2, 3, and 4, respectively.

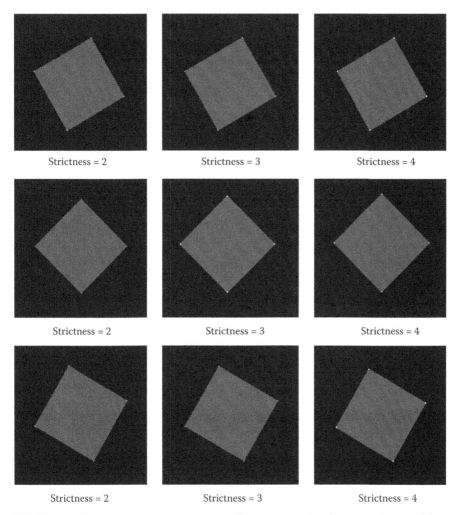

FIGURE 7.18 The corner detection results in different angles with strictness = 2, 3, and 4. The top-, middle-, and bottom-row images are rotated by 30°, 45°, and 60°, respectively.

7.3 Shape Database with Hierarchical Features

A hierarchical shape database includes a shape number, several layers of significant point radii (i.e., Euclidean distances to the boundary), and their x and y coordinates [Shih 1991]. Each local maximum point of the DT is used as a significant point, which has the major contribution in representing features.

TABLE 7.1

Number of Correctly Detected Corners, False Positives, and False Negatives

Test Images	Figure 7.15	Figure 7.16	Figure 7.17
Total number of corners	15	24	12
Laganiere's method			
Correctly detected corners	15	18	8
False positives	0	0	0
False negatives	0	6	4
Our method at strictness = 2			
Correctly detected corners	15	18	7
False positives	0	0	0
False negatives	0	6	5
Our method at strictness = 3			
Correctly detected corners	15	23	12
False positives	0	0	0
False negatives	0	1	0
Our method at strictness = 4			
Correctly detected corners	15	24	12
False positives	0	2	4
False negatives	0	0	0

(a)   (b)   (c)

FIGURE 7.19 The detected corners using a 3 × 3 structuring element at strictness = 2, 3, and 4, respectively.

7.3.1 Shape Number from DT

The shape number is computed from the total distance and the total number of all object points [Danielsson 1978]. It is a measure of shape compactness. Let N be the total number of the pixels belonging to the object, and let X_i be the distance of pixel i to the nearest pixel outside the object. The shape number is mathematically expressed as

$$\text{shape number} = \frac{N^3}{9\pi\left(\sum_{i=1}^{N} X_i\right)^2} \tag{7.14}$$

0	0	0	0	0	0	0	0	0	0	0	0	0
0	0	0	0	0	1	1	1	0	0	0	0	0
0	0	0	1	1	1	1	1	1	1	0	0	0
0	0	1	1	1	1	1	1	1	1	1	0	0
0	0	1	1	1	1	1	1	1	1	1	0	0
0	1	1	1	1	1	1	1	1	1	1	1	0
0	1	1	1	1	1	1	1	1	1	1	1	0
0	1	1	1	1	1	1	1	1	1	1	1	0
0	0	1	1	1	1	1	1	1	1	1	0	0
0	0	1	1	1	1	1	1	1	1	1	0	0
0	0	0	1	1	1	1	1	1	1	0	0	0
0	0	0	0	0	1	1	1	0	0	0	0	0
0	0	0	0	0	0	0	0	0	0	0	0	0

FIGURE 7.20 A 13 × 13 discrete circle.

Note that the constant is chosen such that the shape number of an ideal circle is 1. Hence, the larger the shape number is, the less compact the object shape looks. A 13 × 13 discrete circle is shown in Figure 7.20, where the total number of object pixels is 89. Its city-block DT is shown in Figure 7.21. It is the fact that we have labeled all border pixels with the distance value $x = 1$ (actually, it is 0.5). For precise measurement, each distance value is subtracted by 0.5. The summation of total distance values is $218 - 0.5 \times 89 = 173.5$. Therefore, the shape number is 0.8283. Note that the shape number is not equal to one because of the digitization and the city-block distance used. If we select Euclidean distance, it will increase the computational precision. Figure 7.22 shows the EDT of Figure 7.20. To make the distance summation more accurate, it would be appropriate to decrease every x value by the average error $\sqrt{0.5}/2$.

0	0	0	0	0	0	0	0	0	0	0	0	0
0	0	0	0	0	1	1	1	0	0	0	0	0
0	0	0	1	1	2	2	2	1	1	0	0	0
0	0	1	2	2	3	3	3	2	2	1	0	0
0	0	1	2	3	4	4	4	3	2	1	0	0
0	1	2	3	4	5	5	5	4	3	2	1	0
0	1	2	3	4	5	6	5	4	3	2	1	0
0	1	2	3	4	5	5	5	4	3	2	1	0
0	0	1	2	3	4	4	4	3	2	1	0	0
0	0	1	2	2	3	3	3	2	2	1	0	0
0	0	0	1	1	2	2	2	1	1	0	0	0
0	0	0	0	0	1	1	1	0	0	0	0	0
0	0	0	0	0	0	0	0	0	0	0	0	0

FIGURE 7.21 The city-block DT of Figure 7.20.

0	0	0	0	0	0	0	0	0	0	0	0	0
0	0	0	0	0	1	1	1	0	0	0	0	0
0	0	0	1	1	$\sqrt{2}$	2	$\sqrt{2}$	1	1	0	0	0
0	0	1	$\sqrt{2}$	2	$\sqrt{5}$	$\sqrt{8}$	$\sqrt{5}$	2	$\sqrt{2}$	1	0	0
0	0	1	2	$\sqrt{8}$	$\sqrt{10}$	$\sqrt{13}$	$\sqrt{10}$	$\sqrt{8}$	2	1	0	0
0	1	$\sqrt{2}$	$\sqrt{5}$	$\sqrt{10}$	$\sqrt{17}$	$\sqrt{20}$	$\sqrt{17}$	$\sqrt{10}$	$\sqrt{5}$	$\sqrt{2}$	1	0
0	1	2	$\sqrt{8}$	$\sqrt{13}$	$\sqrt{20}$	$\sqrt{29}$	$\sqrt{20}$	$\sqrt{13}$	$\sqrt{8}$	2	1	0
0	1	$\sqrt{2}$	$\sqrt{5}$	$\sqrt{10}$	$\sqrt{17}$	$\sqrt{20}$	$\sqrt{17}$	$\sqrt{10}$	$\sqrt{5}$	$\sqrt{2}$	1	0
0	0	1	2	$\sqrt{8}$	$\sqrt{10}$	$\sqrt{13}$	$\sqrt{10}$	$\sqrt{8}$	2	1	0	0
0	0	1	$\sqrt{2}$	2	$\sqrt{5}$	$\sqrt{8}$	$\sqrt{5}$	2	$\sqrt{2}$	1	0	0
0	0	0	1	1	$\sqrt{2}$	2	$\sqrt{2}$	1	1	0	0	0
0	0	0	0	0	1	1	1	0	0	0	0	0
0	0	0	0	0	0	0	0	0	0	0	0	0

FIGURE 7.22   The EDT of Figure 7.20.

The summation of total distances is $188.973 - \sqrt{0.5}/2 \times 89 = 157.507$. Hence, the shape number is 1.005, which gives a more accurate measure.

7.3.2 Significant Points Radius and Coordinates

Significant points are defined to be the local maxima of 8-neighborhood in the DT. If the Euclidean distance is used, then the circle drawn with the significant point as the center and the distance as the radius is the maximal inscribed circle in the local region.

Let d be the DT. The significant point extraction algorithm is:

1. Execute $d \oplus_g k$. Here k is a 3×3 structuring element with all zeros.
2. If the object pixel value in the output of step 1 is equal to that in d, then keep the value; otherwise, let this point be zero.
3. If the nonzero output points of step 2 are connected, then represent those points by a single average point.
4. List the values of radii and coordinates of all the nonzero points according to the decreasing order of the radius values.

Figure 7.23 shows a rectangular block with a hole and a slot and the extracted significant points.

7.3.3 Recognition by Matching Database

Let s denote the shape number, and let r_i, x_i, y_i denote the radius and x, y coordinates in the decreasing order of radius, respectively. From the previous section, we already have the shape database registered in the format: $(s, r_1, x_1, y_1, r_2, x_2, y_2, \ldots, r_n, x_n, y_n)$. Compare the unknown object with the

(a)   (b)

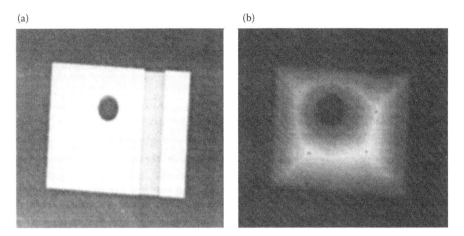

FIGURE 7.23 (a) An industrial part with a hole and a slot. (b) The result of significant points extraction.

registered database and compute the error value using Equation 7.15. If the error is within the tolerance range, then the object is classified as the identical object with the registered object. Otherwise, it is classified as a different object. Let s', r'_i, x'_i, y'_i denote the shape number, radius, and x, y coordinates for the unknown object, respectively. The error is computed as

$$\text{error} = \frac{1}{2}\left(\frac{|s'-s|}{s} + \frac{|r'_1 - r_1| + \cdots + |r'_n - r_n|}{r_1 + \cdots + r_n}\right). \tag{7.15}$$

7.3.4 Localization by Hierarchical Morphological Band-Pass Filter

The problem considered in this section is how to quickly locate an object, which was known and stored in the database. The opening with an image will remove all of the pixels in regions, which are too small to contain the probe. The opposite sequence (closing) will fill in holes and concavities smaller than the probe. Such filters can be used to suppress spatial features or discriminate against objects based on their size distribution. As an example, if a disk-shaped with radius h structuring element is used, then opening with an image is a low-pass filter. The opening residue is a high-pass filter. The difference of two openings of an image with two nonequal radii is a band-pass filter. It can be expressed as follows.

LOW-PASS $= A \circ B^h$,
HIGH-PASS $= A - (A \circ B^h)$,
BAND-PASS $= (A \circ B^{h_1}) - (A \circ B^{h_2})$, where: radii of B, $h_1 < h_2$.

If there are several unknown objects on the workbench and we want to pick up the one specified in the library database, the following algorithm is a simple and rapid in the localization approach.

Localization Algorithm

To locate objects, we change the morphological opening to the erosion operation. Let A denote the binary image, which contains several objects, and r_i denote the significant point radii in the library shape database.

1. Do $(A \ominus_b B^{r_1-1}) - (A \ominus_b B^{r_1+1})$.
2. If the output of step 1 is empty, it means the library object is not on the workbench. If the output has two nonzero points spaced at least a certain interval, it means that at least two objects possibly match the library object. Hence, we need to do erosion with the second radius. Do $(A \ominus_b B^{r_2-1}) - (A \ominus_b B^{r_2+1})$. Apply this procedure recursively until only one point remains. This is the location of the object that matches the library object.

7.4 Corner and Circle Detection

The algorithm of corner and circle detection based on MFs is introduced in this section. A disk structuring element of suitable size is selected. The image is opened with the disk structuring element and the result is subtracted from the original. The opening residue area varies with the angle of the corner. Then two SEs are selected: one is small enough to pass the residue area and the other is too large. The opening residue is eroded with these two SEs and the two outputs are compared. This is equivalent to a shape band-pass filter. Different sized structuring elements are used depending on the corner angles to be located. As the angle decreases, the opening residue area increases, hence the structuring element area increases.

First, the largest area structuring element is selected. If nothing is obtained, then the area of the structuring element is decreased until there is some object pixels are obtained. Thus, the angle information can be obtained.

Corner Detection Algorithm

1. A binary image is dilated by a disk structuring element of a suitable size.
2. The output of step 1 is opened by a disk structuring element of a large size.
3. Subtract the output of step 2 from the output of step 1.

4. The output of step 3 is eroded by two disk structuring elements: One is the same as in step 1 and the other is a little bit larger.

5. Subtraction of two eroded outputs of step 4. This is a band-pass filter. The remaining points are the corner points.

Note that a suitable size of the disk-structuring element in step 1 is determined by the largest area that can be fit inside the opening residue corner (output of Step 3). The corner can be located more accurately by correcting the bias generated in Step 4. Figure 7.24 illustrates an example of applying the corner detection algorithm on a rectangular block with two holes.

The DT algorithm is used to obtain the distance information. According to the shape number, we conclude whether a region is a circle. If it is a circle, the maximum distance is the radius and the location of that value is the center. If there is more than one point, we calculate the average of those points as the center.

Circle Detection Algorithm

1. A grayscale image is thresholded into a binary using a suitable thresholding algorithm.

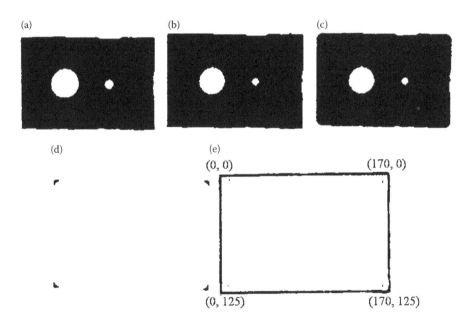

→ There are four rectangular corners in the parts.

Coordinates: (6, 5), (162, 7), (6, 116), (161, 120)

FIGURE 7.24 An example of a corner detector. Use morphology to detect the 90° corners. For clear display, the dark and white regions are reversed. (a) Original image, (b) output of Step 1, (c) output of Step 2, (d) output of Step 3, (e) output of Step 7.

2. Perform a DT algorithm.

3. Use the computed shape number to determine whether a region is a circle.

4. If the region is a circle, detect the maximum distance value in the region and its coordinates.

5. If there is more than one point with the maximum distance value, then select the average of the coordinates as the center and add 0.5 (approximated by a half of pixel unit-length) to the maximum distance as the radius.

The two circular holes in Figure 7.24 can be extracted by the circle detection algorithm and the results are shown in Figure 7.25.

7.5 Size Histogram

In this section, we introduce size histogram, which is a function that gives the frequency of occurrence of each distance value in the DT. It provides useful information pertaining to object shape, and can be used to compute the shape number and convexity and complexity factors. Using the convexity points of size histogram, one can segment an image into object components for further analysis.

For a given image we obtain its DT by the morphological erosion with different shaped SEs and then draw a histogram for each DT. The size histogram is constructed to measure the number of pixels versus each distance value. That is, the distance values on x-axis and the number of pixels with each distance on y-axis.

Size histogram is similar to gray level histogram except that distance values are used instead of intensities. In size histogram, the distance value gives a measure of the radius of a circle at this location to be contained in the object. From it we can also obtain a measure of the compactness of an object. Other shape attributes which can be also measured, such as shape number, which can be computed by finding the convexity points of the curve and local and global shapes.

We call it size histogram because it gives the size (or area) and shape information. The area of the object is the region's size enclosed by the histogram curve. As an illustration, the DT of an image is related with the assigned SEs, which can be regarded as the object shape primitives. It is obvious from the figures, even for a complex shape, that the flavor of distance is preserved corresponding to the structuring element.

The shape of a given image is then determined by applying different SEs and finding their size histograms. Here we present an algorithm to find the size histogram.

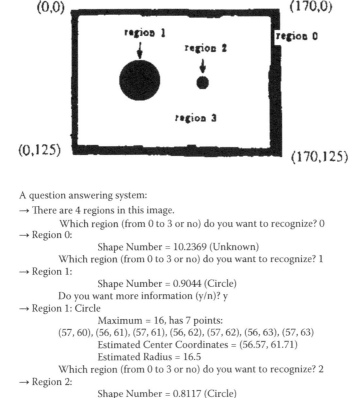

A question answering system:

→ There are 4 regions in this image.

        Which region (from 0 to 3 or no) do you want to recognize? 0

→ Region 0:

        Shape Number = 10.2369 (Unknown)

        Which region (from 0 to 3 or no) do you want to recognize? 1

→ Region 1:

        Shape Number = 0.9044 (Circle)

        Do you want more information (y/n)? y

→ Region 1: Circle

        Maximum = 16, has 7 points:

        (57, 60), (56, 61), (57, 61), (56, 62), (57, 62), (56, 63), (57, 63)

        Estimated Center Coordinates = (56.57, 61.71)

        Estimated Radius = 16.5

        Which region (from 0 to 3 or no) do you want to recognize? 2

→ Region 2:

        Shape Number = 0.8117 (Circle)

        Do you want more information (y/n)? y

→ Region 2: Circle

        Maximum = 5, has 6 points:

        (107, 62), (108, 62), (107, 63), (108, 63), (107, 64), (108, 64)

        Estimated Center Coordinates = (107.50, 63.00)

        Estimated Radius = 7.5

        Which region (from 0 to 3 or no) do you want to recognize? 3

→ Region 3:

        Shape Number = 3.4486 (Unknown)

        Which region (from 0 to 3 or no) do you want to recognize? No

FIGURE 7.25 An example of circle extraction Table 7.1. Number of correctly detected corners, false positives, and false negatives.

Step 1: Compute the DT of the image with the assigned structuring element.

Step 2: Draw the size histogram for the DT.

Step 3: Compute the slopes and compare the slope change for the size histogram.

Step 4: Repeat the procedure for each structure element.

After computing the slopes of size histograms, we choose the structure element whose slopes of the size histogram are nearly the same, as the component that best represents the shape.

Let $y = f(x)$ be any function that defines the size histogram curve, where y is the number of occurrences of each distance value and x is the distance value in the domain $[1, n]$. The slope of the size histogram curve is the first derivative of the function, and the slope change is the second derivative. These values exist only if the function is continuous over the range of x values. As all the distance values have to increase from 1 to n, the function $f(x)$ is continuous and derivable.

Shape Number: Another attribute for shape information is the shape number. If the shape measure has the dimensionless scale-independent quantity and is related to a predetermined ideal shape (usually a circle), then it is conventionally called a shape number. The shape number used here is computed from the DT as referred to Daniellson [1978]. It is a measure of the compactness of the shape. Let A be the total number of pixels (area) which belong to the object, and let d be the *mean distance* which is the distance between an object point and the center of gravity. The shape number is mathematically expressed as

$$\text{Shape Number} = \frac{A}{9\pi \bar{d}^2},\tag{7.16}$$

where $\bar{d}$ can be computed by

$$\bar{d} = \frac{I}{A},\tag{7.17}$$

where I is the moment of inertia which is expressed for any element x

$$I = \iint_A x \, dA.\tag{7.18}$$

Note that $\bar{d}$ can be computed directly from the size histogram $f(x)$ by

$$\bar{d} = \frac{\displaystyle\sum_{x=1}^{n} xf(x)}{\displaystyle\sum_{x=1}^{n} f(x)},\tag{7.19}$$

where n is the largest distance value. The area of the object, which is the total number of occurrences of distances, can be computed by

$$A = \sum_{x=1}^{n} f(x)\tag{7.20}$$

We present a new method of computing shape number using the properties of size histogram. Before deriving the formula, we introduce some definitions regarding the size histogram $f(x)$.

Definition 7.3: The size histogram is *linear* if

$$f(x) = \frac{1}{2}[f(x-1) + f(x+1)] \quad \text{for } 0 < x < n \tag{7.21}$$

Definition 7.4: The size histogram is called *convex* if

$$f(x) > \frac{1}{2}[f(x-1) + f(x+1)] \quad \text{for } 0 < x < n \tag{7.22}$$

Definition 7.5: The size histogram is called *concave* if

$$f(x) < \frac{1}{2}[f(x-1) + f(x+1)] \quad \text{for } 0 < x < n \tag{7.23}$$

If a size histogram contains the combinations of linear, convex, and concave, then the above definitions are applied on the point basis. Based on the above definitions, we propose the following theorems to obtain the mean distance.

Theorem 7.1: If for all x, $f(x)$ is linear, then in the resulting curve (being a triangle) the mean distance d is half of the maximum distance value.

Proof: From Equations 7.16 through 7.18 we have

$$\bar{d} = \frac{\iint_A x \, dA}{A} \tag{7.24}$$

For a linear size histogram, which is a triangle, the mean distance is

$$\bar{d} = \frac{\iint_A x \, dx \, dy}{2A}, \quad \text{where } A = \frac{nf(0)}{2} \tag{7.25}$$

To simplify the computation, we translate the starting point $x = 1$ to be y-axis and the smallest function value, $y = f(n)$, to be x-axis. Hence, integrating the above equation with respect to x from 0 to n and y from 0 to $f(0)$, we have

$$\bar{d} = \frac{1}{4A} \cdot x^2 \big|_0^n \cdot y \big|_0^{f(0)} = \frac{n^2 f(0)}{2n f(0)} = \frac{n}{2} \tag{7.26}$$

Thus, the mean distance $\bar{d}$ is a half of the distance value, which can be computed directly from the histogram.   □

Theorem 7.2: If $0 < m < n$ and $f(m)$ is convex, then the mean distance is computed by

$$\bar{d} = \frac{m^2 f(0) + n^2 f(m)}{2mf(0) + nf(m)} \tag{7.27}$$

The above method will greatly reduce the computational cost of finding mean distance especially for the case of many convex points in size histogram. For the given SEs we compute the shape numbers of all the size histograms. The shape numbers will greatly help in approximating the shape of the object and automate the process of industrial part recognition and inspection based on shape numbers. The other shape attributes can also be extracted from the size histogram. We describe how size histogram is related to shape convexity, elongatedness, and complexity factors.

Convexity properties: From the DT we can obtain the convexity information. We trace the pixels with the same distance and consider them to be lines. For chessboard distance there is no overlapping between lines of different distances. For city-block distance, the overlapping between the lines is more. It is found that the convex edges are nonoverlapping, whereas the concave edges overlap. Similarly, the degree of overlapping using octagonal distance is less than that using city-block distance. These properties will help in finding convexities in the image.

Elongatedness: Elongatedness (or compactness) is another attribute that can be computed from the shape number. We can define elongatedness of a simply connected region as A/t^2, where A is the area and t is the thickness, defined as twice the number of shrinking steps required to make the image disappear. Overall measures of the elongatedness of an image are only of limited usefulness, since it may be partly elongated and partly not, and the elongated parts may have different thicknesses. The only way to detect elongated parts of the image is shrinking, re-expanding, and detecting parts of various thicknesses [Rosenfeld and Kak 1982].

Complexity factors: Other properties like complexity of shape can also be computed. Complexity factor is given by perimeter squared over area and expressed as

$$\frac{P^2}{A} = 4\pi \tag{7.28}$$

The equality holds only for a circle. The complexity factor increases when the shape becomes elongated or irregular. The perimeter of the object is directly obtained from the function value at distance 1 of the size histogram.

In additional to above applications, Zana and Klein [2001] presented an algorithm based on MM and curvature evaluation for the detection of vessel-like patterns in a noisy environment. To define vessel-like patterns, segmentation was performed with respect to a precise model. A vessel is defined as a bright pattern being piecewise connected and locally linear. MM is very well adapted to this description.

References

Agam, G. and Dinstein, I., "Regulated morphological operations," *Pattern Recognition*, vol. 32, no. 6, pp. 947–971, June 1999.

Bouaynaya, N. and Schonfeld, D., "Theoretical foundations of spatially-variant mathematical morphology. Part II: Gray-level images," *IEEE Trans. Pattern Analysis and Machine Intelligence*, vol. 30, no. 5, pp. 837–850, May 2008.

Cheng, F. and Venetsanopoulos, A. N., "An adaptive morphological filter for image processing," *IEEE Trans. Image Processing*, vol. 1, no. 4, pp. 533–539, Oct. 1992.

Danielsson, P. E., "A new shape factor," *Computer Graphics and Image Processing*, vol. 7, no. 2, pp. 292–299, Apr. 1978.

Dinesh, R. and Guru, D. S., "Corner detection using morphological skeleton: An efficient and nonparametric approach," *Lecture Notes in Computer Science*, Springer, Berlin/Heidelberg, pp. 752–760, 2006.

Gauch, J. M., "Image segmentation and analysis via multiscale gradient watershed hierarchies," *IEEE Trans. Image Processing*, vol. 8, no. 1, pp. 69–79, Jan. 1999.

Gu, C. and Lee, M.-C., "Semiautomatic segmentation and tracking of semantic video objects," *IEEE Trans. Image Processing*, vol. 8, no. 5, pp. 572–584, Sep. 1998.

Kitchen, L. and Rosenfeld, A., "Gray-level corner detection," *Pattern Recognition Letters*, vol. 1, no. 2, pp. 95–102, Dec. 1982.

Laganiere, R., "A morphological operator for corner detection," *Pattern Recognition*, vol. 31, no. 11, pp. 1643–1652, Nov. 1998.

Lee, K.-J. and Bien, Z., "A gray-level corner detector using fuzzy logic," *Pattern Recognition Letters*, vol. 17, no. 9, pp. 939–950, Aug. 1996.

Lin, R.-S., Chu, C.-H., and Hsueh, Y.-C., "A modified morphological corner detector," *Pattern Recognition Letters*, vol. 19, no. 3–4, pp. 279–286, Mar. 1998.

Liu, S.-M., Lin, W.-C., and Liang, C.-C., "An iterative edge linking algorithm with noise removal capability," *Proc. International Conference on Pattern Recognition*, vol. 2, pp. 1120–1122, 1988.

Nalwa, V. S. and Pauchon, E., "Edgel aggregation and edge description," *Comput. Vision, Graphics, Image Processing*, vol. 40, no. 1, pp. 79–94, Jan. 1987.

Nevatia, R., "Locating objects boundaries in textured environments," *IEEE Trans. Computers*, vol. 25, no. 11, pp. 1170–1175, Nov. 1976.

Rosenfeld, A. and Kak, A. C., *Digital Picture Processing*, vol. 2, Academic Press, New York, 1982.

Russ, J. C., *The Image Processing Handbook*, CRC Press, Boca Raton, FL, 1992.

Shih, F. Y., "Object representation and recognition using mathematical morphology model," *International Journal of Systems Integration*, vol. 1, no. 2, pp. 235–256, Aug. 1991.

Shih, F. Y. and Cheng, S., "Adaptive mathematical morphology for edge linking," *Information Sciences*, vol. 167, no. 1–4, pp. 9–21, Dec. 2004.

Shih, F. Y., Chuang, C., and Gaddipati, V., "A modified regulated morphological corner detector," *Pattern Recognition Letters*, vol. 26, no. 7, pp. 931–937, June 2005.

Singh, A. and Shneier, M., "Gray level corner detection a generalization and a robust real time implementation," *Comput. Vision Graphics Image Processing*, vol. 51, no. 1, pp. 54–69, July 1990.

Sobania, A., Paul J., and Evans, O., "Morphological corner detector using paired triangular structuring elements," *Pattern Recognition*, vol. 38, no. 7, pp. 1087–1098, July 2005.

Tian, X.-H., Li, Q.-H., and Yan, S.-W., "Regulated morphological operations with weighted structuring element," *Proc. Int. Conf. Machine Learning Cybernat.*, pp. 768–771, Nov. 2002.

Tsai, D.-M., Hou, H.-T., and Su, H.-J., "Boundary-based corner detection using eigenvalues of covariance matrices," *Pattern Recognition Letters*, vol. 20, no. 1, pp. 31–40, Jan. 1999.

Verly, J. G. and Delanoy, R. L., "Adaptive mathematical morphology for range imagery," *IEEE Trans. Image Processing*, vol. 2, no. 2, pp. 272–275, Apr. 1993.

Vincent, L. and Soille, P., "Watersheds in digital spaces: An efficient algorithm based on immersion simulations," *IEEE Trans. on Pattern Analysis and Machine Intelligence*, vol. 13, no. 6, pp. 583–598, June 1991.

Wang, D., "Unsupervised video segmentation based on watersheds and temporal tracking," *IEEE Trans. Circuits and System for Video Technology*, vol. 8, no. 5, pp. 539–546, Sep. 1998.

Yun, I. D., Park, S. H., and Lee, S. U., "Color image segmentation based on 3D clustering: Morphological approach," *Pattern Recognition*, vol. 31, no. 8, pp. 1061–1076, Aug. 1998.

Zana, F. and Klein, J.-C., "Segmentation of vessel-like patterns using mathematical morphology and curvature evaluation," *IEEE Trans. Image Processing*, vol. 10, no. 7, pp. 1010–1019, July 2001.

Zhang, X., and Zhao, D., "A morphological algorithm for detecting dominant points on digital curves," *Proc. SPIE Nonlinear Image Processing*, 2424, pp. 372–383, 1995.

Zheng, Z., Wang, H., and Teoh, E., "Analysis of gray level corner detection," *Pattern Recognition Letters*, vol. 20, no. 2, pp. 149–162, Feb. 1999.

8

Object Representation

Prior to object recognition, one must have an internal representation of it, which is suited for matching its features to image descriptions. Object representation has been playing an important role in many applications, such as image analysis, pattern recognition, computer graphics, and computer animation. There are many approaches to object representation in the literature. This chapter is organized as follows. Section 8.1 introduces object representation and tolerances by MM. Section 8.2 presents the skeletonization or medial axis (MA) transformation. Section 8.3 describes the morphological shape description called geometric spectrum.

8.1 Object Representation and Tolerances

The representation of solid objects must not only possess the nominal (or ideal) geometric shapes, but also reason the geometric inaccuracies (or tolerances) into the locations and shapes of solid objects. Solid modeling is the foundation for computer aided design (CAD) and computer aided manufacturing (CAM) integrations. Its ultimate goal of planning for the automated manufacturing inspections and robotic assembly is to be able to generate a complete process plan automatically, starting from a CAD representation of the mechanical components. Although many solid models have been proposed, only boundary representation and constructive solid geometric (CSG) representation are popularly used as the internal database.

Boundary representation consists of two kinds of information: topological information and geometric information, which represent the vertex coordinates, surface equations, and the connectivity among faces, edges, and vertices.

There are several advantages in boundary representation: large domain, unambiguity, unique, and explicit representation of faces, edges and vertices. There are also several disadvantages: verbose data structure, difficult to create, difficult to check validity, and variational information unavailability.

The idea of CSG representation is to construct a complex part by hierarchically combining simple primitives using Boolean set operations. There are several advantages in CSG representation: large domain, unambiguity, easy to check validity, and easy to create. There are also several disadvantages: nonunique, difficult to edit graphically, input data redundancy, and variational information unavailability.

The mathematical framework presented in this section for modeling solids is MM [Shih 1991; Shih and Gaddipati 2005]. Adopting MM as a tool, the theoretical research aims at studying the representation schemes for the dimension and tolerance of the geometric structure. This section is divided into three parts. The first part defines the representation framework for characterizing dimension and tolerance of solid objects. The second part then adopts the framework to represent several illustrated two-dimensional and three-dimensional objects. The third part describes the added tolerance information to control the quality of the parts and the interchangeability of the parts among assemblies. With the help of variational information, it is known how to manufacture, setup, and inspect to ensure the products within the required tolerance range.

8.1.1 Representation Framework: Formal Languages and MM

There are rules that associate measured entities with features. Measured central planes, axes, and thinned components are associated with symmetric features. The representation framework is formalized as follows.

Let E^N denote the set of all points in the N-dimensional Euclidean space and $p = (x_1, x_2, \ldots, x_N)$ represent a point in E^N. In the following, any object is a subset of E^N. The formal model will be a context-free grammar, G, consisting of a four-tuple [Fu 1982; Ghosh 1988]:

$$G = (V_N, V_T, P, S), \tag{8.1}$$

where V_N is a set of nonterminal symbols, such as complicated shapes; V_T is a set of terminal symbols that contains two sets: one is the decomposed primitive shapes, such as lines and circles, and the other is the shape operators; P is a finite set of rewrite rules or productions denoted by $A \rightarrow \beta$, where $A \in V_N$ and β is a string over $V_N \cup V_T$; S is the start symbol, which is the solid object. The operators we use include morphological dilation and erosion, set union and intersection, and set subtraction.

Note that a production of such a form allows the nonterminal A to be replaced by the string β independent of the context in which the A appears. The grammar G is context-free since for each production in P, the left part is a single nonterminal and the right part is a nonempty string of terminals and nonterminals. The languages generated by context-free grammars are called *context-free languages*. The object representation task may be reviewed as the task of converting a solid shape into a sentence in the language, whereas object recognition is the task of "parsing" a sentence.

There is no general solution for the primitive selection problem at this time. This determination will be largely influenced by the nature of the data, the specific application in question, and the technology available for implementing the system. The following requirements usually serve as a guide for selecting the shape primitives.

1. The primitives should serve as basic shape elements to provide a compact but adequate description of the object shape in terms of the specified structural relations (e.g., the concatenation relation).

2. The primitives should be easily extractable by the existing nonsyntactic (e.g., decision-theoretic) methods, since they are considered to be simple and compact shapes and their structural information is not important.

8.1.2 Dimensional Attributes

Dimension and tolerance are the main components of the variational information. Dimensions are the control parameters that a designer uses to specify the shape of a part without redundancy, and tolerances are essentially associated with the dimensions to specify how accurate the part should be made. For a detailed description of dimension and tolerance refer to American National Standards Institute [1982].

8.1.2.1 The Two-Dimensional Attributes

The commonly used two-dimensional attributes are rectangle, parallelogram, triangle, rhomb, circle, and trapezoid. These can be represented easily in the morphological way, and some of them are illustrated in Figure 8.1. These expressions are not unique, but the evaluation depends on the simplest combination and least computational complexity. The common method is to decompose the attributes into smaller components and apply morphological dilation to "grow up" these components. Most geometric units can be decomposed into thinned small subunits (skeleton) with a certain size structuring element using recursive dilation. In most of the cases, we can decompose the N-dimensional solid object with the morphological operation into one-dimensional lines.

The formal expressions of Figure 8.1 are written as follows:

For Figure 8.1a: Rectangle $\rightarrow \vec{a} \oplus \vec{b}$
For Figure 8.1b: Parallelogram $\rightarrow \vec{a} \oplus \vec{b}$

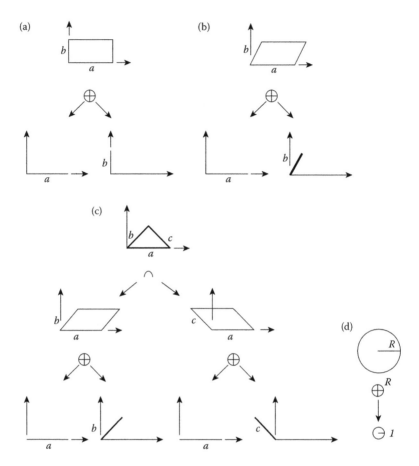

FIGURE 8.1 The decomposition of two-dimensional attributes. (a) Rectangle, (b) parallelogram, (c) triangle, (d) circle.

For Figure 8.1c: Triangle $\rightarrow (\vec{a} \oplus \vec{b}) \cap (\vec{a} \oplus \vec{c})$

For Figure 8.1d: Circle $\rightarrow (\cdots$ (unit-circle $\oplus$ unit-circle) $\oplus \cdots) \oplus$ unit-circle = (unit-circle)$^{\oplus R}$, where the "unit-circle" is a circle with the radius one. This same analogy can be easily extended to three dimensions.

8.1.2.2 The Three-Dimensional Attributes

The basic geometric entities in manufacturing are features. Simple surface features are subsets of an object's boundary that lie in a single surface, which typically is a plane, cylinder, cone, sphere, or torus. Composite surface features are aggregates of simple ones. The three-dimensional attributes, which are similar to two-dimensional attributes except addition of one more dimension, can also apply the similar decomposition method. A representation scheme for a three-dimensional shape model is described in Figure 8.2, and its formal expression is

$$\text{shape} \rightarrow [(\vec{a} \oplus \vec{b}) - (\text{unit\_circle}^{\oplus R})] \oplus \vec{c} \qquad (8.2)$$

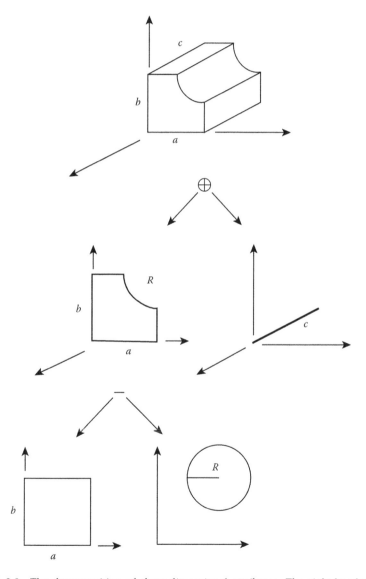

FIGURE 8.2 The decomposition of three-dimensional attributes. The right-handed slide shows the decomposition subunits that construct the attributes through the morphological dilation.

8.1.2.3 Tolerancing Expression

Tolerances constrain an object's features to lie within regions of space called *tolerance zones*. Tolerance zones in Rossignac and Requicha [1985] were constructed by expanding the nominal feature to obtain the region bounded by the outer closed curve, shrinking the nominal feature to obtain the region bounded by the inner curve, and then subtracting the two resulting regions. This procedure is equivalent to the morphological dilation of the offset inner contour with a tolerance-radius disked structuring element. Figure 8.3a shows an annular tolerance zone that corresponds to a circular hole, and Figure 8.3b shows a tolerance zone for an elongated slot. Both could be constructed by dilating the nominal contour with a tolerance-radius disked structuring element whose representation becomes simpler.

Mathematical rules for constructing tolerance zones depend on the types of tolerances, but are independent of the specific geometry of the features and of feature representation methods. The resulting tolerance zones, however, depend on features' geometrics, as shown in Figure 8.4. The tolerance zone for testing the size of a round hole is an annular region lying between two circles with the specified maximal and minimal diameters; the zone corresponding to a form constraint for the hole is also an annulus, defined by two concentric circles whose diameters must differ by a specified amount but are otherwise arbitrary.

The MM does support the conventional limit (±) tolerances on "dimensions" that appear in many engineering drawings. The positive deviation is equivalent to the dilated result and the negative deviation is equivalent to the eroded result. The industrial parts adding tolerance information can be expressed using dilation with a circle (see Figure 8.4).

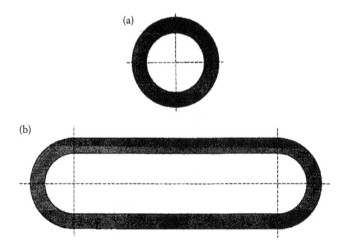

FIGURE 8.3 Tolerance zones. (a) An annular tolerance zone that corresponds to a circular hole, (b) a tolerance zone for an elongated slot.

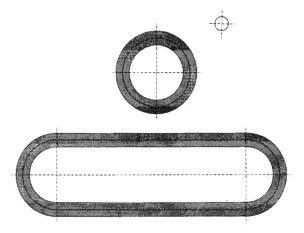

FIGURE 8.4   An example of adding tolerance by morphological dilation.

8.2 Skeletonization or MA Transformation

The studies of skeletonization were motivated from the need of converting a digital image into a linear form in a natural manner. From the skeleton the contour of an object region can be easily regenerated, so that the amount of information involved in an image analysis is the same. Besides, the skeleton emphasizes certain properties of the image; for instance, curvatures of the contour correspond to topological properties of the skeleton. The concept of skeleton was first proposed by Blum [1967].

Skeleton, MA, or symmetrical axis has been extensively used for characterizing objects satisfactorily using the structures composed of line or arc patterns. Bouaynaya, Charif-Chefchaouni, and Schonfeld [2006] presented the spatially-variant morphological restoration and skeleton representation. Kresch and Malah [1998] presented skeleton-based morphological coding of binary images. Trahanias [1992] used the morphological skeleton transform to perform binary shape recognition. Applications include the representation and recognition of handwritten or printed characters, fingerprint ridge patterns, biological cell structures, circuit diagrams, engineering drawings, and the like.

Let us visualize a connected object region as a field of grass and place a fire starting from its contour. Assume that this fire burning spreads uniformly in all directions. The *skeleton* (or *medial axis*, abbreviated MA) is where waves collide with each other in a frontal or circular manner. It receives the name of MA because the pixels are located at midpoints or along local symmetrical axes of the region. The points in MA (or skeleton) bear the same distance from at least two contour points which is the smallest among those computed from all background points. The information on the sizes of local object region is retained by associating each skeleton point with a label representing the aforementioned distance value. Intuitively, the original binary image can be reconstructed by a union of the circular neighborhoods centered at each skeleton point with a radius equal to the associated label.

A skeleton will not appear immediately in the wave front if it is a smooth curve locally. The appearance of a skeleton starts with the minimum radius of curvature in a local contour. The disappearance of a skeleton is encountered when the largest circle can be drawn within a local object region. If the boundary has a regional concave contour, the wave fronts will not collide with each other. Thus, there are no skeleton points located surrounding this area. In fact, the skeleton will originate outside the regional concave contour if the skeletonization is considered to deal with the background of the connected object, that is, the complement of the image.

Many algorithms have been developed to extract the skeleton. The straight-forward procedure to accomplish such a transformation involves an iterative process, which shrinks the object region step-by-step until a one-element thick figure is obtained [O'Gorman 1990]. Naccache and Shinghal [1984] proposed the strategy: visiting all the pixels in the bitmap to iteratively delete the edge points classified as non-safe points (i.e., the points being deleted without the effectness of connectivity). Zhang and Suen [1984] proposed a parallel thinning algorithm that consists of two subiterations: one is to delete the south-east boundary points and the north-west corner points, and the other is to delete the north-west boundary points and the south-east corner points. Xu [2003] presented a generalized discrete morphological skeleton transform.

Several algorithms have aimed to achieve more efficient parallel thinning algorithms for skeletonization. Kwok [1988] proposed a thinning algorithm using contour generation in terms of chain codes. The advantage of this algorithm is that only the contour tracking is performed ineach iteration. Arcelli and Sanniti di Baja [1989] used the city-block DT to detect the skeleton. Rather than peeling, the skeleton is identified as multiple pixels based on the multiplicity defined, and the recursive procedure calls are applied.

An algorithm for generating connected skeletons from approximated Euclidean distances and allowing near reconstruction was developed by Niblack, Capson, and Gibbons [1990]. Jang and Chin [1992] proposed a one-pass parallel thinning algorithm based on a number of criteria including connectivity, unit-width convergence, MA approximation, noise immunity and efficiency. Maragos and Schafer [1986] used morphological operations to extract the skeleton and to optimize the skeleton for the purpose of image coding. Their morphological skeleton is however not connected and more than 1-pixel wide. Jang and Chin [1990] defined digital skeletons based on morphological set transformation and provided the proof of convergence, the condition for 1-pixel thick skeletons, and the connectedness of skeletons.

8.2.1 Medial Axis Transformation by Morphological Dilations

The set of centers and radii (values) of the maximal blocks is called the *MA* (or *symmetric axis*) *transformation*, abbreviated as MAT or SAT. Intuitively, if a block S_P is maximal and is contained in the constant-value region S, it must touch the border of S in at least 2 places; otherwise we could find a neighbor Q of P that was further away than P from the border of S, and then S_Q would contain S_P.

Let the output of the DT be d. A DT converts a binary image, which consists of object (foreground) and nonobject (background) pixels, into an image where every object pixel has a value corresponding to the minimum distance to the background. Consider the slope variation in both X and Y-axes for d. The union of the summit pixels of the slope in both X and Y directions will be the skeleton. From the DT, one can see that distance change between two neighboring pixels is not more than 1. Hence, the four SEs are chosen as follows:

$$h_1 = \begin{bmatrix} 0 \\ 0 \\ \varepsilon \end{bmatrix}, \quad h_2 = \begin{bmatrix} \varepsilon \\ 0 \\ 0 \end{bmatrix}, \quad h_3 = [0 \quad 0 \quad \varepsilon], \quad h_4 = [\varepsilon \quad 0 \quad 0], \tag{8.3}$$

where ε denotes a small positive number less than 1.

MAT Algorithm:

1. Raster scan the distance image d. Do $d \oplus_g h_1$, $d \oplus_g h_2$, $d \oplus_g h_3$, and $d \oplus_g h_4$ at each pixel.

2. If any one of four resulting values is equal to the original gray-value at this pixel, then the gray-value is kept; otherwise, let the gray-value of this pixel be zero. After all pixels are processed, the output is the skeleton with the gray-value indicating the distance.

An example of the EDT and the MA transformation is shown in Figure 8.5.

Reconstruction Algorithm:

Let m denote the grayscale MAT. Using the distance structuring element k, the DT reconstruction from the MAT is achieved by dilating the grayscale MAT by the distance structuring element; that is, $m \oplus_g k$. This reconstruction algorithm is quite simple. The result is exactly the DT. After thresholding at 1, the original binary image can be recovered.

Since the above MAT algorithm produces a skeleton that may be disconnected into several short branches for an object, we present a simple and

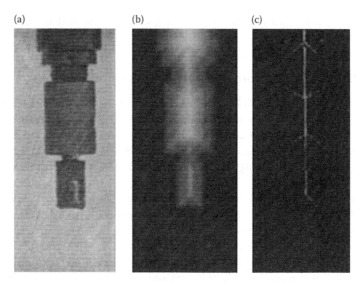

FIGURE 8.5 An example of EDT and skeletonization. (a) Tool picture, (b) DT, (c) skeleton (MA).

efficient algorithm using the maxima tracking approach on EDT to detect skeleton points. The advantages of the skeleton obtained are: (1) connectivity preservation; (2) single-pixel in width; and (3) its locations as close as to the most symmetrical axes.

8.2.2 Thick Skeleton Generation

A grayscale image is first segmented into a binary image, which is then applied by the EDT algorithm. An example of the resulting EDT is shown in Figure 8.6. For any pixel P, we associate a set of disks or circles with various radii centered at P. Let C_p be the largest disk which is completely contained in the object region, and let r_p be the radius of C_p. There may exist another pixel Q such that C_Q contains C_p. If no such a Q exists, C_p is called a *maximal disk*. The set of the centers and radii (or Euclidean distance values) of these maximal inscribed disks is called the skeleton.

8.2.2.1 The Skeleton from Distance Function

The distance function of pixel P in the object region S is defined as the smallest distance of P from all the background pixels S. That is

$$D(P) = \min_{Q \in \bar{S}}\{d(P,Q)\}, \qquad (8.4)$$

```
-  -   -   -   -   -   -   -   -   -   -   -   -   -   -   -   -   -   -   -   -   -   -   -   -   -   -   -   -   -
-  10  10  10  10  10  10  10  10  10  10  10  10  10  10  10  10  10  10  10  10  10  10  10  10  10  10  10  -
-  10  20  20  20  20  20  20  20  20  20  20  20  20  20  20  20  20  20  20  20  20  20  20  20  20  20  10  -   -
-  10  20  30  30  30  30  30  30  30  30  30  30  30  30  30  30  30  30  30  30  30  30  22  14  10  -   -
-  10  20  30  40  36  28  22  20  20  22  28  36  40  40  40  40  40  40  40  40  36  28  20  10  -
-  10  20  30  36  28  22  14  10  10  14  22  28  36  45  50  50  50  50  50  45  36  28  22  14  10  -   -
-  10  20  30  32  22  14  10  -   -   10  14  22  32  41  51  60  60  60  51  41  32  22  14  10  -
-  10  20  30  30  20  10  -   -   -   -   10  20  30  40  50  60  70  60  50  40  30  20  10  -
-  10  20  30  30  20  10  -   -   -   -   10  20  30  40  50  60  70  61  51  41  32  22  14  10  10  10  10
-  10  20  30  32  22  14  10  -   -   10  14  22  32  41  51  60  60  60  54  45  36  28  22  20  20  20  10  -
-  10  20  30  36  28  22  14  10  10  14  22  28  36  45  50  50  50  50  50  45  36  28  22  20  20  20  10  -
-  10  20  30  40  36  28  22  20  20  22  28  36  40  40  40  40  40  40  40  40  32  22  14  10  10  10  10  -
-  10  20  30  30  30  30  30  30  30  30  30  30  30  30  30  30  30  30  30  30  20  10  -
-  10  20  20  20  20  20  20  20  20  20  20  20  20  20  20  20  20  20  20  20  20  10  -
-  10  10  10  10  10  10  10  10  10  10  10  10  10  10  10  10  10  10  10  10  10  10  -   -   -   -   -
```

FIGURE 8.6 The Euclidean distance function with the ridge points underlined. For expressional simplicity, we use ten times of the Euclidean distance in rounded numbers.

where $d(P, Q)$ is the Euclidean distance between P and Q. If we visualize the distance values as the altitude on a surface, then the 'ridges' of the surface constitute the skeleton in which a tangent vector cannot be uniquely defined. It can be easily derived that if and only if a point P belongs to the skeleton of an object region S, the maximal disk with the radius $D(P)$ hits the contour of S at least two places. Let us define the set

$$A(P) = \{Q \mid d(P, Q) = D(P), Q \in \overline{S}\}. \tag{8.5}$$

If the set $A(P)$ contains more than one element, then P is a skeleton point.

8.2.2.2 Detection of Ridge Points

Considering two neighbors of P in any of the horizontal, vertical, 45°-diagonal, and 135°-diagonal directions, if the altitude of P is higher than one neighbor and not lower than the other, then P is called the *ridge point*. Figure 8.6 shows the EDT with ridge points underlined. Note that the resulting skeleton on the right hand side is disconnected from the major skeleton.

8.2.2.3 Trivial Uphill Generation

The trivial uphill of a point P is the set of all the neighbors with a higher altitude. Figure 8.7 shows the resulting skeleton if we add the uphill of ridge points and continuously add the uphill of the new skeleton points until no further uphill can be generated. Finally, a connected skeleton can be obtained, but a thick one. To make the skeleton to be single-pixel wide, we need get rid of the loose definition of ridge points and take into account the directional-neighborhood.

```
- -   -   -   -   -   -   -   -   -   -   -   -   -   -   -   -   -   -   -   -   -   -   -   -   -   -   -   -
- 10  10  10  10  10  10  10  10  10  10  10  10  10  10  10  10  10  10  10  10  10  10  10  10  10  10  10  -
- 10  20  20  20  20  20  20  20  20  20  20  20  20  20  20  20  20  20  20  20  20  20  20  20  20  10  -   -
- 10  20  30  30  30  30  30  30  30  30  30  30  30  30  30  30  30  30  30  30  30  30  22  14  10  -   -   -
- 10  20  30  40  36  28  22  20  20  22  28  36  40  40  40  40  40  40  40  40  36  28  20  10  -   -   -   -
- 10  20  30  36  28  22  14  10  10  14  22  28  36  45  50  50  50  50  50  45  36  28  22  14  10  -   -   -
- 10  20  30  32  22  14  10  -   -   10  14  22  32  41  51  60  60  60  51  41  32  22  14  10  -   -   -   -
- 10  20  30  30  20  10  -   -   -   -   10  20  30  40  50  60  70  60  50  40  30  20  10  -   -   -   -   -
- 10  20  30  30  20  10  -   -   -   -   10  20  30  40  50  60  70  61  51  41  32  22  14  10  10  10  10  -
- 10  20  30  32  22  14  10  -   -   10  14  22  32  41  51  60  60  60  54  45  36  28  22  20  20  20  10  -
- 10  20  30  36  28  22  14  10  10  14  22  28  36  45  50  50  50  50  50  45  36  28  22  20  20  20  10  -
- 10  20  30  40  36  28  22  20  20  22  28  36  40  40  40  40  40  40  40  32  22  14  10  10  10  10  -
- 10  20  30  30  30  30  30  30  30  30  30  30  30  30  30  30  30  30  30  30  20  10  -   -   -   -   -
- 10  20  20  20  20  20  20  20  20  20  20  20  20  20  20  20  20  20  20  20  10  -   -   -   -   -
- 10  10  10  10  10  10  10  10  10  10  10  10  10  10  10  10  10  10  10  10  10  -   -   -   -   -
- -   -   -   -   -   -   -   -   -   -   -   -   -   -   -   -   -   -   -   -   -   -   -   -   -   -   -   -
```

FIGURE 8.7   The ridges points and their trivial uphill generation are underlined. It is indeed connected but too thick.

8.2.3 Basic Definitions

Prior to the discussion of our algorithm, the basic definitions of base point, apex point, directional-uphill generation, and directional-downhill generation are introduced below.

8.2.3.1 Base Point

The base point is defined as a corner point, which has the distance value 1 and is surrounded by a majority of background points. It belongs to one of the following three configurations as well as their variations up to eight 45°-rotations, respectively:

$$
\begin{array}{ccc}
1\ 0\ 0 & 1\ 1\ 0 & 1\ 1\ 1 \\
1\ \underline{1}\ 0, & 1\ \underline{1}\ 0,\ \text{and} & 1\ \underline{1}\ 0, \\
0\ 0\ 0 & 0\ 0\ 0 & 0\ 0\ 0
\end{array}
$$

where the center pixels underlined represent the 45°, 90°, and 135° corner points, respectively. In a digital image, only the above three degrees are considered in a local 3×3 window.

If all the three configurations are considered as base points, then more nonsignificant short skeletal branches are produced. In other words, an approach based on the skeletal branches originating from all corner points in a 3×3 window would lead to unmanageable complexity in the skeletal structure. Therefore, an appropriate shape-informative skeleton should reach a compromise among the representativity of the connected object structure, the required re-constructivity, and the cost of deleting nonsignificant branches.

We consider sharper convexities as more significant. In detail, if the amplitude of the angle formed by two intersecting wavefronts of the fireline is viewed as the parameter characterizing the sharpness, the values smaller than or equal to 90° are identified as suitable convexities. More strictly, only 45°-convexities are detected as base points. By using these points as the source to grow up the skeleton, the remaining procedures are acquired to preserve the skeletal connectedness.

8.2.3.2 Apex Point

The apex point is the pixel being the local maximum in its 3×3 neighborhood. Note that the local-maximum pixels only construct a small portion of a disconnected skeleton. They occur as 45°-corner points or interior elements, which have the highest altitude locally. The 45°-corner points also serve as base points. The base points and apex points are considered as sources or elementary cells of the skeleton, which grows up emitting from them. Figure 8.8 shows the Euclidean distance function with the base points labeled "B" and the apex points underlined.

8.2.3.3 Directional-Uphill Generation

The set of points $\{P_i\}$ is called the *directional-neighborhood* of P, denoted by D_P, if they are in the 8-neighborhood of P and located within $\pm 45°$ slope changes from the current MA orientation of P. For example, using the 8-neighbors labeled $P_1, P_2, \ldots,$ and P_8 counterclockwise from the positive x-axis of P, if P_7 and P are the skeleton points, the points $P_2, P_3,$ and P_4 are the directional-neighbors of P; that is, $D_P = \{P_2, P_3, P_4\}$. Several cases are illustrated below.

$$\begin{bmatrix} P_4 & P_3 & P_2 \\ \cdot & P & \cdot \\ \cdot & P_7 & \cdot \end{bmatrix}, \begin{bmatrix} \cdot & \cdot & P_2 \\ P_5 & P & P_1 \\ \cdot & \cdot & P_8 \end{bmatrix}, \begin{bmatrix} P_4 & P_3 & \cdot \\ P_5 & P & \cdot \\ \cdot & \cdot & P_8 \end{bmatrix}, \text{ and } \begin{bmatrix} \cdot & P_3 & P_2 \\ \cdot & P & P_1 \\ P_6 & \cdot & \cdot \end{bmatrix}$$

```
-  -   -   -   -   -   -   -   -   -   -   -   -   -   -   -   -   -   -   -   -   -   -   -   -   -   -   -   -
-  B  10  10  10  10  10  10  10  10  10  10  10  10  10  10  10  10  10  10  10  10  10  10  10  10  10   B  -
- 10  20  20  20  20  20  20  20  20  20  20  20  20  20  20  20  20  20  20  20  20  20  20  20  10   -   -
- 10  20  30  30  30  30  30  30  30  30  30  30  30  30  30  30  30  30  30  30  30  30  30  22  14  10   -   -
- 10  20  30  40  36  28  22  20  20  22  28  36  40  40  40  40  40  40  40  40  36  28  20  10   -   -   -
- 10  20  30  36  28  22  14  10  10  14  22  28  36  45  50  50  50  50  50  45  36  28  22  14  10   -   -   -
- 10  20  30  32  22  14  10   -   -  10  14  22  32  41  51  60  60  60  51  41  32  22  14  10   -   -   -   -
- 10  20  30  30  20  10   -   -   -   -  10  20  30  40  50  60  70  60  50  40  30  20  10   -   -   -   -   -
- 10  20  30  30  20  10   -   -   -   -  10  20  30  40  50  60  70  61  51  41  32  22  14  10  10  10   B  -
- 10  20  30  32  22  14  10   -   -  10  14  22  32  41  51  60  60  60  54  45  36  28  22  20  20  20  10  -
- 10  20  30  36  28  22  14  10  10  14  22  28  36  45  50  50  50  50  50  45  36  28  22  20  20  20  10  -
- 10  20  30  40  36  28  22  20  20  22  28  36  40  40  40  40  40  40  40  32  22  14  10  10  10   B  -
- 10  20  30  30  30  30  30  30  30  30  30  30  30  30  30  30  30  30  30  20  10   -   -   -   -   -
- 10  20  20  20  20  20  20  20  20  20  20  20  20  20  20  20  20  20  20  20  10   -   -   -   -   -
-  B  10  10  10  10  10  10  10  10  10  10  10  10  10  10  10  10  10  10  10  10   B   -   -   -   -   -
-  -   -   -   -   -   -   -   -   -   -   -   -   -   -   -   -   -   -   -   -   -   -   -   -   -   -   -   -
```

FIGURE 8.8 The Euclidean distance function with the base points labeled "B" and the apex points underlined.

```
-  -  -  -  -  -  -  -  -  -  -  -  -  -  -  -  -  -  -  -  -  -  -  -  -  -  -  -  -
-  10 10 10 10 10 10 10 10 10 10 10 10 10 10 10 10 10 10 10 10 10 10 10 10 10 10 10 -
-  10 20 20 20 20 20 20 20 20 20 20 20 20 20 20 20 20 20 20 20 20 20 20 20 20 10 -  -
-  10 20 30 30 30 30 30 30 30 30 30 30 30 30 30 30 30 30 30 30 30 30 22 14 10 -  -  -
-  10 20 30 40 36 28 22 20 20 22 28 36 40 40 40 40 40 40 40 40 40 36 28 20 10 -  -  -
-  10 20 30 36 28 22 14 10 10 14 22 28 36 45 50 50 50 50 50 45 36 28 22 14 10 -  -  -
-  10 20 30 32 22 14 10 -  -  10 14 22 32 41 51 60 60 60 51 41 32 22 14 10 -  -  -  -
-  10 20 30 30 20 10 -  -  -  -  10 20 30 40 50 60 70 60 50 40 30 20 10 -  -  -  -  -
-  10 20 30 20 10 -  -  -  -  10 20 30 40 50 60 70 61 51 41 32 22 14 10 10 10 10 -
-  10 20 30 32 22 14 10 -  -  10 14 22 32 41 51 60 60 60 54 45 36 28 22 20 20 20 10 -
-  10 20 30 40 36 28 22 20 20 22 28 36 40 40 40 40 40 40 40 40 32 22 14 10 10 10 10 -
-  10 20 30 30 30 30 30 30 30 30 30 30 30 30 30 30 30 30 30 30 20 10 -  -  -  -  -
-  10 20 20 20 20 20 20 20 20 20 20 20 20 20 20 20 20 20 20 20 10 -  -  -  -  -  -
-  10 10 10 10 10 10 10 10 10 10 10 10 10 10 10 10 10 10 10 10 10 -  -  -  -  -  -
-  -  -  -  -  -  -  -  -  -  -  -  -  -  -  -  -  -  -  -  -  -  -  -  -  -  -  -  -
```

FIGURE 8.9   The results of directional-uphill generation of Figure 8.8.

Note that the set of directional neighbors always contains three elements. The directional-uphill generation adds the point, which is the maximum of P, and its directional-neighborhood. That is to say

$$P^U_{\text{next}} = \max_{P_i \in D_P \cup \{P\}} \{P_i\}. \tag{8.6}$$

Figure 8.9 shows the result of directional-uphill generation of Figure 8.8.

8.2.3.4 Directional-Downhill Generation

From Figure 8.9, we observe that there should exist a vertical path connecting two apex points of value "40" underlined on the left hand side, in which the altitude changes are not always increasing; instead, they are a mixture of decreasing and increasing values. The directional-downhill generation, which is similar to the directional-uphill generation except the maxima tracking of the set excluding the central point P, is used to produce this type of skeletal branch. That is to say

$$P^D_{\text{next}} = \max_{P_i \in D_P} \{P_i\}. \tag{8.7}$$

The directional-downhill generation is initialized from the apex points, which cannot be further tracked by the directional-uphill generation. Hence, the altitude of the next directional-downhill should be lower in the beginning. However, the tracking procedure is continued without taking into account the comparison of neighbors with P. The next directional-downhill altitude could be lower or even higher until a skeleton point appears in the directional neighborhood. Figure 8.10 shows the result of directional-downhill generation of Figure 8.9. The skeleton is now connected and possesses single-pixel in width except for two pixels having the same local maximum of Euclidean distance.

```
-  -   -   -   -   -   -   -   -   -   -   -   -   -   -   -   -   -   -   -   -   -   -   -   -   -   -   -   -
-  10  10  10  10  10  10  10  10  10  10  10  10  10  10  10  10  10  10  10  10  10  10  10  10  10  10  10  -
-  10  20  20  20  20  20  20  20  20  20  20  20  20  20  20  20  20  20  20  20  20  20  20  20  20  10  -   -
-  10  20  30  30  30  30  30  30  30  30  30  30  30  30  30  30  30  30  30  30  30  22  14  10  -   -   -
-  10  20  30  40  36  28  22  20  20  22  28  36  40  40  40  40  40  40  40  40  36  28  20  10  -   -   -
-  10  20  30  36  28  22  14  10  10  14  22  28  36  45  50  50  50  50  50  45  36  28  22  14  10  -   -   -
-  10  20  30  32  22  14  10  -   -   10  14  22  32  41  51  60  60  60  51  41  32  22  14  10  -   -   -   -
-  10  20  30  30  20  10  -   -   -   -   10  20  30  40  50  60  70  60  50  40  30  20  10  -   -   -   -   -
-  10  20  30  30  20  10  -   -   -   10  20  30  40  50  60  70  61  51  41  32  22  14  10  10  10  10  -
-  10  20  30  32  22  14  10  -   -   10  14  22  32  41  51  60  60  60  54  45  36  28  22  20  20  20  10  -
-  10  20  30  36  28  22  14  10  10  14  22  28  36  45  50  50  50  50  50  45  36  28  22  20  20  20  10  -
-  10  20  30  40  36  28  22  20  20  22  28  36  40  40  40  40  40  40  40  40  36  28  14  10  10  10  10  -
-  10  20  30  30  30  30  30  30  30  30  30  30  30  30  30  30  30  30  30  20  10  -   -   -   -   -
-  10  20  20  20  20  20  20  20  20  20  20  20  20  20  20  20  20  20  20  10  -   -   -   -   -
-  10  10  10  10  10  10  10  10  10  10  10  10  10  10  10  10  10  10  10  10  -   -   -   -   -   -
```

FIGURE 8.10   The results of directional-downhill generation of Figure 8.9.

8.2.4 The Skeletonization Algorithm and Connectivity Properties

The skeletonization algorithm traces the skeleton points by choosing the local maxima on the EDT and takes into consideration of the least slope changes in medial axes. The algorithm is described as follows:

1. The base points and apex points are detected as the initial skeleton points.

2. Starting with these skeleton points, the directional-uphill generation in Equation 8.6 is used to add more skeleton points, and continuously add the directional-uphill of the new skeleton points until no further point can be generated. The points which cannot be further tracked are marked.

3. Starting with the marked points, the directional downhill generation in Equation 8.7 is used to complete the skeleton tracking.

Figure 8.11 illustrates the skeleton of a rectangle with holes and notches obtained by the skeletonization algorithm, where M indicates the skeleton pixel.

```
-  -  -  -  -  -  -  -  -  -  -  -  -  -  -  -  -  -  -  -  -  -  -  -  -  -  -  -  -  -  -
-  M  O  O  O  O  O  O  O  O  O  O  O  O  O  O  O  O  O  O  O  O  O  O  O  O  M  M  M  -
-  O  M  O  O  O  O  O  O  O  O  O  O  O  O  O  O  O  O  O  O  O  O  O  O  M  O  -  -  -
-  O  O  M  O  O  M  M  M  M  M  M  O  O  O  O  O  O  O  O  O  O  O  M  M  O  O  -  -
-  O  O  O  M  M  O  O  O  O  O  O  M  M  O  O  O  O  O  O  O  M  M  O  O  -  -  -
-  O  O  O  M  O  O  O  O  O  O  O  O  M  O  O  O  O  O  M  O  O  O  O  O  -  -  -
-  O  O  O  M  O  O  O  -  -  O  O  O  O  O  M  O  O  O  M  O  O  O  O  -  -  -  -
-  O  O  O  M  O  O  -  -  -  -  O  O  O  O  O  M  M  M  O  O  O  O  O  -  -  -  -
-  O  O  O  M  O  O  -  -  -  O  O  O  O  O  M  M  O  O  O  O  O  -  -  -  -  -
-  O  O  O  M  O  O  O  -  -  O  O  O  O  O  M  O  O  M  M  M  M  M  M  M  M  O  -
-  O  O  O  M  O  O  O  O  O  O  O  O  O  O  M  O  O  O  O  O  O  O  O  M  M  O  -
-  O  O  O  M  O  O  O  O  O  O  O  O  M  M  O  O  O  O  O  O  O  O  O  O  M  -
-  O  O  M  O  O  M  M  M  M  M  M  O  O  O  O  O  O  O  O  M  O  O  -  -  -
-  O  M  O  O  O  O  O  O  O  O  O  O  O  O  O  O  O  O  O  M  O  -  -  -  -
-  M  O  O  O  O  O  O  O  O  O  O  O  O  O  O  O  O  O  O  M  -  -  -  -  -
```

FIGURE 8.11   The resulting skeleton on a rectangle with holes and notches.

In the experiment, the 8-connectedness is applied for foreground and the 4-connectedness for background. The skeletonization algorithm possesses the following three connectivity properties:

(C1) After skeletonization, an originally connected object will not be separated into two or more sub-objects.

(C2) After skeletonization, a connected object will not disappear at all.

(C3) After skeletonization, the originally disconnected background components are not 4-connected.

The proofs of the above three properties are given below.

Proof of C1. Induction hypothesis: For any m $(m < n)$ marked apex points of an object, the skeleton obtained will preserve the property C1, where n could be any number.

The base case is $m = 1$. The algorithm starts tracking from each of the base and the apex points, and it will stop when the current point is connected with another skeleton point or touches with the point marked. That means every sub-skeleton branch starting from the base point will connect with some skeleton point or meet with another branch at the marked apex point. When $m = 2$, it is regarded as two sub-objects, each sub-object contains a marked apex point. As discussed above, each sub-object is 8-connected. The algorithm will trace from each marked-apex and stop when it is connected with some skeleton points contained in the other skeleton subset. The reason why it would not connect with a skeleton point, which belongs to the same skeleton subset, is the directional-neighbors that we use. Using directional-neighbors will enforce the tracking direction to go towards the region, which has not been tracked up to now. Finally, it will lead tracking toward another skeleton subset.

As the induction hypothesis claim, when $m = n$, the skeleton obtained is 8-connected. Now consider the case when $m = n + 1$. Those $n + 1$ marked-apex points can be regarded as 2 sub-objects: One contains n marked-apex points and the other contains only a marked-apex (i.e., the $(n + 1)$th point). As discussed above, the one with n marked-apex points is 8-connected, and the other with a marked-apex is also 8-connected. Tracking from the $(n + 1)$th marked-apex, it will lead the tracking to go towards the skeleton subset with n marked-apex points. That means the whole object is 8-connected. □

Proof of C2. Tracking starts from each base and apex points. The algorithm marks base points as the skeleton points. For any size of an object, there must be at least one base point or at least one apex point, and the skeleton obtained must have at least one point. For the ideal case of a circle, there is no base point but one apex point is present, which is the local maximum, representing the center of the circle. That is after skeletonization, an 8-connected object will not disappear at all, and the skeleton contains at least a pixel. □

Proof of C3. According to the definition of the apex points, there exists at least one apex in the object between any two background components. As the algorithm is performed, it will trace from each apex point, which is not 8-connected to any skeleton point. After tracking, there will be one skeleton branch that will make these two background components disconnected. Therefore, the skeleton obtained will not allow the originally disconnected background components to be 4-connected. □

The algorithm including three procedures and two functions is described in pseudo-codes below.

```
Procedure SKEPIK
  /* trace starting from each base and apex points */
  for i, j in 1 ... N, 1 ... N loop
    if BASE_APEX (p(i, j)) then
      Uphill-Generation (p(i, j), p(i, j));
    end if
  end loop
  /* trace from each marked apex point in the procedure
  UpHill-Generation */
  while (marked_pixel)
    DownHill-Generation (marked_apex, marked_apex);
  end while
end SKEPIK
function APEX (p): Boolean
/* apex point is the local maximum point */
  if (p is local maxima) then
    return TRUE
  else
    return FALSE;
  end APEX
function BASE_APEX (p): Boolean
/* Base point is the point with distance 1 and has 4 or more
   zeros in its 8-neighborhood */
  if (distance of p = 1) then
    find 8-neighbors of p;
    if ((number of 8-neighbors with distance 0) > 4) then
      return TRUE;
    else
      if APEX (p) then
        return TRUE;
      else
        return FALSE;
      end if
    end if
  end if
end BASE_APEX
procedure Uphill-Generation (current-skeleton-pixel, previous-
skeleton-pixel)
  if (number of maximum in 8-neighbors > 1 and
```

```
    distance of maximum≥distance of current-skeleton-pixel) then
        maximum-pixel = the maximum of the directional-neighbors;
    UpHill-Generation (maximum-pixel, current-skeleton-pixel);
  else
    if (number of maximum in 8-neighbors = 1 and
      distance of maximum>distance of current-skeleton-pixel)
        then
      maximum-pixel = the maximum;
      UpHill-Generation (maximum-pixel, current-skeleton-pixel);
    else
      mark-apex; /*mark current-skeleton-pixel for later
                 processing */
    end if
  end if
end UpHill-Generation
procedure DownHill-Generation (current-skeleton-pixel,
  previous-skeleton-pixel)
  maximum-pixel = the maximum of the directional-neighbors;
  if (maximum-pixel is not a skeleton point) then
    DownHill-Generation (maximum-pixel, current-skeleton-pixel);
  end if
end DownHill-Generation
```

8.2.5 A Modified Algorithm

The modified maxima tracking algorithm will extract the skeleton by eliminating nonsignificant short skeletal branches, which touch the object boundary at corners. It is different from the previous algorithm that detects the base and apex points as initial skeleton points. Instead, the maxima tracking starts from the apex points only. The algorithm will recursively repeat the same procedure which selects the maxima in the directional-neighborhood as the next skeleton point until another apex points are reached.

The modified maxima tracking algorithm is given as follows [Shih and Pu 1995]:

1. The apex points are detected as the initial skeleton points.
2. Starting with each apex point, we use the directional-uphill generation to generate the skeleton points. Recursively repeat this procedure until an apex point is reached, and then the apex point is marked.
3. Starting with these marked apex points, the directional-downhill generation is used to track the new skeleton points.

Figure 8.12 illustrates the resulting skeleton by using the modified algorithm, where there is a closed curve indicating an inside hole.

Figure 8.13 illustrates that the character "e" and its rotations by 30°, 45°, and 90° using the modified algorithm will produce the identical rotated skeleton provided the digitization error is disregarded.

```
-  -  -  -  -  -  -  -  -  -  -  -  -  -  -  -  -  -  -  -  -  -  -  -  -  -  -  -  -  -  -
-  O  O  O  O  O  O  O  O  O  O  O  O  O  O  O  O  O  O  O  O  O  O  O  O  O  M  M  -
-  O  O  O  O  O  O  O  O  O  O  O  O  O  O  O  O  O  O  O  O  O  O  O  O  M  O  -  -
-  O  O  O  O  M  M  M  M  M  O  O  O  O  O  O  O  O  O  O  O  O  M  M  O  O  -  -
-  O  O  O  M  M  O  O  O  O  O  M  M  O  O  O  O  O  O  M  M  O  O  O  -  -  -
-  O  O  O  M  O  O  O  O  O  O  O  M  O  O  O  O  O  M  O  O  O  O  O  -  -  -
-  O  O  O  M  O  O  -  -  -  -  O  O  O  O  O  M  O  O  O  M  O  O  O  O  -  -  -  -
-  O  O  O  M  O  O  -  -  -  -  O  O  O  O  O  M  M  M  O  O  O  O  O  -  -  -  -  -
-  O  O  O  M  O  O  O  -  -  O  O  O  O  O  M  M  O  O  O  O  O  O  O  O  O  -
-  O  O  O  M  O  O  O  -  -  O  O  O  O  O  M  O  O  M  M  M  M  M  M  M  M  O  -
-  O  O  O  M  O  O  O  O  O  O  O  O  M  O  O  O  O  M  O  O  O  O  O  M  M  O  -
-  O  O  O  M  O  O  O  O  O  O  O  M  O  O  O  O  O  O  O  O  O  O  O  O  O  -
-  O  O  O  O  O  M  M  M  M  M  M  O  O  O  O  O  O  O  O  O  O  O  -  -  -  -  -
-  O  O  O  O  O  O  O  O  O  O  O  O  O  O  O  O  O  O  O  O  O  O  O  -  -  -  -  -
-  O  O  O  O  O  O  O  O  O  O  O  O  O  O  O  O  O  O  O  O  O  O  O  -  -  -  -  -
```

FIGURE 8.12 The resulting skeleton using the modified algorithm on a rectangle with holes and notches.

(a)   (b)   (c)   (d)

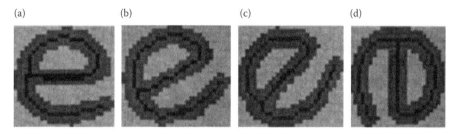

FIGURE 8.13 An example of the modified skeleton for a set of the character 'e' in various rotaions. (a) 0°, (b) 30°, (c) 45°, and (d) 90°.

8.3 Morphological Shape Description

A useful morphological shape description tool, called *geometric spectrum* or *G-spectrum*, is presented for quantifying the geometric features on multi-dimensional binary images. The basis of this tool relies upon the cardinality of a set of non-overlapping segments in an image using morphological operations. The G-spectrum preserves the translation invariance property. With a chosen set of isotropic SEs, the G-spectrum also preserves rotation invariance. After the normalization procedure, the G-spectrum can also preserve scaling invariance. These properties and proofs of the G-spectrum are discussed.

8.3.1 Introduction

Shape description describes the object shape according to its geometric features. The shape of an object refers to its profile and physical structure. These characteristics can be represented by boundary, region, moment, and structural representations. These representations can be used for matching shapes, recognizing objects, or making measurements on the shape characteristics. Therefore, shape description is a very active and important issue in image processing, computer vision, and pattern recognition during recent decades.

Many algorithms have been proposed to represent shape. The skeleton representation and shape decomposition are the two most important categories. The idea of measuring the successive results of morphological openings on an image by different sized SEs was initialized by Matheron and Serra. Matheron [1975] explored the properties of these successive openings for size distributions called *granulometries*. Intuitively, a binary image is treated as a collection of grains (or particles), and the grains are sieved through a filter of increasing mesh sizes. After each filter is applied, the total number of remaining pixels is counted, and the distribution of these counts reflects the grain size distribution. Serra [1982] applied the size distribution for the continuous size and the shape description on two-dimensional binary images. Xu [2001] presented the morphological representation of two-dimensional binary shapes using rectangular components. Loncaric and Dhawan [1993] presented a morphological signature transform for shape description. Yu and Wang [2005] presented a shape representation algorithm based on MM. It consists of two steps. First, an input shape is decomposed into a union of meaningful convex subparts by a recursive scheme. Second, the shape of each subpart is approximated by a morphological dilation of basic SEs.

The notations are introduced as follows. Let X be a set in a multidimensional Euclidean space, which represents a binary image, and let $Card(X)$ be the cardinality of X (i.e., the total number of elements in the set X). The cardinality of a one-dimensional set is the length, of a two-dimensional set is the area, of a three-dimensional set is the volume, and so forth. Let $\{B(n) \mid n = 0, 1, \ldots, N-1\}$ be a sequence of SEs, such that the origin $(0, 0) \in B(n)$, and $B(n)$ contains at least one element, where

$$N = \max\{n \mid X \ominus A(n) \neq \phi\} \qquad (8.8)$$

and

$$A(0) = (0, 0), \qquad (8.9)$$

$$A(n + 1) = A(n) \oplus B(n) \quad \text{for } n = 0, 1, \ldots, N - 1. \qquad (8.10)$$

More rigorously, the set $\{X \circ A(n) \mid n = 0, 1, \ldots, N\}$ provides a great deal of shape and size information of a given image X. Due to the anti-extensive property of the morphological opening, the following is obtained:

$$X \circ A(0) \supseteq X \circ A(1) \supseteq \cdots \supseteq X \circ A(N), \qquad (8.11)$$

Maragos [1989b] proposed a shape-size descriptor called *pattern spectrum*. The discrete version of pattern spectrum is a useful quantity for shape analysis, which is given by

$$PS_X(n, B(n)) = Card(X \circ A(N) - X \circ A(N + 1)) \quad \text{for } n = 0, 1, \dots, N - 1, \quad (8.12)$$

where $A(n + 1)$ is computed by Equation 8.10. Maragos [1989a] also presented a representation theory for morphological image and signal processing.

Bronskill and Venetsanopoulos [1988] proposed *pecstrum* of which the discrete version is

$$P_n(X) = \frac{PS_X(n, B(n))}{Card(X)} \quad \text{for } n = 0, 1, \dots, N - 1 \quad (8.13)$$

Another variety called *probability distribution function* [Dougherty and Pelz 1991] is defined as

$$\Phi(n) = 1 - \frac{Card(X \circ A(n))}{Card(X)} \quad \text{for } n = 0, 1, \dots, N - 1 \quad (8.14)$$

We present a useful morphological shape description tool, called *geometric spectrum* or *G-spectrum*, for quantifying geometric features on multidimensional binary images. The G-spectrum is a shape descriptor superior to the above descriptors in Equations 8.12 through 8.14 because of its less redundancy property. The G-spectrum, based on the morphological erosion and set transformation, is defined and some examples are given in the next section.

8.3.2 G-Spectrum

The G-spectrum is a measurement for quantifying the geometric shape of discrete multidimensional images. From Equation 8.10, we know that

$$X \ominus A(n + 1) = (X \ominus A(n)) \ominus B(n) \subseteq X \ominus A(n) \quad \text{for } n = 0, 1, \dots, N \quad (8.15)$$

Let $\{\Psi_n \mid n = 0, 1, \dots, N\}$ be a sequence of set transformations satisfying

$$X \ominus A(n + 1) \subseteq \Psi_n[X \ominus A(n + 1)] \subseteq X \ominus A(n) \quad \text{for } n = 0, 1, \dots, N \quad (8.16)$$

Some widely-used examples of the Ψ_n transformation [Goutsias and Schonfeld 1991] are given as follows.

Example 8.1: $\Psi_n(X) = X \oplus B(n) \quad \text{for } n = 0, 1, \dots, N.$

Example 8.2: $\Psi_n(X) = X \oplus (B(n) \bullet A(n)) \quad \text{for } n = 0, 1, \dots, N.$

Example 8.3: $\Psi_n(X) = (X \oplus B(n)) \bullet A(n) \quad \text{for } n = 0, 1, \dots, N.$

We introduce the G-spectrum as a shape descriptor, which is based on a set of size distributions. The formal definition of G-spectrum is given as follows.

Definition 8.1: The G-spectrum is a set of values defined by

$$\text{G-spectrum} = \{G_0(X), G_1(X), \ldots, G_N(X)\} \tag{8.17}$$

where

$$G_n(X) = (Card(X \ominus A(n)) - Card(\Psi_n[X \ominus A(n+1)]))/Card(X) \tag{8.18}$$

If we can find a sequence of transformations $\{\Psi_n | n = 0, 1, \ldots, N\}$ satisfying Equation 8.16 then

$$X \ominus A(n) - \Psi_n[X \ominus A(n+1)] \subseteq X \ominus A(n) - X \ominus A(n+1) \tag{8.19}$$

It is clear that

$$Card(X \ominus A(n) - \Psi_n[X \ominus A(n+1)]) \leq Card(X \ominus A(n) - X \ominus A(n+1)) \tag{8.20}$$

According to Equations 8.18 and 8.20, we can observe that the defined G-spectrum is less redundant than $R_n(X)$; that is

$$G_n(X) \leq (Card(X \ominus A(n)) - Card(\Psi_n[X \ominus A(n+1)]))/Card(X) = R_n(X) \tag{8.21}$$

It has been proved [Goutsias and Schonfeld 1991] that the upper-bound of the set transformations $\{\Psi_n\}$ satisfying Equation 8.16 is $X \ominus A(n) \circ B(n) \bullet A(n)$, and the following equation is satisfied by

$$X \circ A(k) = \bigcup_{n=k}^{N} [(X \ominus A(n) - \Psi_n[X \ominus A(n+1)]) \oplus A(n)] \tag{8.22}$$

The difference between two successive openings is

$$X \circ A(n) - X \circ A(n+1) = (X \ominus A(n) - \Psi_n[X \ominus A(n+1)]) \oplus A(n)$$
$$\supseteq X \ominus A(n) - \Psi_n[X \ominus A(n+1)] \tag{8.23}$$

It is obvious that

$$Card(X \circ A(n) - X \circ A(n+1)) \geq Card(X \ominus A(n) - \Psi_n[X \ominus A(n+1)]) \tag{8.24}$$

implies $G_n(X) \leq P_n(X)$ according to Equations 8.12, 8.13, and 8.18, where $P_n(X)$ is the nth element of "pecstrum." Hence, the defined G-spectrum has the least redundant size distribution, such that $G_n(X) \leq R_n(X)$ and $G_n(X) \leq P_n(X)$.

8.3.3 The Properties of G-Spectrum

The properties of G-spectrum are presented and discussed in this section.

Proposition 8.1: For a given image X, each element of G-spectrum is a positive valued function. That is

$$G_n(X) \geq 0 \quad \text{for } n = 0, 1, \ldots, N \tag{8.25}$$

Proof: From Equation 8.16, we know that

$$X \ominus A(n) \supseteq \Psi_n[X \ominus A(n+1)] \tag{8.26}$$

By applying the cardinality to both sides yields

$$Card(X \ominus A(n)) \geq Card(\Psi_n[X \ominus A(n+1)]) \tag{8.27}$$

Because $Card(X) \geq 0$, we have

$$(Card(X \ominus A(n)) - Card(\Psi_n[X \ominus A(n+1)]))/Card(X) \geq 0 \tag{8.28}$$

According to Equation 8.18, the result in Equation 8.25 is obtained. □

Definition 8.2: The redundant reduction rate (RRT) of a given image X is defined as

$$RRT(X) = \frac{1}{Card(X)} \sum_{n=0}^{N} Card(\Psi_n[X * A(n+1)] - X * A(n+1)). \tag{8.29}$$

As stated in Proposition 8.1, the G-spectrum is a set of positive values, which gives the quantitative feature of an image based upon geometry. The RRT is an indicator of how much redundant information can be reduced by the G-spectrum. It is found that RRT can also be used in the matching procedure of object recognition.

Proposition 8.2: With a compact region of support, the summation of G-spectrum is equal to 1 minus the RRT(). That is

$$\sum_{n=0}^{N} G_n(X) = 1 - RRT(X) \tag{8.30}$$

Proof: From the definition of G-spectrum in Equation 8.18, we have

$$\sum_{n=0}^{N} G_n(X) = \frac{1}{Card(X)} \sum_{n=0}^{N} (Card(X * A(n)) - Card(\Psi_n[X * A(n+1)])) \tag{8.31}$$

$$\sum_{n=0}^{N} G_n(X) = \frac{1}{Card(X)} \sum_{n=0}^{N} Card(X * A(n) - \Psi_n[X * A(n+1)]) \qquad (8.32)$$

$$\sum_{n=0}^{N} G_n(X) = \frac{1}{Card(X)} \sum_{n=0}^{N} Card(X * A(n) - X * A(n+1)$$
$$- (\Psi_n[X * A(n+1)] - X * A(n+1))) \qquad (8.33)$$

$$\sum_{n=0}^{N} G_n(X) = \frac{1}{Card(X)} \sum_{n=0}^{N} Card(X * A(n) - X * A(n+1)$$
$$- \frac{1}{Card(X)} \sum_{n=0}^{N} Card(\Psi_n[X * A(n+1)] - X * A(n+1)) \qquad (8.34)$$

Because $X \ominus A(0) = X$ and $X \ominus A(N+1) = \phi$, we have

$$\sum_{n=0}^{N} Card(X * A(n) - X * A(n+1)) = Card(X * A(0) - X * A(n+1))$$
$$= Card(X) \qquad (8.35)$$

Therefore,

$$\sum_{n=0}^{N} G_n(X) = \frac{Card(X)}{Card(X)} - \frac{1}{Card(X)} \sum_{n=0}^{N} Card(\Psi_n[X * A(n+1)] - X * A(n+1))\mathcal{E}$$
$$= 1 - RRT(X) \qquad (8.36)$$

The summation of G-spectrum is used to determine the degree of redundancy for image representation. If $\sum_{n=0}^{N} G_n(X)$ is smaller, it means that the more redundant information is removed from the image. For an image X, the $RRT(X)$ will be varied with respect to the different sets of transformations. By employing the above concept, we are able to select a suitable transformation, which leads to the best performance on image coding.

Proposition 8.3: If $\Psi_n[X \oplus \{z\}] = \Psi_n[X] \oplus \{z\}$, for $n = 0, 1, \ldots, N$, then the G-spectrum is translation invariant. That is

$$G_n(X \oplus \{z\}) = G_n(X) \quad \text{for } n = 0, 1, \ldots, N \qquad (8.37)$$

Proof: The proof is derived from the fact of the translation invariance of erosion and cardinality, and from the assumption of the translation invariance of Ψ_n. That is

$$G_n(X \% \{z\}) = \frac{Card((X \% \{z\}) \ominus A(n)) - Card(\Psi_n[(X \% \{z\}) \ominus A(n+1)])}{Card(X \% \{z\})} \tag{8.38}$$

$$G_n(X \% \{z\}) = \frac{Card((X \ominus A(n)) \% \{z\}) - Card(\Psi_n[X \ominus A(n+1)] \% \{z\})}{Card(X \% \{z\})} \tag{8.39}$$

$$G_n(X \% \{z\}) = \frac{Card(X \ominus A(n)) - Card(\Psi_n[X \ominus A(n+1)])}{Card(X)} = G_n(X) \tag{8.40}$$

$\square$

The translation invariance property of G-spectrum is useful in object recognition. Given any two G-spectrums $\{G_n(X_1) \mid n = 0, 1, \ldots, N\}$ and $\{G_n(X_2) \mid n = 0, 1, \ldots, N\}$, if both are the same or differ within a tolerance range, we say these two objects are matched. Although the G-spectrum is not scaling invariant, we can normalize the object into a pre-defined size. After normalization, the G-spectrum can preserve scaling invariance.

Proposition 8.4: The G-spectrum is scaling invariant if the set X is normalized. That is

$$G_n(Nr(\xi X)) = G_n(Nr(X)) \quad \text{for } n = 0, 1, \ldots, N, \tag{8.41}$$

where ξ is an unknown scaling factor and $Nr(X)$ is a normalization function which is defined as

$$Nr(X) = \frac{\tau X}{Card(X)}, \tag{8.42}$$

where τ is a pre-defined value.

Proof: Because

$$Nr(\xi X) = \frac{\tau \xi X}{Card(\xi X)} = \frac{\tau \xi X}{\xi Card(X)} = Nr(X), \tag{8.43}$$

we have

$$G_n(Nr(\xi X)) = \frac{Card(Nr(\xi X) \ominus A(n)) - Card(\Psi_n[Nr(\xi X) \ominus A(n+1)])}{Card(Nr(\xi X))} \tag{8.44}$$

$$G_n(Nr(\xi X)) = \frac{Card(Nr(X) \ominus A(n)) - Card(\Psi_n[Nr(X) \ominus A(n+1)])}{Card(Nr(X))}$$

$$= G_n(Nr(X)) \tag{8.45}$$

$\square$

From Proposition 8.4, if we perform the normalization (note that *Card* $(Nr(X)) = \tau$) on the images with various scaling factors ξ, the G-spectrums of ξX and X are the same. This implies that the normalization according to a pre-defined value τ can produce the scaling invariant version of the G-spectrum.

Proposition 8.5: The first k elements of G-spectrum are zeros. That is

$$G_n(X) = 0 \quad \text{for } n = 0, 1, \ldots, k-1 \tag{8.46}$$

if and only if the following equations are satisfied:

$$X \circ A(k) = X \tag{8.47}$$

$$\Psi_n[X \ominus A(n+1)] = X \ominus A(n) \circ B(n) \bullet A(n) \tag{8.48}$$

Proof: Case 1: Assuming that Equation 8.46 is true, we prove that Equations 8.47 and 8.48 are also true. According to Equation 8.18, we have

$$Card(X \ominus A(n)) = Card(\Psi_n[X \ominus A(n+1)]). \tag{8.49}$$

This implies that

$$X \ominus A(n) = \Psi_n[X \ominus A(n+1)]. \tag{8.50}$$

We replace X by $X \circ A(n+1)$ on the left-hand side of Equation 8.50 and obtain

$$(X \circ A(n+1)) \ominus A(n) = (X \ominus A(n+1) \oplus A(n+1) \ominus A(n) \tag{8.51}$$

$$= (X \ominus (A(n) \oplus B(n)) \oplus (A(n) \oplus B(n))) \ominus A(n) \tag{8.52}$$

$$= X \ominus A(n) \ominus B(n) \oplus B(n) \oplus A(n) \ominus A(n) \tag{8.53}$$

$$= X \ominus A(n) \circ B(n) \bullet A(n) \tag{8.54}$$

We replace X by $X \circ A(n+1)$ on the right-hand side of Equation 8.50 and obtain

$$\Psi_n[(X \circ A(n+1)) \ominus A(n+1)] = \Psi_n[(X \ominus A(n+1)) \oplus A(n+1) \ominus A(n+1)] \tag{8.55}$$

$$= \Psi_n[(X \ominus A(n+1)) \bullet A(n+1)] \tag{8.56}$$

According to the MM property, a set, which is eroded by a structuring element, is the same as the eroded result followed by a closing with the same structuring element.

We then have

$$\Psi_n[(X \circ A(n + 1)) \ominus A(n + 1)] = \Psi_n[X \ominus A(n + 1)] \tag{8.57}$$

From Equations 8.54 and 8.57, we can obtain

$$\Psi_n[X \ominus A(n + 1)] = X \ominus A(n) \circ B(n) \bullet A(n) \tag{8.58}$$

From Equations 8.22 and 8.50, we have

$$X \circ A(k) = \bigcup_{n=0}^{N} [X * A(n) - \Psi_n[X * A(n + 1)] \% A(n)]$$

$$- \bigcup_{n=0}^{k-1} [X * A(n) - \Psi_n[X * A(n + 1)] \% A(n)] \tag{8.59}$$

$$= X \circ A(0) - \phi \tag{8.60}$$

$$= X \tag{8.61}$$

Case 2: Assuming that Equations 8.47 and 8.48 are true, we prove that Equation 8.46 is also true.

$$\Psi_n[X \ominus A(n + 1)] = X \ominus A(n) \circ B(n) \bullet A(n) \tag{8.62}$$

$$= X \ominus A(n) \ominus B(n) \oplus B(n) \oplus A(n) \ominus A(n) \tag{8.63}$$

$$= (X \ominus (A(n) \oplus B(n)) \oplus (A(n) \oplus B(n))) \ominus A(n) \tag{8.64}$$

By applying Equation 8.10, we obtain

$$\Psi_n[X \ominus A(n + 1)] = [X \circ A(n + 1)] \ominus A(n) \tag{8.65}$$

According to Equations 8.11 and 8.47, we obtain

$$X \circ A(n + 1) = X \quad \text{for } n = 0, 1, \dots, k - 1 \tag{8.66}$$

Hence,

$$\Psi_n[X \ominus A(n + 1)] = X \ominus A(n) \quad \text{for } n = 0, 1, \dots, k - 1 \tag{8.67}$$

From the definition of $G_n(X)$ in Equation 8.18, we obtain

$$G_n(X) = \frac{Card(X \ominus A(n)) - Card(\Psi_n[X \ominus A(n + 1)])}{Card(X)} \tag{8.68}$$

$$= \frac{Card(X \ominus A(n)) - Card(X \ominus A(n))}{Card(X)} = 0 \quad \text{for } n = 0, 1, \dots, k - 1 \tag{8.69}$$

$\square$

If we can find a sequence of the set $\{A(n) \mid n = 0, 1, \ldots, N\}$ which satisfies Equations 8.47 and 8.48, the recognition problem can be simplified by matching only $N - k + 1$ elements of the G-spectrum. It means, if G-spectrums of two sets X_1 and X_2 satisfy

$$G_n(X_1) - G_n(X_2) \le \sigma \quad \text{for } n = k, k + 1, \ldots, N, \tag{8.70}$$

these two sets are regarded as the same.

Proposition 8.6: If the set of SEs is chosen to be isotropic, the G-spectrum can be regarded as rotation invariance.

The proof is straightforward and skipped. Since the SEs are isotropic, the G-spectrum defined as erosions by them, is naturally rotation-invariant. There is a relationship between pattern spectrum and G-spectrum. The next proposition will explore the relationship and prove that under some constraints, the elements of the G-spectrum and those of the pattern spectrum are equal to zero.

Proposition 8.7: There exists that for some n

$$G_n(X) = 0 \tag{8.71}$$

if and only if, the following equations are satisfied:

$$\Psi_n[X \ominus A(n + 1)] = X \ominus A(n) \circ B(n) \bullet A(n) \tag{8.72}$$

and

$$PS_X(n, B(n)) = 0 \tag{8.73}$$

Proof: Case 1: Assuming that Equation 8.71 is true, we prove that Equations 8.72 and 8.73 are also true. According to Equations 8.18 and 8.71, we have $Card(X \ominus A(n)) = Card(\Psi_n[X \ominus A(n + 1)])$. Since both sides are operated as the erosion on the same set X, we obtain

$$X \ominus A(n) = \Psi_n[X \ominus A(n + 1)]. \tag{8.74}$$

The proof of Equation 8.72 is the same as the proof of Equation 8.48 in Proposition 8.5. To derive Equation 8.73, we first give

$$X \circ A(n) = X \ominus A(n) \oplus A(n) \tag{8.75}$$

$$= (\Psi_n[X \ominus A(n + 1)]) \oplus A(n). \tag{8.76}$$

From Equation 8.72 we have

$$X \circ A(n) = (X \ominus A(n) \circ B(n) \bullet A(n)) \oplus A(n) \tag{8.77}$$

$$= X \ominus A(n) \ominus B(n) \oplus B(n) \oplus A(n) \ominus A(n) \oplus A(n) \tag{8.78}$$

$$= X \ominus (A(n) \oplus B(n)) \oplus (A(n) \oplus B(n)) \ominus A(n) \oplus A(n) \tag{8.79}$$

$$= X \ominus A(n+1) \oplus A(n+1) \circ A(n) \tag{8.80}$$

$$= X \circ A(n+1) \circ A(n) \tag{8.81}$$

$$\subseteq X \circ A(n+1) \tag{8.82}$$

From Equation 8.11, we know $X \circ A(n) \supseteq X \circ A(n+1)$. Hence, $X \circ A(n) = X \circ A(n+1)$. From Equation 8.12, we conclude $PS_X(n, B(n)) = 0$.

Case 2: Assuming that equations. 8.72 and 8.73 are true, we prove that Equation 8.71 is also true. From Equations 8.12 and 8.73 we have

$$X \circ A(n) = X \circ A(n+1). \tag{8.83}$$

From the MM property we obtain

$$X \ominus A(n) = (X \circ A(n)) \ominus A(n) \tag{8.84}$$

$$= (X \circ A(n+1)) \ominus A(n) \tag{8.85}$$

$$= X \ominus A(n) \circ B(n) \bullet A(n). \tag{8.86}$$

From Equations 8.18, 8.72, and 8.86, we have

$$G_n(X) = \frac{Card(X \ominus A(n) \circ B(n) \bullet A(n)) - Card(X \ominus A(n) \circ B(n) \bullet A(n))}{Card(X)} = 0. \tag{8.87}$$
$\square$

Proposition 8.7 tells us that if the transformation Ψ_n is constrained by Equation 8.72 and the nth element of the pattern spectrum is equal to zero, then the nth element of G-spectrum will be equal to zero or vice versa.

As many quantitative measures with respect to shape such as turning angles, length of sides, area, perimeter, radial lengths, and boundary coordinates have been proposed, it is difficult to standardize a minimum set of the shape descriptor to adequately quantify various object forms. The shape description techniques can be broadly classified into *external* and *internal*. The external shape description is based on the contour of a region under consideration, whereas the internal description deals with the region under consideration as an enclosed space.

The skeleton is one of the most important internal shape descriptors. The idea of transforming a binary image into an object skeleton and using the skeleton, as a shape descriptor was first introduced by Blum [1967]. The skeleton sets can be used as the base for the shape description, classification and decomposition [Maragos and Schafer 1986; Shih and Pu 1990; 1992].

The G-spectrum $\{G_n(X) \mid n = 0, 1, \ldots, N\}$ can be used as an internal shape descriptor, which is based on the quantified geometric features instead of a sequence of discrete or connected pixels (e.g. the skeleton representation). Although the original image cannot be reconstructed from the G-spectrum (i.e., G-spectrum is not an information-preserving descriptor), the G-spectrum is more useful than the skeleton in the shape recognition. It is not straightforward to apply a set of skeletons in solving the matching problem during shape recognition; however, it is easier when a set of quantified geometric features, such as G-spectrum, is used. In other words, the G-spectrum is not only a shape descriptor, but also a tool for object recognition.

In addition, a general-purpose technique based on multiscale MM for object recognition was presented in Jalba, Wilkinson, and Roerdink [2006], where a connected operator similar to the morphological hat transform is defined, and two scale-space representations are built, using the curvature function as the underlying one-dimensional signal.

References

American National Standards Institute (ANSI), *Dimensioning and Tolerancing*, ANSI Standard Y14.5M, ASME, New York, 1982.

Arcelli, C. and Sanniti di Baja, G., "A one-pass two-operation process to detect the skeletal pixels on the 4-distance transform," *IEEE Trans. Pattern Analysis and Machine Intelligence*, vol. 11, no. 4, pp. 411–414, Apr. 1989.

Blum, H., "A transformation for extracting new descriptors of shape," in W. Dunn (Ed), *Models for the Perception of Speech and Visual Forms*, Proc. of Meeting held November 1964, MIT Press, Massachusetts, pp. 362–380, 1967.

Bouaynaya, N., Charif-Chefchaouni, M., and Schonfeld, D., "Spatially-variant morphological restoration and skeleton representation," *IEEE Trans. Image Processing*, vol. 15, no. 11, pp. 3579–3591, Nov. 2006.

Bronskill, J. and Venetsanopoulos, A., "Multidimensional shape description and recognition using mathematical morphology," *Journal of Intell. and Robotics Sys.*, vol. 1, no. 2, pp. 117–143, June 1988.

Dougherty, E. and Pelz, J., "Morphological granulometric and analysis of electrophotographic images—size distribution statistics for process control," *Optical Engineering*, vol. 30, no. 4, pp. 438–445, Apr. 1991.

Fu, K. S., *Syntactic Pattern Recognition and Applications*, Prentice-Hall, 1982.

Ghosh, P. K., "A mathematical model for shape description using Minkowski operators," *Computer Vision, Graphics, and Image Processing*, vol. 44, no. 3, pp. 239–269, Dec. 1988.

Goutsias, J. and Schonfeld, D., "Morphological representation of discrete and binary images," *IEEE Trans. Signal Processing*, vol. 39, no. 6, pp. 1369–1379, June 1991.

Jalba, A. C. Wilkinson, M. H. F., and Roerdink, J. B. T. M., "Shape representation and recognition through morphological curvature scale spaces," *IEEE Trans. Image Processing*, vol. 15, no. 2, pp. 331–341, Feb. 2006.

Jang, B. K. and Chin, R. T., "Analysis of thinning algorithm using mathematical morphology," *IEEE Trans. Pattern Analysis and Machine Intelligence*, vol. 12, no. 6, pp. 541–551, June 1990.

Jang, B. K. and Chin, R. T., "One-pass parallel thinning: Analysis, property, and quantitative evaluation," *IEEE Trans. Pattern Analysis and Machine Intelligence*, vol. 14, no. 11, pp. 1129–1140, Nov. 1992.

Kresch, R. and Malah, D., "Skeleton-based morphological coding of binary images," *IEEE Trans. Image Processing*, vol. 7, no. 10, pp. 1387–1399, Oct. 1998.

Kwok, P. C. K., "A thinning algorithm by contour generation," *Comm. ACM*, vol. 31, no. 11, pp. 1314–1324, Nov. 1988.

Loncaric, S. and Dhawan, A. P., "A morphological signature transform for shape description," *Pattern Recognition*, vol. 26, no. 7, pp. 1029–1037, July 1993.

Margos, P. A. and Schafer, R. W., "Morphological skeleton representation and coding of binary images," *IEEE Trans. Acoustics, Speech and Signal Processing*, vol. 34, no. 5, pp. 1228–1244, Oct. 1986.

Maragos, P. A., "A representation theory for morphological image and signal processing," *IEEE Trans. Pattern Analysis and Machine Intelligence*, vol. 11, no. 6, pp. 586–599, June 1989a.

Maragos, P. A., "Pattern spectrum and multiscale shape representation," *IEEE Trans. Patt. Anal. Mach. Intell.*, vol. 11, no. 7, pp. 701–716, July 1989b.

Matheron, G., *Random Sets and Integral Geometry*, New York: Wiley, 1975.

Niblack, C. W., Capson, D. W., and Gibbons, P. B., "Generating skeletons and center-lines from the MA transform," *Proc. Intl. Conf. Pattern Recognition*, Atlantic City, New Jersey, pp. 881–885, June 1990.

Naccache, N. J. and Shinghal, R., "SPTA: A proposed algorithm for thinning binary patterns," *IEEE Trans. Systems, Man and Cybernetics*, vol. 14, no. 3, pp. 409–418, May 1984.

O'Gorman, L. "K × K thinning," *Computer Vision, Graphics and Image Processing*, vol. 51, no. 2, pp. 195–215, Aug. 1990.

Rossignac, J. R. and Requicha, A. A. G., "Offsetting operations in solid modelling," Production Automation Project, Univ. Rochester, Rochester, NY Tech. Memo, 53, June 1985.

Serra, J., *Image Analysis and Mathematical Morphology*, New York: Academic, 1982.

Shih, F. Y., "Object representation and recognition using mathematical morphology model," *Inter. Journal of Systems Integration*, vol. 1, no. 2, pp. 235–256, Aug. 1991.

Shih, F. Y. and Gaddipati, V., "Geometric modeling and representation based on sweep mathematical morphology," *Information Sciences*, vol. 171, no. 3, pp. 213–231, March 2005.

Shih, F. Y. and Pu, C. C., "Medial axis transformation with single-pixel and connectivity preservation using Euclidean distance computation," *Proc. IEEE Intl. Conf. on Pattern Recognition*, Atlantic City, New Jersey, pp. 723–725, 1990.

Shih, F. Y. and Pu, C. C., "Morphological shape description using geometric spectrum on multidimensional binary images," *Pattern Recognition*, vol. 25, no. 9, pp. 921–928, Sep. 1992.

Shih, F. Y. and Pu, C. C., "A skeletonization algorithm by maxima tracking on Euclidean distance transform," *Pattern Recognition*, vol. 28, no. 3, pp. 331–341, March 1995.

Trahanias, P., "Binary shape recognition using the morphological skeleton transform," *Pattern Recognition*, vol. 25, no. 11, pp. 1277–288, Nov. 1992.

Xu, J., "Morphological representation of 2-D binary shapes using rectangular components," *Pattern Recognition*, vol. 34, no. 2, pp. 277–286, Feb. 2001.

Xu, J., "A generalized discrete morphological skeleton transform with multiple structuring elements for the extraction of structural shape components," *IEEE Trans. Image Processing*, vol. 12, no. 12, pp. 1677–1686, Dec. 2003.

Yu, L. and Wang, R., "Shape representation based on mathematical morphology," *Pattern Recognition Letters*, vol. 26, no. 9, pp. 1354–1362, July 2005.

Zhang, T. Y. and Suen, C. Y., "A fast thinning algorithm for thinning digital patterns," *Comm. ACM*, vol. 27, no. 3, pp. 236–239, Mar. 1984.

9

Decomposition of Morphological Structuring Elements

Mathematical morphology has been widely used for many applications in image processing and analysis. Most image processing architectures adapted to morphological operations use structuring elements (SEs) of a limited size. Therefore, difficulties arise when we deal with a large sized structuring element. Various algorithms have been developed for decomposing a large sized structuring element into dilations of small structuring components. In this chapter, we present the techniques of dealing with the decomposition of morphological SEs.

This chapter is organized as follows. Section 9.1 presents the decomposition of geometric-shaped SEs. Section 9.2 presents the decomposition of binary SEs. Section 9.3 presents the decomposition of grayscale SEs.

9.1 Decomposition of Geometric-Shaped SEs

Several widely used geometric-shaped morphological SEs in MM can be used to explore the shape characteristics of an object. In this section, we present a technique of simplifying the decomposition of various types of big geometric-shaped SEs into dilations of smaller structuring components by the use of a mathematical transformation. Hence, the desired morphological erosion and dilation are equivalent to a simple inverse transformation over the result of operations on the transformed decomposable SEs. We also present a strategy to decompose a large cyclic cosine structuring element.

The technique of decomposing a two-dimensional convex structuring element into one-dimensional structuring components is also developed.

We first introduce the definitions concerning the types of morphological SEs and the MM properties related to the decomposition. We then present the decomposition technique for one-dimensional and two-dimensional geometric-shaped SEs. A new mathematical transformation that is applied on a nondecreasing cosine structuring element to become decreasing is presented. A strategy for dealing with the decomposition of two-dimensional SEs into one-dimensional elements is described.

9.1.1 Definitions of Types of SEs

The definitions related to the types of SEs are given. For expressional simplicity, these definitions are described in one dimension. The extension to two dimensions can be similarly derived. Let the origin of a structuring element $k_{(2n+1)}(x)$ of the odd size $2n + 1$ be located at the central point as shown in Figure 9.1, such that the x-coordinate of the structuring element is: $x = -n, \ldots, -1, 0, 1, \ldots, n$. Let m_i denote the slope of the line segment i; that is, $m_i = k(i) - k(i - 1)$.

Definition 9.1: A structuring element $k(x)$ is called *eccentrically decreasing* if

$$\begin{cases} k(i+1) \geq k(i) & \text{for } i = -n, \ldots, -1 \\ k(i-1) \geq k(i) & \text{for } i = 1, \ldots, n \end{cases} \tag{9.1}$$

Otherwise, $k(x)$ is called *eccentrically nondecreasing*. If $k(i + 1) = k(i)$ for all $i = -n, \ldots, -1, 0, 1, \ldots, n - 1$, it is called *flat*.

Note that we are only interested in the morphological operations by the eccentrically decreasing SEs because the morphological operations by a nondecreasing structuring element will induce nonsignificant results. Figure 9.2 illustrates that the dilated result of a binary image f with two gray levels, zero indicating background and C indicating foreground, by a large cyclic cosine structuring element k (it is certainly nondecreasing) is a

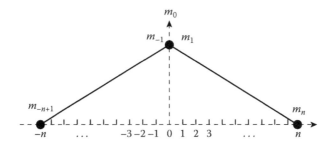

FIGURE 9.1 A linearly-sloped structuring element.

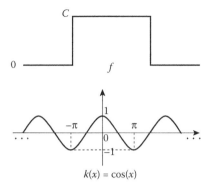

$$k(x) = \cos(x)$$

FIGURE 9.2 A binary image f with two gray levels, 0 indicating background and C indicating foreground, and a large cyclic cosine structuring element k.

constant image with the gray level of the highest value $C + 1$ in the calculation, and the eroded result is a constant image with the gray level of the lowest value -1.

Definition 9.2: A structuring element $k(x)$ is called *linearly-sloped* if $k(x)$ is eccentrically decreasing and

$$\begin{cases} m_i = c_1 & \text{for } i = -(n-1), \ldots, -1, 0 \\ m_i = c_2 & \text{for } i = 1, \ldots, n \end{cases} \tag{9.2}$$

where c_1 and c_2 are constants. An example of a linearly-sloped structuring element with $c_1 = c_2$ is shown in Figure 9.1.

Definition 9.3: A structuring element $k(x)$ is called *convex* or *eccentrically convex*, as shown in Figure 9.3, if $k(x)$ is eccentrically decreasing and

$$\begin{cases} m_i \leq m_{i-1} & \text{for } i = -(n-1), \ldots, -1, 0 \\ -m_i \leq -m_{i+1} & \text{for } i = 1, \ldots, n \end{cases} \tag{9.3}$$

Definition 9.4: A structuring element $k(x)$ is called *concave* or *eccentrically concave*, as shown in Figure 9.4, if $k(x)$ is eccentrically decreasing and

$$\begin{cases} m_i \geq m_{i-1} & \text{for } i = -(n-1), \ldots, -1, 0 \\ -m_i \geq -m_{i+1} & \text{for } i = 1, \ldots, n \end{cases} \tag{9.4}$$

When working on the two-dimensional SEs, we need the following definitions.

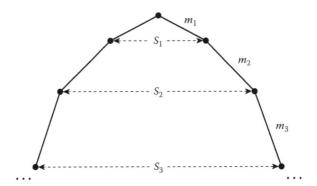

FIGURE 9.3   A convex structuring element.

Definition 9.5: A two-dimensional grayscale structuring element can be viewed as a three-dimensional geometric surface with its heights corresponding to the element's gray values. Mathematically, either a parametric or a nonparametric form can be used to represent a surface. A nonparametric representation is either implicit or explicit. For a three-dimensional geometric surface, an implicit, non-parametric form is given by

$$f(x,y,z) = 0 \tag{9.5}$$

In this form, for each x- and y-values the multiple z-values are allowed to exist. The representation of a two-dimensional grayscale structuring element can be expressed in an explicit, nonparametric form of

$$k(x,y) = \{z \mid z = g(x,y), (x,y) \in \text{geometric shape domain}\} \tag{9.6}$$

In this form, for each of x- and y-values only a z-value is obtained. Hence, the two-dimensional grayscale structuring element can be assigned a unique

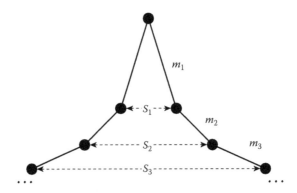

FIGURE 9.4   A concave structuring element.

datum at each (x, y) location. Some often-used two-dimensional geometric shapes are sphere, cone, ellipsoid, hyperboloid, Gaussian, exponential, damped sine, and damped cosine.

Definition 9.6: A structuring element $k(x, y)$ is called *additively-separable* if

$$k(x, y) = p(x) + q(y), \tag{9.7}$$

where $p(x)$ and $q(y)$ are the functions of only x and y, respectively.

9.1.2 Decomposition Properties

The decomposition properties presented in this section can be suited for any binary (with two gray levels, 0 indicating background and C indicating foreground) or grayscale image f, but only limited to the decreasing gray-scale SEs k. Note that the value C must be larger enough than the height of the structuring element.

Property 9.1: If a structuring element k can be decomposed into sequential dilations of several smaller structuring components, then a dilation (or erosion) of an image f by k is equal to successively applying dilations (or erosions) of the previous stage output by those structuring components. All convex SEs can be decomposed using this technique. As shown in Figure 9.3, let $k_1, k_2, \dots,$ and k_n denote the segmented smaller linearly-sloped structuring components with the sizes $s_1, (s_2 - s_1), \dots,$ and $(s_n - s_{n-1})$, respectively. That is

$$\text{if } k = k_1 \oplus k_2 \oplus \cdots \oplus k_n, \quad \text{then } f \oplus k = (\cdots ((f \oplus k_1) \oplus k_2) \oplus \cdots) \oplus k_n$$
$$\text{and } f \ominus k = (\cdots ((f \ominus k_1) \ominus k_2) \ominus \cdots) \ominus k_n \tag{9.8}$$

Example 9.1: Assume a linearly-sloped structuring element is

$$k = \{-4 \quad -2 \quad 0 \quad -1 \quad -2\}.$$

It can be decomposed as

$$k = k_1 \oplus k_2 = \{-2 \quad 0 \quad -1\} \oplus \{-2 \quad 0 \quad -1\}.$$

Example 9.2: Given a paraboloid structuring element: $k(x) = -6x^2$, where $-2 \le x \le 2$. It is

$$k = \{-24 \quad -6 \quad 0 \quad -6 \quad -24\}.$$

It can be decomposed as

$$k = k_1 \oplus k_2 = \{-6 \quad 0 \quad -6\} \oplus \{-18 \quad 0 \quad -18\}.$$

Property 9.2: If a structuring element k can be decomposed into a maximum selection over several structuring components, then a dilation (or erosion) of an image f by k is equal to finding a maximum (or minimum) of each applied dilation (or erosion) of the image by these structuring components. All concave SEs can be decomposed using this technique. As shown in Figure 9.4, let $k_1, k_2, \ldots$, and k_n denote the segmented structuring components with the sizes $s_1, s_2, \ldots$, and s_n, respectively. That is

$$\text{if } k = \max(k_1, k_2, \ldots, k_n), \quad \text{then } f \oplus k = \max(f \oplus k_1, f \oplus k_2, \ldots, f \oplus k_n)$$
$$\text{and } f \ominus k = \min(f \ominus k_1, f \ominus k_2, \ldots, f \ominus k_n) \tag{9.9}$$

Example 9.3: Given an exponential structuring element: $k(x) = 100(e^{-|x|} - 1)$, where $-2 \leq x \leq 2$. It is expressed as

$$k = \{-86 \quad -63 \quad 0 \quad -63 \quad -86\}.$$

Select the small SEs k_1 and k_2 as

$$k = \max\{k_1, k_2\} = \max(\{-63 \quad 0 \quad -63\}, \{-86 \quad -63 \quad -63 \quad -63 \quad -86\})$$
$$= \max(\{-63 \quad 0 \quad -63\}, (\{0 \quad 0 \quad 0\} \oplus \{-23 \quad 0 \quad -23\}) + (-\bar{6}\bar{3}))$$

Example 9.4: Given a concave structuring element $k(x) = \{-15 \quad -9 \quad 0 \quad -10 \quad -14\}$.
Select the small structuring element k_1 and k_2 as

$$k = \max\{k_1, k_2\} = \max(\{-9 \quad 0 \quad -10\}, \{-15 \quad -10 \quad -10 \quad -10 \quad 14\})$$
$$= \max(\{-9 \quad 0 \quad -10\}, (\{0 \quad 0 \quad 0\} \oplus \{-5 \quad 0 \quad -4\}) + (\bar{1}\bar{0}))$$

The computational complexity increases much more for the concave decomposition because further decompositions are required to perform the dilations (or erosions) by the structuring components $k_2, k_3, \ldots, k_n$, which have larger sizes than those in the convex decomposition and also an additional maximum (or minimum) operation.

A mix-typed structuring element is the combination of linearly-sloped, convex, and concave SEs, which can be decomposed by using the linearly-sloped, convex, and concave decomposition algorithms, respectively (see Figure 9.5). The decomposition algorithm is described as follows. A combined structuring element k is segmented into a maximum selection of its convex structuring components, $k_1, k_2, \ldots, k_n$. Each convex structuring component is decomposed into sequential dilations of its segmented linearly-sloped structuring components.

Example 9.5: Given a three-sloped combination of linearly-sloped, convex, and concave SEs: $k(x) = -6|x|$ when $0 \leq |x| \leq 2$, $k(x) = -|x| - 10$ when $2 \leq |x| \leq 4$, and $k(x) = -2|x| - 6$ when $4 \leq |x| \leq 6$. It can be expressed as

$$k = \{-18 \quad -16 \quad -14 \quad -13 \quad -12 \quad -6 \quad 0 \quad -6 \quad -12 \quad -13 \quad -14 \quad -16 \quad -18\}.$$

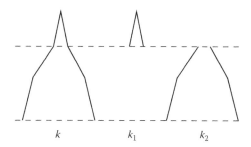

FIGURE 9.5 Decomposition of a mix-typed structuring element.

Select the small structuring components k_1, k_2, and k_3 as

$$k_1 = \{-6 \quad 0 \quad -6\}, \quad k_2 = \{-1 \quad 0 \quad -1\}, \quad k_3 = \{-2 \quad 0 \quad -2\}$$

Hence,

$$k = \max\{k_1 \oplus k_1, (((\{0 \quad 0 \quad 0 \quad 0 \quad 0\} \oplus k_2) \oplus k_2) \oplus k_3) \oplus k_3) - \vec{12}\}.$$

Property 9.3: Let $f : F \rightarrow E$ and $k : K \rightarrow E$. Let $x \in (F \oplus K) \cap (F \ominus \hat{K})$ be given. The relationship between dilation and erosion is

$$f \ominus_g k = -((-f) \oplus_g \hat{k}), \text{ where } \hat{k}(x) = k(-x), \text{ reflection of } k \qquad (9.10)$$

This implies that we may calculate erosion using dilation, or vice versa.

Property 9.4: For computational simplicity in decomposition, the gray-value at the origin (i.e., the highest value) of a decreasing structuring element can be suppressed to zero. In other words, we could subtract a constant $\mathbf{P}_c$, which is equal to the highest value, from the structuring element. The dilation and erosion then become

$$f \oplus (k - \mathbf{P}_c) = (f \oplus k) - \mathbf{P}_c \qquad (9.11)$$

$$f \ominus (k - \mathbf{P}_c) = (f \ominus k) + \mathbf{P}_c \qquad (9.12)$$

Property 9.5: All the linearly-sloped and one-dimensional convex SEs can be decomposed into dilations of smaller structuring components. For two-dimensional convex SEs, if each value $k(x, y)$ can be expressed as: $k(x, y) = p(x) = q(y)$, that is, two-dimensional additively-separable, then it can be decomposed into dilations of smaller structuring components. The other types of convex and all the concave SEs can be decomposed using a maximum selection over several segmented structuring components. Interested readers should refer to Shih and Mitchell [1991] for details.

9.1.3 One-Dimensional Geometric-Shaped SEs

9.1.3.1 Semicircle, Semiellipse, Gaussian, Parabola, Semihyperbola, Cosine, and Sine

A one-dimensional geometric-shaped structuring element $k(x)$ is actually the geometric curve: $f(x,y) = 0$, in a two-dimensional Euclidean space. We assign each point in the structuring element by the negated distance from this point to the horizontal curve tangent to the top of the curve to ensure that $k(x)$ is decreasing. An origin-centered semicircle can be represented implicitly by

$$f(x,y) = x^2 + y^2 - r^2 = 0, \qquad (9.13)$$

where $|x| \le r, 0 \le y \le r$, and r denotes the radius. Equation 9.13 can be rewritten in the explicit form as

$$y = \sqrt{r^2 - x^2} \qquad (9.14)$$

For computational simplicity, a constant $-r$ is added to suppress the y-value at the origin to zero. Hence,

$$k(x) = \sqrt{r^2 - x^2} - r \qquad (9.15)$$

Assuming that $r = 3$, the structuring element $k(x) = \sqrt{9 - x^2} - 3$, is numerically expressed as

$$k(x) = [-3 \quad \sqrt{5} - 3 \quad \sqrt{8} - 3 \quad 0 \quad \sqrt{8} - 3 \quad \sqrt{5} - 3 \quad -3]$$

This is a convex structuring element because the slopes $|m_i|$ is increasing eccentrically from the center. The similar method can be also applied to other geometric-shaped SEs as listed in Table 9.1. Note that the cosine and sine SEs both contain a combination of convex and concave components.

9.1.3.2 Decomposition Strategy

The following propositions are based on the assumption that the input image f is binary with two gray values: C indicating foreground and 0 indicating background, where C is a large number. The reason for setting the object points to a sufficiently large number is that the minimum selection in erosion or the maximum selection in dilation will not be affected by the object's gray value. The erosion of f by a geometric-shaped structuring element results in the output image reflecting the minimum distance (i.e., the predefined values in the structuring element) from each object pixel to all the background pixels.

Proposition 9.1: A one-dimensional eccentrically decreasing, geometric-shaped structuring element $k(x)$ can be represented as $k(x) = g(x) + c$, where

TABLE 9.1

One-Dimensional Geometric-Shaped SEs

Name	Representation	Structuring Element
Semiellipse	$\frac{x^2}{a^2} + \frac{y^2}{b^2} = 1$	$b\left[1 - \frac{x^2}{a^2}\right]^{1/2} - b$
Gaussian	$y = \frac{1}{\sqrt{2\pi}} e^{-x^2/2}$	$\frac{1}{\sqrt{2\pi}} e^{-x^2/2} - \frac{1}{\sqrt{2\pi}}$
Parabola	$x^2 = -4ay$	$-x^2/4a$
Semihyperbola	$\frac{y^2}{b^2} - \frac{x^2}{a^2} = 1$	$-b\left[1 + \frac{x^2}{a^2}\right]^{1/2} + b$
Sine	$y = a\sin(x)$	$a\sin(x) - a$
Cosine	$y = a\cos(x)$	$a\cos(x) - a$

$k(x) \leq 0$, $g(x)$ represents a geometric-shaped function, and c is a constant. It satisfies that for all $0 < x_1 < x_2$ and $x_2 < x_1 < 0$, there exists $g(x_1) > g(x_2)$. If a transformation Ψ satisfies that for all $0 < x_1 < x_2$ and $x_2 < x_1 < 0$, there exits $0 \leq \Psi[|k(x_1)|] < \Psi[|k(x_2)|]$, then we can obtain a new structuring element k^*, in which $k^*(x) = -\Psi[|k(x)|]$ where $k^*(x) \leq 0$. Let Ψ^{-1} is the inverse transformation of Ψ. The $k^*(x)$ can be used in the morphological erosion on a binary image f, so that the desired result $(f \ominus k)(x)$ is obtained by

$$(f \ominus k)(x) = \Psi^{-1}[(f \ominus k^*)(x)] \tag{9.16}$$

for all nonzero values in $(f \ominus k^*)(x)$. Those zeros in $(f \ominus k^*)(x)$ still remain zeros in $(f \ominus k)(x)$.

Proof: Since the binary image f is represented by two values C and 0, we obtain

$$(f \ominus k)(x) = \min\{C - k(z), 0 - k(z')\} \tag{9.17}$$

where all z's satisfy that $x + z$ is located within the foreground and all values of z' satisfy that $x + z'$ is located within background. Note that C has been selected to be large enough to ensure that $C - k(z)$ is always greater than $0 - k(z')$. Hence, we have

$$(f \ominus k)(x) = \min\{-k(z')\} \tag{9.18}$$

Similarly, if the new transformed structuring element k^* is used for erosion, we obtain

$$(f \ominus k^*)(x) = \min\{-k^*(z_1')\}, \tag{9.19}$$

where all values of z_1' satisfy that the pixel $x + z_1'$ is located within the background. Because

$$g(x_1) > g(x_2) \rightarrow k(x_1) > k(x_2) \tag{9.20}$$

and

$$\Psi[|k(x_1)|] < \Psi[|k(x_2)|] \rightarrow k^*(x_1) > k^*(x_2) \tag{9.21}$$

for all $0 < x_1 < x_2$ and $x_2 < x_1 < 0$, we have $k(x_1) > k(x_2)$ and $k^*(x_1) > k^*(x_2)$. This ensures that each location z' in Equation 9.18 is the same as the location z_1' in Equation 9.19 for the same input image f. Because $k^*(x) = -\Psi[|k(x)|]$, we have $k^*(z_1') = -\Psi[|k(z')|]$. Then $k(z')$ can be obtained by $|k(z')| = \Psi^{-1}[-k^*(z_1')]$. Hence, $-k(z') = \Psi^{-1}[-k^*(z_1')]$. Because the transformation Ψ preserves the monotonically increasing property, we obtain

$$\min\{-k(z')\} = \Psi^{-1}[\min\{-k^*(z_1')\}] \tag{9.22}$$

Combining Equations 9.18, 9.19, and 9.22, we can easily obtain the result in Equation 9.16.   □

Proposition 9.2: If $k(x)$, Ψ, and $k^*(x)$ all satisfy the conditions given in Proposition 9.1, then the $k^*(x)$ can be used in the morphological dilation on a binary image f. The desired result $(f \oplus k)(x)$ can be obtained by

$$(f \oplus k)(x) = C - \Psi^{-1}[C - (f \oplus k^*)(x)] \tag{9.23}$$

for all nonzero values in $(f \oplus k^*)(x)$, where C represents the foreground value. Those zeros in $(f \oplus k^*)(x)$, hence, remain zeros in $(f \oplus k)(x)$. The proof, which is similar to the proof in Proposition 9.1, is skipped.

Proposition 9.3: The result of a binary image f eroded by a geometric-shaped structuring element k is an inverted geometric shape of k within the foreground domain. As shown in Figure 9.6, those slopes eccentrically from the center in a convex structuring element k are $|m_1|$, $|m_2|$, and $|m_3|$. After erosion, its shape is inverted to become concave; hence the slopes eccentrically from the center become $|m_3|$, $|m_2|$, and $|m_1|$.

Proof: From Equation 9.18, the erosion is the minimum of negated values of the structuring element. The weights in the geometric-shaped structuring element are designed as the negative of the heights counting from the point to the horizontal line which is tangent to the toppest location (i.e. the center) of the structuring element. Since k is decreasing and has values all negative except the center zero, the result of minimum selection will be the coordinate z' in $k(z')$ such that $x + z'$ is located in the background and z' is the closest to the center of the structuring element. Hence, if we trace the resulting erosion curve from the boundary toward the center of an object, it is equivalent to trace the geometric curve of the structuring element from its center outwards. In other words, the erosion curve is an inverted geometric shape of the structuring element.   □

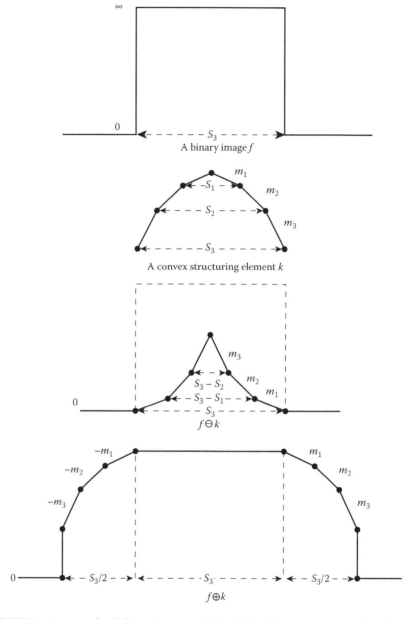

FIGURE 9.6 An example of a binary image eroded and dilated by a convex structuring element.

Proposition 9.4: The result of a binary image f dilated by a geometric-shaped structuring element k is the geometric shape of k located in the background initiating from the object boundary toward background. As shown in Figure 9.6, those slopes eccentrically from the center in a convex structuring element k are $|m_1|$, $|m_2|$, and $|m_3|$. After dilation, its shape starting from the edge of the object falls down by following the same sequence of slopes $|m_1|$, $|m_2|$, and $|m_3|$ correspondingly. The proof is similar to the proof in Proposition 9.3 is skipped.

Proposition 9.5: Morphological operations by a one-dimensional eccentrically decreasing, geometric-shaped structuring element can be implemented by the operations by a linearly-sloped structuring element.

Proof: Select $\Psi = |k^{-1}|$ in Proposition 9.1. Then $k^*(x) = -\Psi[|k(x)|] = -|k^{-1}|$ $(|k(x)|) = -x$, which is a linearly-sloped structuring element. Hence, the morphological erosion and dilation can apply Equations 9.16 and 9.23. $\square$

Example 9.6: Let a cosine curve be $y = \cos(x)$, where $|x| \le \pi$. We quantize the cosine function in an interval of $\pi/4$. When x is

$$\left[-\pi \quad -\frac{3}{4}\pi \quad -\frac{1}{2}\pi \quad -\frac{1}{4}\pi \quad 0 \quad \frac{1}{4}\pi \quad \frac{1}{2}\pi \quad \frac{3}{4}\pi \quad \pi \right]$$

y is equal to

$$[-1 \quad -0.7 \quad 0 \quad 0.7 \quad 1 \quad 0.7 \quad 0 \quad -0.7 \quad -1]$$

It is a combination of convex and concave structuring components, such that when $|x| \le \pi/2$, it is convex, and when $\pi/2 \le |x| \le \pi$, it is concave. The structuring element constructed by $k(x) = \cos(x) - 1$, can be expressed as

$$k: [-2 \quad -1.7 \quad -1 \quad -0.3 \quad 0 \quad -0.3 \quad -1 \quad -1.7 \quad -2]$$

The mathematical transformation defined as $\Psi[|k(x)|] = 10^{|k(x)|} - 1$ can satisfy the conditions in Proposition 9.1. By applying $k^* = -\Psi[|k(x)|]$, we obtain

$$k^*: [-99 \quad -49 \quad -9 \quad -1 \quad 0 \quad -1 \quad -9 \quad -49 \quad -99]$$

It is observed that k^* is convex. Hence, k^* can be decomposed using dilations of smaller structuring components, such that $k^* = k_1 \oplus k_2 \oplus k_3 \oplus k_4$, where

$$k_1 = [-1 \quad 0 \quad -1], \quad k_2 = [-8 \quad 0 \quad -8], \quad k_3 = [-40 \quad 0 \quad -40], \quad \text{and}$$
$$k_4 = [-50 \quad 0 \quad -50]$$

Hence,

$$f \ominus k^* = (((f \ominus k_1) \ominus k_2) \ominus k_3) \ominus k_4 \tag{9.24}$$

The result of $f \ominus k$ can be obtained by

$$f \ominus k = \Psi^{-1}[f \ominus k^*] = \log_{10}[(f \ominus k^*) + 1] \qquad (9.25)$$

for all nonzero values in $f \ominus k^*$.

Example 9.7: Let a parabolic structuring element be $k(x) = -x^2$, where a is equal to ¼. We quantize the parabolic function. When x is

$$[-4 \quad -3 \quad -2 \quad -1 \quad 0 \quad 1 \quad 2 \quad 3 \quad 4]$$

y is equal to

$$[-16 \quad -9 \quad -4 \quad -1 \quad 0 \quad -1 \quad -4 \quad -9 \quad -16]$$

It is clearly a convex structuring element. The mathematical transformation is defined as $\Psi[|k(x)|] = |k(x)|^{1/2}$. By applying $k^* = -\Psi[|k(x)|] = -x$, we obtain

$$k^* = [-4 \quad -3 \quad -2 \quad -1 \quad 0 \quad -1 \quad -2 \quad -3 \quad -4]$$

The structuring element k^* is linearly-sloped. Hence k^* can be decomposed using iterative dilations of a structuring component of size 3, such that $k^* = k_1 \oplus k_1 \oplus k_1 \oplus k_1$, where

$$k_1 = [-1 \quad 0 \quad -1]$$

Hence,

$$f \ominus k^* = (((f \ominus k_1) \ominus k_1) \ominus k_1) \ominus k_1 \qquad (9.26)$$

The result of $f \ominus k$ can be obtained

$$f \ominus k = \Psi^{-1}[f \ominus k^*] = (f \ominus k^*)^2 \qquad (9.27)$$

for all nonzero values in $f \ominus k^*$.

9.1.4 Two-Dimensional Geometric-Shaped SEs

9.1.4.1 Hemisphere, Hemiellipsoid, Gaussian, Elliptic Paraboloid, and Hemihyperboloid

A two-dimensional geometric-shaped structuring element $k(x, y)$ is actually the geometric surface: $f(x, y, z) = 0$ in a three-dimensional Euclidean space. We assign each point in the structuring element by the negated distance from this point to the horizontal plane tangent to the top of the surface to ensure that $k(x,y)$ is decreasing. A hemisphere can be represented implicitly by

$$f(x,y,z) = x^2 + y^2 + z^2 - r^2 = 0 \tag{9.28}$$

where $|x| \le r$, $|y| \le r$, $0 \le z \le r$, and r denotes the radius. It can be rewritten in the explicit form as

$$z = \sqrt{r^2 - x^2 - y^2} \tag{9.29}$$

It is known that the hemisphere is an eccentrically decreasing structuring element. For computational simplicity, a constant $-r$ is added to suppress the z-value at the origin to zero.
Hence,

$$k(x,y) = \sqrt{r^2 - x^2 - y^2} - r \tag{9.30}$$

The similar method can be also applied to other geometric-shaped SEs as listed in Table 9.2.

The aforementioned two-dimensional geometric-shaped SEs are all convex except that the Gaussian shape is a combined convex and concave structuring element. Since it is not additively-separable, it can only be decomposed using the maximum selection of several segmented structuring components [Shih and Mitchell 1991], which is inefficient in calculation. In the following section, we present an improved strategy to decompose it into dilations of smaller structuring components.

9.1.4.2 Decomposition Strategy

Proposition 9.6: A two-dimensional eccentrically decreasing, geometric-shaped structuring element $k(x,y)$ can be represented by $k(x,y) = g(p(x) + q(y)) + c$, where $g(\ )$ represents a geometric-shaped function, $p(x)$ and $q(y)$ are functions of x and y, respectively, and c is a constant. If the function $g(\ )$ and

TABLE 9.2

Two-Dimensional Geometric-Shaped SEs

Name	Representation	Structuring Element
Hemiellipsoid	$\dfrac{x^2}{a^2} + \dfrac{y^2}{b^2} + \dfrac{z^2}{c^2} = 1$	$c\left[1 - \dfrac{x^2}{a^2} - \dfrac{y^2}{b^2}\right]^{1/2} - c$
Gaussian surface	$z = \dfrac{1}{\sqrt{2\pi}} e^{-(x^2+y^2)/2}$	$\dfrac{1}{\sqrt{2\pi}} e^{-(x^2+y^2)/2} - \dfrac{1}{\sqrt{2\pi}}$
Elliptic paraboloid	$\dfrac{x^2}{a^2} + \dfrac{y^2}{b^2} + cz = 0$	$-\dfrac{x^2}{ca^2} - \dfrac{y^2}{cb^2}$
Hemihyperboloid	$\dfrac{z^2}{c^2} - \dfrac{x^2}{a^2} - \dfrac{y^2}{b^2} = 1$	$-c\left[\dfrac{x^2}{a^2} - \dfrac{y^2}{b^2} + 1\right]^{1/2} + c$

a transformation Ψ satisfy that for all $0 < x_1 < x_2$ and $x_2 < x_1 < 0$, there exist $g(p(x_1) + q(y)) > g(p(x_2) + q(y))$ and $0 \le \Psi[|k(x_1, y)|] < \Psi[|k(x_2, y)|]$, and for all $0 < y_1 < y_2$ and $y_2 < y_1 < 0$, there exists $g(p(x) + q(y_1)) > g(p(x) + q(y_2))$ and $0 \le \Psi[|k(x, y_1)|] < \Psi[|k(x, y_2)|]$. Then we can obtain a new structuring element k^*, in which $k^*(x, y) = -\Psi[|k(x, y)|]$, where $k^*(x, y) \le 0$. The $k^*(x, y)$ can be used in the morphological erosion and dilation on a binary image f, so that the desired result $(f \ominus k)(x, y)$ can be obtained by

$$(f \ominus k)(x, y) = \Psi^{-1}[(f \ominus k^*)(x, y)] \tag{9.31}$$

for all nonzero values in $(f \ominus k^*)(x, y)$. The dilation $(f \oplus k)(x, y)$ can be obtained by

$$(f \oplus k)(x, y) = C - \Psi^{-1}[C - (f \oplus k^*)(x, y)] \tag{9.32}$$

for all nonzero values in $(f \oplus k^*)(x, y)$. The proof can be similarly derived from the proof in Proposition 9.1.

Proposition 9.7: If a two-dimensional eccentrically decreasing, geometric-shaped structuring element $k(x, y)$ satisfies that $k(x, y) = g(p(x) + q(y)) + c$, where $g(\ )$ represents a geometric-shaped function, $p(x)$ and $q(y)$ are convex and in the same sign, and c is a constant, then k can be decomposed using dilations of smaller structuring components through a mathematical transformation.

Proof: Select a transformation $\Psi = |g^{-1}|$ in Proposition 9.6. We have

$$k^*(x, y) = -\Psi[|k(x, y)|] = -\Psi[|g(p(x) + q(y) + c|] = -|p(x) + q(y) + c| \tag{9.33}$$

Since $p(x)$ and $q(y)$ are convex and in the same sign, we obtain

$$k^*(x, y) = -|p(x)| - |q(y)| \pm c \tag{9.34}$$

It is clearly that $k^*(x, y)$ is an additively-separable convex structuring element, hence it can be decomposed using dilations of smaller structuring components. □

An example of the decomposition of a hemispheric-shaped structuring element is given below.

Example 9.8: A three-dimensional hemispheric surface is given by placing $r = 4$ in Equation 9.29 as

$$z = \sqrt{16 - x^2 - y^2} \tag{9.35}$$

The hemispheric-shaped structuring element can be constructed as

$$k(x, y) = \sqrt{16 - x^2 - y^2} - 4 = -(4 - \sqrt{16 - x^2 - y^2}) \tag{9.36}$$

The mathematical transformation defined as $\Psi[\,|\,k(x,y)\,|\,] = -((-(\,|\,k(x,y)\,|\, - 4))^2 - 16)$ can satisfy the conditions in Proposition 9.6. By applying $k^* = -\Psi[\,|\,k(x,y)\,|\,]$, we obtain

$$k^*(x,y) = -x^2 - y^2 \tag{9.37}$$

The new structuring element k^* is additively-separable and convex and it can be numerically represented as

$$
k^*:\quad
\begin{array}{rrrrrrr}
 & -13 & -10 & -9 & -10 & -13 & \\
-13 & -8 & -5 & -4 & -5 & -8 & -13 \\
-10 & -5 & -2 & -1 & -2 & -5 & -10 \\
-9 & -4 & -1 & 0 & -1 & -4 & -9 \\
-10 & -5 & -2 & -1 & -2 & -5 & -10 \\
-13 & -8 & -5 & -4 & -5 & -8 & -13 \\
 & -13 & -10 & -9 & -10 & -13 &
\end{array}
$$

Hence, k^* can be decomposed as: $k^* = k_1 \oplus k_2 \oplus k_3$, where

$$
k_1 = \begin{bmatrix} -2 & -1 & -2 \\ -1 & 0 & -1 \\ -2 & -1 & -2 \end{bmatrix}, \quad
k_2 = \begin{bmatrix} -6 & -3 & -6 \\ -3 & 0 & -3 \\ -6 & -3 & -6 \end{bmatrix}, \quad
k_3 = \begin{bmatrix} & -5 & \\ -5 & 0 & -5 \\ & -5 & \end{bmatrix},
$$

Therefore, the result of $f \ominus k$ can be obtained by

$$f \ominus k = \Psi^{-1}[(f \ominus k^*)] = -(-(f \ominus k^*) + 16)^{1/2} + 4 \tag{9.38}$$

9.1.5 Decomposition of a Large Transformed Cyclic Cosine Structuring Element

We discussed in Figure 9.2 that a large cyclic cosine structuring element (i.e. nondecreasing) has no significant meaning, because its morphological operations will result in a constant image. However, in image processing and computer graphics, it is sometimes desirable to form geometric-shaped surfaces onto a xy-plane. By applying a mathematical transformation, the nondecreasing structuring element can be converted into a decreasing element. Therefore, the morphological erosion by the new transformed structuring element followed by an inverse transformation may yield such a geometric-shaped surface.

A two-dimensional cyclic cosine structuring element $k(x,y)$ is given by

$$k(x,y) = \cos(\sqrt{x^2 + y^2}) - 1 \tag{9.39}$$

When the domain of $\sqrt{x^2 + y^2}$ exceeds a cycle 2π, $k(x,y)$ will become nondecreasing. By applying a transformation $\Psi[|k(x,y)|] = (\cos^{-1}(-(|k(x,y)| - 1)))^2$, a new structuring element is constructed, such that $k^*(x,y) = -\Psi[|k(x,y)|] = -x^2 - y^2$ which is decreasing and convex. The inverse transformed morphological erosion $\Psi^{-1}[f \ominus k^*]$, as illustrated in Figure 9.7, is equal to

$$\Psi^{-1}[f \ominus k^*] = -\cos(\theta(f \ominus k^*)^{1/2}) + 1 \tag{9.40}$$

where θ denotes an angle. When $x = 0$ in the cosine function, it represents $x = 1$ in k^*. In the following decomposition, we describe the procedure by illustrating a 17×17 two-dimensional cyclic cosine structuring element.

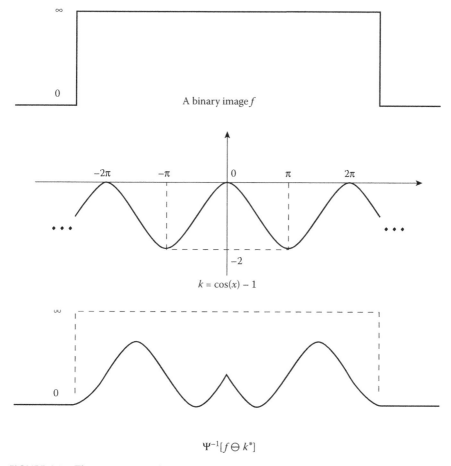

FIGURE 9.7 The inverse transformation of the eroded result of f by k^*, such that k^* is the new transformed cyclic structuring element of k.

Many large-sized one-dimensional or two-dimensional cyclic geometric-shaped SEs can be obtained by applying the similar procedure.

The new structuring element $k^*(x,y) = -(x^2 + y^2)$, which is symmetric to its center, can be numerically expressed for simplicity by its upper-right quadrant using the unit $\theta = \pi/4$ in the cosine function as

$$
\begin{array}{rrrrrrrrr}
-64 & -65 & -68 & -73 & -80 & -89 & -100 & -113 & -128 \\
-49 & -50 & -53 & -58 & -65 & -74 & -85 & -98 & -113 \\
-36 & -37 & -40 & -45 & -52 & -61 & -72 & -85 & -100 \\
-25 & -26 & -29 & -34 & -41 & -50 & -61 & -74 & -89 \\
-16 & -17 & -20 & -25 & -32 & -41 & -52 & -65 & -80 \\
-9 & -10 & -13 & -18 & -25 & -34 & -45 & -58 & -73 \\
-4 & -5 & -8 & -13 & -20 & -29 & -40 & -53 & -68 \\
-1 & -2 & -5 & -10 & -17 & -26 & -37 & -50 & -65 \\
0 & -1 & -4 & -9 & -16 & -25 & -36 & -49 & -64
\end{array}
$$

The erosion of a binary image f by the structuring element k^* is

$$f \ominus k^* = ((((((f \ominus k_1) \ominus k_2) \ominus k_3) \ominus k_4) \ominus k_5) \ominus k_6) \ominus k_7) \ominus k_8 \tag{9.41}$$

where

$$
k_1 = \begin{bmatrix} -2 & -1 & -2 \\ -1 & 0 & -1 \\ -2 & -1 & -2 \end{bmatrix}, \quad
k_2 = \begin{bmatrix} -6 & -3 & -6 \\ -3 & 0 & -3 \\ -6 & -3 & -6 \end{bmatrix}, \quad
k_3 = \begin{bmatrix} -10 & -5 & -10 \\ -5 & 0 & -5 \\ -10 & -5 & -10 \end{bmatrix},
$$

$$
k_4 = \begin{bmatrix} -14 & -7 & -14 \\ -7 & 0 & -7 \\ -14 & -7 & -14 \end{bmatrix}, \quad
k_5 = \begin{bmatrix} -18 & -9 & -18 \\ -9 & 0 & -9 \\ -18 & -9 & -18 \end{bmatrix}, \quad
k_6 = \begin{bmatrix} -22 & -11 & -22 \\ -11 & 0 & -11 \\ -22 & -11 & -22 \end{bmatrix},
$$

$$
k_7 = \begin{bmatrix} -26 & -13 & -26 \\ -13 & 0 & -13 \\ -26 & -13 & -26 \end{bmatrix}, \quad \text{and} \quad
k_8 = \begin{bmatrix} -30 & -15 & -30 \\ -15 & 0 & -15 \\ -30 & -15 & -30 \end{bmatrix}.
$$

Hence,

$$\Psi^{-1}[f \ominus k^*] = -\cos\left[\frac{\pi}{4}(f \ominus k^*)^{1/2}\right] + 1 \tag{9.42}$$

9.1.6 Decomposition of Two-Dimensional SEs into One-Dimensional Elements

According to Proposition 9.6, we observe that many two-dimensional geometric-shaped SEs in the form of $k(x, y) = g(p(x) + q(y)) + c$, can be transformed into $k^*(x, y)$ which is additively-separable and convex. This kind of structuring element $k^*(x, y)$ can be further decomposed into a dilation of two one-dimensional structuring components, one in x-direction and the other in y-direction, and both are convex. Furthermore, these one-dimensional convex SEs can be again decomposed into a series of dilations of structuring subcomponents of size 3.

Example 9.9: Let $k^*(x, y)$ be

$$k^* = \begin{bmatrix} -8 & -5 & -4 & -5 & -8 \\ -5 & -2 & -1 & -2 & -5 \\ -4 & -1 & 0 & -1 & -4 \\ -5 & -2 & -1 & -2 & -5 \\ -8 & -5 & -4 & -5 & -8 \end{bmatrix}$$

The $k^*(x, y)$ can be decomposed as $k^*(x, y) = k_1(x) \oplus k_2(y)$, where

$$k_1(x) = \begin{bmatrix} -4 & -1 & 0 & -1 & -4 \end{bmatrix}, \quad k_2(y) = \begin{bmatrix} -4 \\ -1 \\ 0 \\ -1 \\ -4 \end{bmatrix}$$

The $k_1(x)$ and $k_2(y)$ can be further decomposed into

$$k_1(x) = \begin{bmatrix} -1 & 0 & -1 \end{bmatrix} \oplus \begin{bmatrix} -3 & 0 & -3 \end{bmatrix}$$

$$k_2(y) = \begin{bmatrix} -1 \\ 0 \\ -1 \end{bmatrix} \oplus \begin{bmatrix} -3 \\ 0 \\ -3 \end{bmatrix}$$

9.2 Decomposition of Binary SEs

MM, which is based on set-theoretic concept, provides an efficient tool to image processing and analysis [Serra 1982; Haralick, Sternberg, and Zhuang

1987; Shih and Mitchell 1989; Shih and Puttagunta 1995]. Most image processing architectures adapted to morphological operations use SEs of a limited size. Implementation becomes difficult when a large sized structuring element is used. Many techniques have been proposed for the decomposition of binary SEs. Zhuang and Haralick [1986] presented a tree-search algorithm for decomposing binary SEs. Park and Chin [1994, 1995] proposed algorithms for decomposing convex SEs with an extension to arbitrary shape, but they are complicated and have indecomposable cases. Hashimoto and Barrera [2002] improved that with decomposition on some indecomposable SEs. Richardson and Schafer [1991] presented a lower bound for structuring element decompositions. Anelli, Broggi, and Destri [1998] developed a method for decomposing arbitrarily shaped binary SEs using genetic algorithms (GAs). However, its disadvantages are that it takes too much time to obtain a stochastic solution and the result is not optimal. Yang and Lee [2005] presented the decomposition of morphological SEs with integer linear programming.

We present an efficient technique for decomposing arbitrary binary SEs using GAs to obtain better results. We first present the overview of decomposition using GAs, and points out the advantages of structuring element decomposition. We then present the decomposition technique using GAs and some experimental results.

9.2.1 Overview of Decomposition Using GAs

To decompose arbitrarily shaped binary SEs efficiently, we adapt Anellis's method [1998]. We use Park's 13 prime factors [Park and Chin 1994] as the initial population in GAs and recursively reduce sizes. We only discuss the decomposition of odd-sized SEs. Even-sized SEs can be changed to be odd-sized by adding a column and a row.

Let $S_{N \times N}$ denote a structuring element of size $N \times N$; for notation simplicity, we use S_{NN}. If $N \geq 5$ and the structuring element is decomposable by dilations, we can obtain two factors, $S_{(N-2)(N-2)}$ and S_{33}, in the first iteration; otherwise, a third factor S_{NN}^{*} is generated such that $S_{NN} = (S_{(N-2)(N-2)} \oplus S_{33}) \cup S_{NN}^{*}$. Note that in the computational ordering, morphological operators receive a higher priority than set operators. Furthermore, S_{NN}^{*} can be decomposed into union of factors of 3×3. Then, $S_{(N-2)(N-2)}$ can be recursively decomposed until every factor is of size 3×3. The size of 3×3 is often used as the elementary structuring component for decomposition. The decomposition algorithm is presented below and its flowchart is shown in Figure 9.8. Note that the prime component, denoted by S_{33}^{PC}, is obtained from Park's 13 prime factors. The overview of the decomposition algorithm is presented as follows.

1. Generate the initial population, $S_{(N-2)(N-2)}$ and S_{33}^{PC}, using Park's 13 prime factors 2.

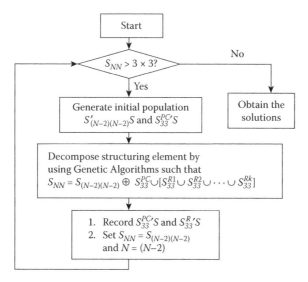

FIGURE 9.8 The flowchart of the decomposition algorithm.

2. Use the fitness functions in GAs to proceed the decomposition, such that

$$S_{NN} = S_{(N-2)(N-2)} \oplus S_{33}^{PC} \cup (S_{33}^{R1} \cup S_{33}^{R2} \cup \cdots \cup S_{33}^{Rk}). \quad (9.43)$$

3. Repeat steps 1 and 2 for further decomposition until all factors are of size 3×3.

Note that when using the morphological property, we can further reduce the decomposition into the form of unions of dilations of smaller SEs as

$$S^1 \oplus (S^2 \cup S^3) = (S^1 \oplus S^2) \cup (S^1 \oplus S^3). \quad (9.44)$$

Therefore, the decomposition result is in the form of $S_{NN} - \cup_{i=1}^{m} B_i$, where $B_i = SE_{33}^{PC\_1} \oplus SE_{33}^{PC\_2} \oplus \cdots \oplus SE_{33}^{PC\_k} \oplus SE_{33}^{R}$, $k \geq 0$, and m and k are integers.

9.2.2 Advantages of Structuring Element Decomposition

Let A denote a binary image and S denote a structuring element. If we decompose S into $S_1 \oplus S_2 \oplus \cdots \oplus S_k$, the dilation and erosion become

Dilation:

$$A \oplus S = A \oplus (S_1 \oplus S_2 \oplus \cdots \oplus S_k) = (\cdots ((A \oplus S_1) \oplus S_2) \oplus \cdots) \oplus S_k, \quad (9.45)$$

Erosion:

$$A \ominus S = A \ominus (S_1 \oplus S_2 \oplus \cdots \oplus S_k) = (\cdots ((A \ominus S_1) \ominus S_2) \ominus \cdots) \ominus S_k. \quad (9.46)$$

Suppose that sizes of A and S are $N \times N$ and $M \times M$, respectively. The time complexity for the dilation or erosion operator is $N^2 \times M^2$. For example, if we decompose a structuring element of size 13×13 into six small structuring components of size 3×3, the computational cost will decrease from $169N^2$ to $54N^2$.

For a parallel pipelined architecture, we decompose the structuring element S using the union together with the dilation. For example, S is decomposed as

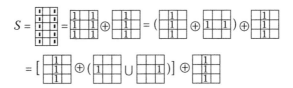

We further decompose the above decomposition as follows, and its parallel pipelined architecture is shown in Figure 9.9.

9.2.3 The Decomposition Technique Using GAs

GAs, introduced by Hollard [1975] in his seminal work, are commonly used as adaptive approaches to provide randomized, parallel, and global searching

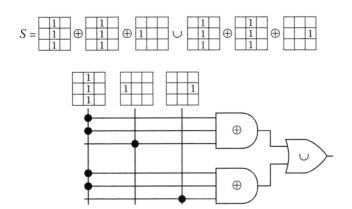

FIGURE 9.9 An example showing parallel implementation.

based on the mechanics of natural selection and natural genetics in order to find solutions of a problem. There are four distinctions between normal optimization and search procedures [Herrera, Lozano, and Verdegay 1994]: (a) GAs work with a coded parameter set, not the parameters themselves; (b) GAs search from randomly selected points, not from a single point; (c) GAs use the objective function information; (d) GAs use probabilistic transition rules, not deterministic ones.

Although GAs have been developed in many kinds [Ho, Chen, and Shu 1999; Tang et al. 1998], the fundamental of GAs is based on the Simple Genetic Algorithm (SGA) [Holland 1992]. In general, GAs start with some randomly selected population, called the first generation. Each individual, called chromosome, in the population corresponds to a solution in the problem domain. An objective, called fitness function is used to evaluate the quality of each chromosome. Next generation will be generated from some chromosomes whose fitness values are high. Reproduction, crossover, and mutation are the three basic operators used to reproduce chromosomes to obtain the best solution. The process will repeat for many times until a predefined condition is satisfied or a constant number of iterations is reached.

To apply GAs in decomposition efficiently, the dynamic varying chromosomes are developed. That is, the length of chromosomes gradually decreases as the size of SEs decreases. Each chromosome consists of two sub-chromosomes: One, ξ^F, of fixed size and the other, ξ^V, of variable size. We scan a two-dimensional structuring element into a one-dimensional string of chromosome. For example, if a structuring element is of size 7×7, we choose a chromosome ξ consisting of ξ^V and ξ^F of sizes 5^2 and 3^2 genes, respectively.

$$\xi = \xi^V \ || \ \xi^F, \ \xi^V = g_0 g_1 \cdots g_{24}, \quad \text{and} \quad \xi^F = g_0 g_1 \cdots g_8$$

where "$||$" indicates concatenation.

Figures 9.10a and b illustrate the numbering scheme for $\xi^V(g_i(0 \le i \le 24))$ and $\xi^F(g_i(0 \le i \le 8))$. Their corresponding SEs are shown in Figures 9.10c and d. The one-dimensional string of chromosomes ξ^V and ξ^F are $\xi^V = 1101010100010001001110010$ and $\xi^F = 010111010$, respectively.

(a)

0	1	2	3	4
5	6	7	8	9
10	11	12	13	14
15	16	17	18	19
20	21	22	23	24

(b)

0	1	2
3	4	5
6	7	8

(c)

1	1	0	1	0
1	0	1	0	0
0	1	0	0	0
1	0	0	1	1
1	0	0	1	0

(d)

0	1	0
1	1	1
0	1	0

FIGURE 9.10 The positions correspond to the genes.

For general-purpose serial computers, the computational complexity of a dilation of two SEs is related to the total number of non-zero points in the elements. To minimize the complexity for serial and parallel implementations, two fitness functions are used. Suppose that a structuring element S is decomposed into three parts as $S = S' \oplus S^F \cup S^R$, the first fitness function is defined as

$$Evaluation\_1(\xi) = \sum_{i=0}^{all\ pixels} (S_i' + S_i^F + S_i^R), \qquad (9.47)$$

where ξ is a chromosome, S' and S^F are the decomposed SEs, and S^R is the difference between S and $S' \oplus S^F$. Note that if $S = S' \oplus S^F$, all pixels in S^R are 0. The first fitness function, which counts all non-zero points in the decomposed SEs, is aimed at minimization. The second fitness function is defined as

$$Evaluation\_2(\xi) = \sum_{i=0}^{all\ pixels} (S_i - S_i^{New}), \qquad (9.48)$$

where S^{New} is the dilation of the two decomposed SEs, that is, $S' \oplus S^F$. Similarly, the second fitness function, which counts the difference between the original S and the current $S' \oplus S^F$, is also aimed at minimization. The threshold value for determining whether ξ is good or not can be automatically selected. In the first iteration, the minimum fitness value is used as the threshold. In the following iterations, if there exists a smaller fitness value, the average of this value and the current threshold will be used as the new threshold. The good chromosomes will be adopted for further operations in reproduction, crossover, and mutation.

Reproduction:

$$Reproduction(\xi) = \{\xi_i \mid Evaluation(\xi_i) \leq \Omega \text{ and } \xi_i \in \Psi\}, \qquad (9.49)$$

where Ω is used to sieve chromosomes and Ψ is the population. We reproduce those good chromosomes below Ω from population.

Crossover:

$$Crossover(\xi) = \{\xi_i \, \varpi \, \xi_j \mid \xi_i, \xi_j \in \Psi\}, \qquad (9.50)$$

where ϖ denotes the operation that reproduces ξ by exchanging genes from their parents, ξ_i and ξ_j. The often-used crossovers are one-point, two-point, and multi-point crossovers. Figure 9.11 provides an example of one-point crossover in a 3×3 structuring element.

	Chromosome			Structuring element	
				SE-A	SE-B
Before crossover	Chromosome A	10000	1110	1 0 0 ↔ 0 1 0 / 0 0 1 1 1 0 / 1 1 0 0 1 1	
	Chromosome B	01011	0011		
				SE-A	SE-B
After crossover	Chromosome A	01011	1110	0 1 0 1 0 0 / 1 1 1 0 0 0 / 1 1 0 0 1 1	
	Chromosome B	10000	0011		

FIGURE 9.11 The one-point crossover in a 3×3 structuring element.

Mutation:

$$Mutation(\xi) = \xi_i \, \vartheta \, j \mid 0 \leq j \leq Max\_Length(\xi_i) \text{ and } \xi_i \in \Psi\}, \qquad (9.51)$$

where ϑ is the operation by randomly selecting chromosome ξ_i from Ψ and randomly changing bits from ξ_i. The often-used mutations are one-point, two-point, and multi-point mutations. Figure 9.12 illustrates a one-point mutation in a 3×3 structuring element.

The Decomposition Algorithm:

1. Check the size of the structuring element S. If it is 3×3, set $S^F = S$ and go to step 7.
2. Generate the initial population using Park's 13 prime factors [Park and Chin 1994].
3. Evaluate two fitness values for each chromosome and update the threshold values to obtain better chromosomes.
4. Reproduce new chromosomes using crossover and mutation.

	Chromosome		Structuring element
Before mutation	Chromosome A	1000 0 1110 ▲	1 0 0 / 0 0 1 / 1 1 0
After mutation	Chromosome A	1000 1 1110 ▲	1 0 0 / 0 1 1 / 1 1 0

FIGURE 9.12 One-point and two-point mutations.

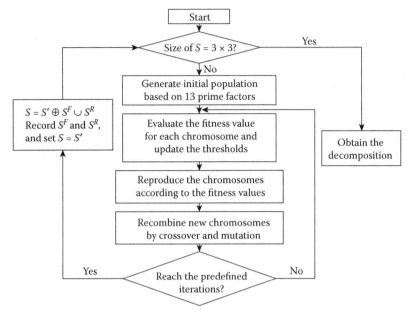

FIGURE 9.13   The flowchart of the decomposition algorithm.

5. If the preset limited number of iterations is not reached yet, go to step 3.

6. Record the decomposition such that $S = S' \oplus S^F \cup S^R$, record the decomposed SEs S^F and S^R, and set $S = S'$. Go to step 1.

7. Output all of the decomposed SEs S^Fs and S^Rs.

The flowchart of the decomposition algorithm is shown in Figure 9.13.

9.2.4 Experimental Results

We have successfully tested the decomposition algorithm on many examples. This section shows some results and the comparisons with the Hashimoto and Barrera [2002] and Anelli, Broggi, and Destri [1998] methods.

Example 9.10: The structuring element shown in Figure 9.14 is indecomposable in Park's algorithm [Park and Chin 1995, Example 3]. Using the decomposition algorithm, we can find the same decomposition as in Hashimoto's algorithm [Hashimoto and Barrera 2002], as shown in Figure 9.14a. Moreover, we can obtain a better decomposition, as shown in Figure 9.14b, which is suited for parallel implementation. In Figures 9.14a and b, the original 21 non-zero points in the structuring element are reduced to 10. Note that the origin of the structuring element is placed at the center. In Figure 9.14b, a structuring element is shifted by (−1, −2). An illustration of translation

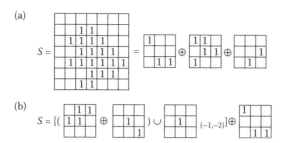

FIGURE 9.14   An example of decomposing a structuring element.

is shown in Figure 9.15, where: (a) denotes the geometric coordinates; (b) an extension to 7×7; and (c) and (d) are, respectively, a translation by (2, 2) and (–2, 1).

Example 9.11: Figure 9.16 shows the same example as used in the Anelli, Broggi, and Destri [1998] paper, where they demonstrated the decomposition of using 22 points in the structuring element and 8 unions. We run the decomposition algorithm and obtain the following result, which needs 18 points in the structuring element and 3 unions.

Example 9.12: Figure 9.17 shows the decomposition of a 17×17 structuring element. Anelli, Broggi, and Destri [1998] presented the decomposition of a 16×16 structuring element on a two processors HP 9000 with 128 megabytes of RAM, which took about six hours of CPU time. We run the proposed decomposition algorithm on a Pentium III/866 MHz processor with 256 megabytes of RAM, and it takes about 20 minutes of CPU time. The decomposition requires 39 points and 8 unions.

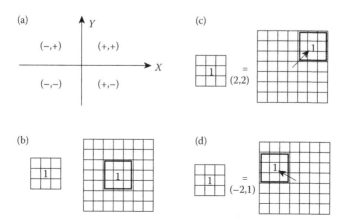

FIGURE 9.15   The translation scheme.

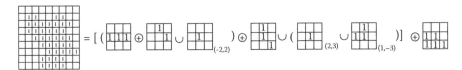

FIGURE 9.16 The same example as used in the Anelli, Broggi, and Destri [1998] paper.

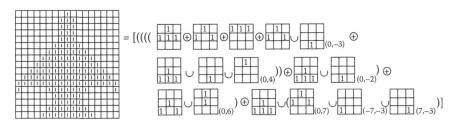

FIGURE 9.17 The decomposition of a 17 × 17 structuring element.

Example 9.13: Figure 9.18 shows six examples of different shapes, such as dark, ship, car, fish, lamp and telephone. Note that the telephone with a hollow illustrates a difficult case in decomposition. Table 9.3 lists the computational cost comparisons of the original and decomposed SEs.

9.3 Decomposition of Grayscale SEs

Let $f: F \rightarrow E$ and $k: K \rightarrow E$ denote the grayscale image and the grayscale structuring element, respectively. The symbols, E, F, and K, respectively, represent Euclidean space, bounded domain for f, and bounded domain for k. The *grayscale dilation* of f by k is given by

$$(f \oplus k)(x) = \max_{x-z \in F, z \in K} [f(x-z) + k(z)]. \tag{9.52}$$

The *grayscale erosion* of f by k is given by

$$(f * k)(x) = \min_{x+z \in F, z \in K} [f(x+z) - k(z)]. \tag{9.53}$$

Most image processing architectures adapted to morphological operations use the SEs of a small size. One advantage of the structuring element decomposition is to allow the freedom of choosing any size of SEs in implementing morphological algorithms on any image processing architecture. For example,

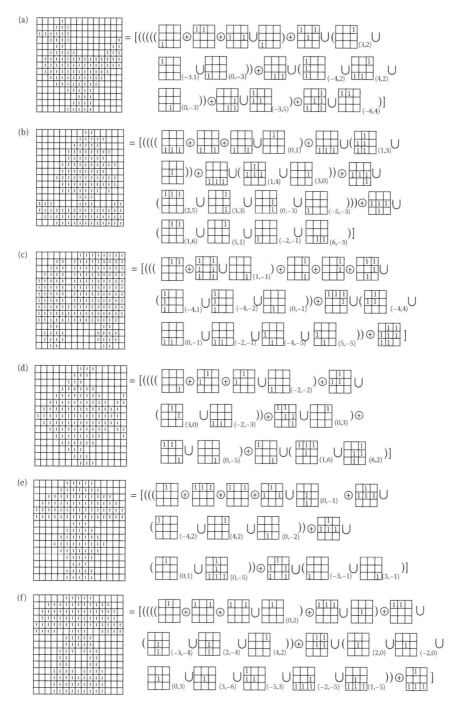

FIGURE 9.18 Examples of different shapes of (a) dark, (b) ship, (c) car, (d) fish, (e) lamp, and (f) telephone.

TABLE 9.3

Cost Comparisons of the Original and Decomposed SEs

	(a) Dark	(b) Ship	(c) Car	(d) Fish	(e) Lamp	(f) Telephone
Original points	102	125	168	112	115	127
Original union	0	0	0	0	0	0
Decomposed points	33	47	46	33	35	42
Decomposed union	9	13	9	7	8	12

the Cytocomputer cannot process any template of size larger than 3×3 on each pipeline stage. Besides, the small-sized structuring element will simplify the design of the image processing architectures. Another advantage is the reduction of computational time. For example, if we decompose a structuring element of size $(p + q - 1) \times (p + q - 1)$ into a dilation of two smaller structuring components of sizes $p \times p$ and $q \times q$, the number of calculations required in morphological operations will be reduced from $(p + q - 1)^2$ to $p^2 + q^2$. Furthermore, if a very big structuring element is recursively decomposed into a sequence of small ones, the computational time will be reduced dramatically. For example, if a structuring element of size $n \times n$ (where n is an odd number) is decomposed into smaller ones of size 3×3, the calculations required in morphological operations will be reduced from n^2 to $3^2(n - 1)/2$.

In this section, we focus on the grayscale structuring element decomposition. Camps, Kanungo, and Haralick [1996] developed decomposition of grayscale SEs by the tree search approach with Umbra and Top operations. Gader [1991] presented the separable decomposition and the approximation of grayscale SEs into horizontal and vertical SEs. If a structuring element is not separable, the approximation of separable decomposition is adopted. Shih and Mitchell [1991] introduced a strategy for decomposing certain types of grayscale SEs into dilations or maximum selections of small components. Shih and Wu [1992] developed the decomposition of geometric-shaped SEs using morphological transformations. However, their method can only be used to decompose those SEs with specific geometric features. Sussner and Ritter [1997, 2000] presented the decomposition of grayscale morphological templates using the rank method. Their algorithm has to guess an outer product representation, which may cause a failure if a wrong guess were made. Furthermore, the computation of intensive search procedure is time-consuming to affect the performance of their algorithm. Sussner, Pardalos, and Ritter [1997] utilized an integer programming approach to prove that the decomposition of grayscale SEs is an NP-completeness problem. Shih and Mitchell [1992] applied the strategy of decomposing the Euclidean structuring element for computing EDT efficiently.

We present the decomposition algorithms for arbitrary grayscale SEs without going through additional pre-processes, such as checking the type of SEs and the decomposition rules. It is organized as follows. We first introduce the general properties of structuring element decomposition. Then we describe the one-dimensional decomposition and the two-dimensional decomposition. We also provide the complexity analysis.

9.3.1 General Properties of Structuring Element Decomposition

The structuring element decomposition properties described in this section are suited for binary (with two values, "0" indicating background and "1" indicating foreground) and grayscale images.

Property 9.6: If a structuring element k can be decomposed into sequential dilations of smaller structuring components, then a dilation (erosion, respectively) of an image f by k is equal to successively applied dilations (erosions, respectively) of the previous stage output by these structuring components. That is, if $k = k_1 \oplus k_2 \oplus \cdots \oplus k_n$, then

$$f \oplus k = (\cdots ((f \oplus k_1) \oplus k_2) \oplus \cdots) \oplus k_n \quad \text{and}$$
$$f \ominus k = (\cdots ((f \ominus k_1) \ominus k_2) \ominus \cdots) \ominus k_n. \tag{9.54}$$

Property 9.7: If a structuring element k can be decomposed into a maximum selection over several structuring components, then a dilation (erosion, respectively) of an image f by k is equal to finding a maximum (minimum, respectively) over each applied dilation (erosion, respectively) of the image by these structuring components. That is, if $k = \max(k_1, k_2, \ldots, k_n)$, then

$$f \oplus k = \max(f \oplus k_1, f \oplus k_2, \ldots, f \oplus k_n) \quad \text{and}$$
$$f \ominus k = \min(f \ominus k_1, f \ominus k_2, \ldots f \ominus k_n). \tag{9.55}$$

Property 9.8: Let $f_{(x)}$ denote an image f being shifted by x. The dilation and erosion satisfy the translation invariance property as

$$f_{(x)} \oplus k = (f \oplus k)_{(x)} \quad \text{and} \quad f_{(x)} \ominus k = (f \ominus k)_{(x)}. \tag{9.56}$$

Property 9.9: Let u denote a constant. The dilation and erosion satisfy

$$f \oplus k = (f(x) - u) \oplus (k(z) + u) \quad \text{and} \quad f \ominus k = (f(x) - u) \ominus (k(z) - u). \tag{9.57}$$

Example 9.14: Let $S = [8, 9, 10, 11, 10]$. If we decompose S into $[8, 9, 10] \oplus [0, 1, 0]$, according to Property 9.9, the decomposed results could be $[4, 5, 6] \oplus [4, 5, 4]$, $[10, 11, 12] \oplus [-2, -1, -2]$, and so on.

9.3.2 The One-Dimensional Arbitrary Grayscale Structuring Element Decomposition

The decomposition of certain types of SEs such as linearly-sloped, eccentrically convex, and eccentrically concave, was developed by Shih and Mitchell

[1991]. Furthermore, the decomposition of certain geometric-shaped SEs such as sine, cosine, semiellipse, parabola, semihyperbola, and Gaussian was presented by Shih and Wu [1992]. In this section, we introduce a general decomposition strategy to decompose one-dimensional arbitrary grayscale SEs.

Let the origin of the structuring element be placed at the leftmost point. The decomposition of a structuring element of odd size n can be expressed as

$$[a_1, a_2, a_3, \ldots, a_n] = [b_1^1, b_2^1, b_3^1] \oplus [b_1^2, b_2^2, b_3^2] \oplus \cdots \oplus [b_1^{(n-1)/2}, b_2^{(n-1)/2}, b_3^{(n-1)/2}],$$

$$(9.58)$$

and the decomposition of a structuring element of even size m can be expressed as

$$[a_1, a_2, a_3, \ldots, a_m] = [c_1, c_2] \oplus [b_1^1, b_2^1, b_3^1] \oplus [b_1^2, b_2^2, b_3^2]$$
$$\oplus \cdots \oplus [b_1^{(m-2)/2}, b_2^{(m-2)/2}, b_3^{(m-2)/2}], \quad (9.59)$$

where a_i, b_j^k, and c_l denote the factors in the SEs. Note that, b_j^k indicates the jth factor in the kth structuring element, where j ($1 \leq j \leq 3$) and k ($1 \leq k \leq (n-1)/2$) are integers. For example, a structuring element of size 7 will be decomposed into dilations of three small SEs of size 3, and b_2^1 will be the second factor in the first small structuring element.

Proposition 9.8: If a structuring element satisfies the following two conditions, it can be decomposed by Equations 9.58 and 9.59. Let $S_N = [a_1, a_2, a_3, \ldots, a_c, \ldots, a_N]$, where a_c denotes an apex element, and N denotes an odd number. The two conditions are:

(a) $a_1 \leq a_2 \leq a_3 \leq \cdots \leq a_c$ and $a_c \geq a_{c+1} \geq \cdots \geq a_N$

(b) $a_2 - a_1 \geq a_3 - a_2 \geq \cdots \geq a_c - a_{c-1}$ and $a_c - a_{c+1} \leq a_{c+1} - a_{c+2} \leq \cdots \leq a_{N-1} - a_N$.

Proof: Let S_N be decomposed into S_{N-2} and S_3, where $S_{N-2} = [a_1, a_2, a_3, \ldots, a_{N-2}]$ obtained by duplicating the first $N-2$ elements in S_N, and $S_3 = [c_1, c_2, c_3]$ is computed using $c_1 = 0$, $c_2 = a_{N-1} - a_{N-2}$ and $c_3 = a_N - a_{N-2}$. We will prove that by choosing S_{N-2} and S_3 in this way, the equality of $S_N = S_{N-2} \oplus S_3$ holds. Let a_{y-2}, a_{y-1}, and a_y denote any three adjacent elements in S_{N-2}. By looking at the location of a_y in calculating $S_{N-2} \oplus S_3$ as shown in Figure 9.19, we obtain the three values: a_y, $(a_{N-1} - a_{N-2}) + a_{y-1}$, and $(a_N - a_{N-2}) + a_{y-2}$. We need to prove that a_y is the maximum value. By subtracting a_y from the three values, we obtain 0, $(a_{N-1} - a_{N-2}) + (a_{y-1} - a_y)$, and $(a_N - a_{N-2}) + (a_{y-2} - a_y)$. Therefore, we will prove that $(a_{N-1} - a_{N-2}) + (a_{y-1} - a_y) \leq 0$ and $(a_N - a_{N-2}) + (a_{y-2} - a_y) \leq 0$ in the following three cases:

1. If a_{y-2} is on the right hand side of a_c (i.e., a_{N-2} is on the right hand side of a_c): From condition (a), we know $a_c \geq a_{y-2} \geq a_{y-1} \geq a_y \geq \cdots \geq a_{N-2}$

$$
\begin{array}{ccc}
& C_3 & C_2 \quad C_1 \\
S_3 = [\; \boxed{a_N - a_{N-2}} & , \; \boxed{a_{N-1} - a_{N-2}} , \; \boxed{0}\;]
\end{array}
$$

(a) In element a_y: $S_{N-2} = [\; a_1, a_2, \ldots, \; \boxed{a_{y-2}} \; , \; \boxed{a_{y-1}} , \; \boxed{a_y} , \ldots, a_{N-2}]$

(b) Dilation result: $\max \left\{ \begin{array}{l} a_{y-2} + (a_N - a_{N-2}), \\ a_{y-1} + (a_{N-1} - a_{N-2}) \\ a_y + 0 \end{array} \right\} = \max \left\{ \begin{array}{l} (a_{y-2} - a_y) + (a_N - a_{N-2}), \\ (a_{y-1} - a_y) + (a_{N-1} - a_{N-2}) \\ 0 \end{array} \right\}$

FIGURE 9.19 Dilation of structuring elements to illustrate Proposition 9.8.

$\geq a_{N-1} \geq a_N$. By combining with condition (b), we obtain $a_{y-1} - a_y \leq a_{N-2} - a_{N-1}$ and $a_{y-2} - a_{y-1} \leq a_{N-1} - a_N$. The summation of the above two inequalities gives $a_{y-2} - a_y \leq a_{N-2} - a_N$. Therefore, we conclude that $(a_{N-1} - a_{N-2}) + (a_{y-1} - a_y) \leq 0$ and $(a_N - a_{N-2}) + (a_{y-2} - a_y) \leq 0$.

2. If a_y and a_N are both on the left hand side of a_c: We obtain $a_{y-1} - a_y \leq 0$ and $a_{y-2} - a_y \leq 0$ because $a_1 \leq \cdots \leq a_{y-2} \leq a_{y-1} \leq a_y \leq a_c$. We obtain $a_N - a_{N-1} \leq a_y - a_{y-1}$ and $a_N - a_{N-2} \leq a_y - a_{y-2}$ because $a_2 - a_1 \geq a_3 - a_2 \geq \cdots \geq a_c - a_{c-1}$ and $a_1 \leq \cdots \leq a_{y-2} \leq a_{y-1} \leq a_y \leq \cdots \leq a_{N-2} \leq a_{N-1} \leq a_N \leq a_c$. Therefore, we conclude that $(a_{N-1} - a_{N-2}) + (a_{y-1} - a_y) \leq 0$ and $(a_N - a_{N-2}) + (a_{y-2} - a_y) \leq 0$.

3. If a_y is on the left hand side of a_c, but a_{N-2} is on the right hand side of a_c: We obtain $a_{N-1} - a_{N-2} \leq 0$ and $a_N - a_{N-2} \leq 0$ because $a_c \geq a_{c+1} \geq \cdots \geq a_N$. We obtain $a_{y-1} - a_y \leq 0$ and $a_{y-2} - a_y \leq 0$ because $a_1 \leq \ldots \leq a_{y-2} \leq a_{y-1} \leq a_y \leq a_c$. Therefore, we conclude that $(a_{N-1} - a_{N-2}) + (a_{y-1} - a_y) \leq 0$ and $(a_N - a_{N-2}) + (a_{y-2} - a_y) \leq 0$.

We continuously decompose S_{N-2} using the same scheme until the decomposed size is 3. ◻

Note that the two conditions listed in Proposition 9.8 are called *convex eccentrically decreasing* in Shih and Wu [1992]. If a structuring element belongs to this type, it can be decomposed by Equations 9.58 and 9.59. However, the decomposition strategy in general cases for arbitrary grayscale SEs is different from the one in Shih and Wu [1992] which is designed for geometric-shaped SEs.

We can decompose an arbitrary grayscale structuring element into the formats as shown in Equations 9.60 and 9.61, where n and m denote odd and even numbers, respectively, and × denotes don't care. Note that × is equivalent to $-\infty$ in MM. The assumptions (a) and (b) are a particular case in the general decomposition of arbitrary grayscale SEs. The general algorithm does not use assumptions (a) and (b). By applying the presented algorithm, if a structuring element satisfies assumptions (a) and (b), then the dilated result of the two decomposed components will be the same as the original structuring element, so the maximum operation will not be needed. As a result, the decomposed components are equivalent to those of using Equations 9.58 and 9.59.

$$
[a_1, a_2, a_3, \ldots, a_n]
$$

$$
= \max \begin{cases}
([b_1^{(1,1)}, b_2^{(1,1)}, b_3^{(1,1)}] \oplus [b_1^{(1,2)}, b_2^{(1,2)}, b_3^{(1,2)}] \oplus \cdots \oplus [b_1^{(1,(n-1)/2)}, b_2^{(1,(n-1)/2)}, b_3^{(1,(n-1)/2)}])_{(0)}, \\
([b_1^{(2,1)}, b_2^{(2,1)}, b_3^{(2,1)}] \oplus [b_1^{(2,2)}, b_2^{(2,2)}, b_3^{(2,2)}] \oplus \cdots \oplus [b_1^{(2,(n-3)/2)}, b_2^{(2,(n-3)/2)}, b_3^{(2,(n-3)/2)}])_{(2)}, \\
\vdots \\
([b_1^{((n-1)/2,1)}, b_2^{((n-1)/2,1)}, b_3^{((n-1)/2,1)}])_{(n-3)}
\end{cases}
$$

$$(9.60)$$

$$
[a_1, a_2, a_3, \cdots, a_m]
$$

$$
= \max \begin{cases}
([c_1, c_2] \oplus ([b_1^{(1,1)}, b_2^{(1,1)}, b_3^{(1,1)}] \oplus [b_1^{(1,2)}, b_2^{(1,2)}, b_3^{(1,2)}] \oplus \cdots \oplus [b_1^{(1,(m-2)/2)}, b_2^{(1,(m-2)/2)}, b_3^{(1,(n-2)/2)}])_{(0)}, \\
([b_1^{(2,1)}, b_2^{(2,1)}, b_3^{(2,1)}] \oplus [b_1^{(2,2)}, b_2^{(2,2)}, b_3^{(2,2)}] \oplus \cdots \oplus [b_1^{(2,(m-2)/2)}, b_2^{(2,(m-2)/2)}, b_3^{(2,(m-2)/2)}])_{(0)}, \\
([b_1^{(3,1)}, b_2^{(3,1)}, b_3^{(3,1)}] \oplus [b_1^{(3,2)}, b_2^{(3,2)}, b_3^{(3,2)}] \oplus \cdots \oplus [b_1^{(3,(m-4)/2)}, b_2^{(3,(m-4)/2)}, b_3^{(3,(m-4)/2)}])_{(2)}, \\
\vdots \\
([b_1^{(m/2,1)}, b_2^{(m/2,1)}, b_3^{(m/2,1)}])_{(m-4)}, \\
([\times, \times, e])_{(m-3)}
\end{cases}
$$

$$(9.61)$$

where e, c_1, c_2, and $b_j^{(t,k)}$ denote the factors in the SEs. Note that, $b_j^{(t,k)}$ indicates the jth factor in the kth structuring element of the tth decomposition.

Let S_N denote a structuring element of size N. The algorithm for the one-dimensional structuring element decomposition is presented below, and its flowchart is shown in Figure 9.20.

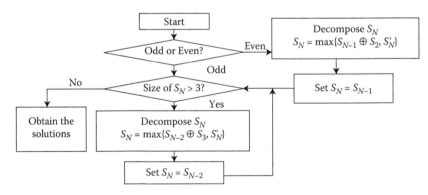

FIGURE 9.20 The flowchart of the one-dimensional structuring element decomposition.

The One-Dimensional Decomposition Algorithm:

1. If S_N is odd sized, go to step 5.

2. Decompose $S_N = [a_1, a_2, a_3, \ldots, a_N]$ into S_{N-1} and S_2, where $S_{N-1} = [a_1, a_2, a_3, \ldots, a_{N-1}]$ is obtained by duplicating the $N - 1$ elements in S_N, and $S_2 = [d_1, d_2]$ is computed using $d_1 = 0$ and $d_2 = a_N - a_{N-1}$.

3. Compare S_N and $S_{N-1} \oplus S_2$. If they are equal, we obtain $S_N = S_{N-1} \oplus S_2$; otherwise, $S_N = \max\{S_{N-1} \oplus [0, \times], [\times, a_N]_{(N-2)}\}$. Note that, unlike some existing algorithms, the presented algorithm does not need to check the type of the structuring element before decomposing the structuring element. In fact, the result of comparing S_N and $S_{N-1} \oplus S_2$ implies whether a structuring element is decomposable or not.

4. Set $S_N = S_{N-1}$.

5. If $N \leq 3$, go to step 9.

6. Decompose $S_N = [a_1, a_2, a_3, \ldots, a_N]$ into S_{N-2} and S_3, where $S_{N-2} = [a_1, a_2, a_3, \ldots, a_{N-2}]$ is obtained by duplicating the $N - 2$ elements in S_N, and $S_3 = [c_1, c_2, c_3]$ is computed using $c_1 = 0$, $c_2 = a_{N-1} - a_{N-2}$, and $c_3 = a_N - a_{N-2}$.

7. Compare S_N and $S_{N-2} \oplus S_3$. If they are equal, we obtain $S_N = S_{N-2} \oplus S_3$; otherwise, $S_N = \max\{S_{N-2} \oplus [0, 0, \times], [\times, a_{N-1}, a_{N-2}]_{(N-3)}\}$.

8. Set $S_N = S_{N-2}$, and go to step 5.

9. Obtain the result of the decomposition. Let m and k be integers and $1 \leq k \leq m$. For even-sized SEs, $S_N = \max\{Q_k, P_k\}$, and for odd-sized, $S_N = \max\{P_k\}$, where $P_k = S_3^1 \oplus S_3^2 \oplus \cdots \oplus S_3^k$ and $Q_k = S_2 \oplus S_3^1 \oplus S_3^2 \oplus \cdots \oplus S_3^k$.

Example 9.15: Let $S = [0, 2, 4, 6, 5, 4, 3]$. Since S is a linearly-sloped structuring element, we apply the decomposition algorithm in Shih and Mitchell [1991] to obtain $S = [0, 2, 1] \oplus [0, 2, 1] \oplus [0, 2, 1]$. By using the decomposition algorithm, we can obtain an alternative solution as

$$S = [b_1^1, b_2^1, b_3^1] \oplus [b_1^2, b_2^2, b_3^2] \oplus [b_1^3, b_2^3, b_3^3] = [0, 2, 4] \oplus [0, 2, 1] \oplus [0, -1, -2].$$

Example 9.16: Let $S = [3, 6, 8, 9, 8, 6, 3]$. Since S is a convex shaped structuring element, we apply the decomposition algorithm in Shih and Mitchell [1991] to obtain $S = [3, 4, 3] \oplus [0, 2, 0] \oplus [0, 3, 0]$. By using the decomposition algorithm, we can obtain an alternative solution as

$$S = [3, 6, 8] \oplus [0, 1, 0] \oplus [0, -2, -5].$$

Example 9.17: Let $S = [9, 7, 3, 10, 8, 2, 6]$. Since S is neither linearly-sloped nor convex-shaped, it cannot be decomposed using Shih and Mitchell [1991]. By using the decomposition algorithm, we can obtain the result as

$$S = \max \left\{ \begin{array}{l} [b_1^{(1,1)}, b_2^{(1,1)}, b_3^{(1,1)}] \oplus [b_1^{(1,2)}, b_2^{(1,2)}, b_3^{(1,2)}] \oplus [b_1^{(1,3)}, b_2^{(1,3)}, b_3^{(1,3)}] \\ [b_1^{(2,1)}, b_2^{(2,1)}, b_3^{(2,1)}]_{(2)} \oplus [b_1^{(2,2)}, b_2^{(2,2)}, b_3^{(2,2)}] \\ [b_1^{(3,1)}, b_2^{(3,1)}, b_3^{(3,1)}]_{(4)} \end{array} \right\}$$

$$= \max \{ [9, 7, 3, 10, 8] \oplus [0, -6, \times], [\times, \times, 6]_{(4)} \}$$

$$= \max \{ \max \{ [9, 7, 3] \oplus [0, \times, \times], [\times, 10, 8]_{(2)} \} \oplus [0, -6, \times], [\times, \times, 6]_{(4)} \}$$

$$= \max \left\{ \begin{array}{l} [9, 7, 3] \oplus [0, \times, \times] \oplus [0, -6, \times] \\ [\times, 10, 8]_{(2)} \oplus [0, -6, \times] \\ [\times, \times, 6]_{(4)} \end{array} \right\}$$

Example 9.18: Let an even-sized structuring element be $S = [a_1, a_2, a_3, a_4, a_5, a_6] = [8, 9, 10, 11, 10, 3]$. It is decomposed into $E = [e_1, e_2, e_3, e_4, e_5]$ and $D = [d_1, d_2]$ in the first step, where $d_1 = 0$ and $d_2 = a_6 - a_5 = -7$, and then E is decomposed into $B = [b_1, b_2, b_3]$ and $C = [c_1, c_2, c_3]$, where $[b_1, b_2, b_3] = [a_1, a_2, a_3]$, $c_1 = 0$, $c_2 = a_4 - a_3 = 1$, and $c_3 = a_5 - a_3 = 0$. Therefore, the result is $S = [8, 9, 10] \oplus [0, 1, 0] \oplus [0, -7]$.

9.3.3 The Two-Dimensional Arbitrary Grayscale Structuring Element Decomposition

We can decompose the two-dimensional arbitrary grayscale SEs by two methods. One method is to decompose them into a set of one-dimensional row or column SEs, which can be continuously decomposed using the afore-mentioned technique in the previous section. Figure 9.21 illustrates the row decomposition, where the row SEs are shifted by (x, y). The translation is explained in Figure 9.22, where (a) describes the coordinate system, (b) is a translation by $(2, 0)$, and (c) is a translation by $(0, 2)$.

Another method, called the *combined row-and-column* approach, for the decomposition of two-dimensional grayscale SEs is presented below. It itera-tively decomposes a two-dimensional grayscale structuring element into a dilation of one-dimensional row and one-dimensional column structuring components until the original structuring element can be reconstructed by the union of all dilations of pairs of row and column one-dimensional grayscale structuring components.

Its flowchart is shown in Figure 9.23. Let $S_{N \times N}$ denote a two-dimensional structuring element of size $N \times N$, and let $B_{N \times 1}$ and $C_{1 \times N}$ denote the one-dimensional row and column structuring components of size N, respectively.

The Two-Dimensional Combined Row-and-Column Decomposition Algorithm:

1. Choose either row-major or column-major and perform the decompo-sition as $S_{N \times N} = \max \{ B_{N \times 1} \oplus C_{1 \times N}, S'_{N \times N} \}$. We only show the row-major

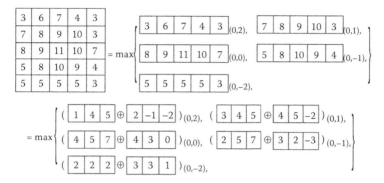

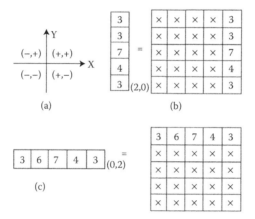

FIGURE 9.21 The row decomposition of a two-dimensional structuring element.

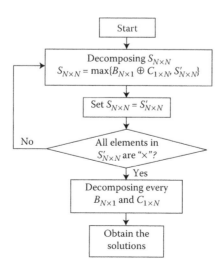

FIGURE 9.22 The representation of the translation.

FIGURE 9.23 The flowchart of the combined row-and-column decomposition.

decomposition. The column-major decomposition can be derived similarly. We select one row, say $B_{N\times1}$, which is the maximum over the sum of the row elements in $S_{N\times N}$. The $C_{1\times N}$ can be obtained by selecting the row-wise minimum of the matrix generated from subtracting each row of $S_{N\times N}$ by $B_{N\times1}$. For illustration, if selecting

$$B_{3\times1} = [5,7,8] \quad \text{in} \quad S_{3\times3} = \begin{bmatrix} 5 & 7 & 8 \\ 3 & 2 & 9 \\ 6 & 3 & 5 \end{bmatrix}, \quad \text{we have}$$

$$C_{1\times3} = \min_{\text{row-wise}} \left\{ \begin{bmatrix} 5 & 7 & 8 \\ 3 & 2 & 9 \\ 6 & 3 & 5 \end{bmatrix} - \begin{bmatrix} 5 & 7 & 8 \\ 5 & 7 & 8 \\ 5 & 7 & 8 \end{bmatrix} \right\} = \begin{bmatrix} 0 \\ -5 \\ -4 \end{bmatrix}.$$

Note that it will cause errors if the row-wise minimum is not selected. For example, if we randomly select the value from each row of the differences as

$$C_{1\times3} = \text{random}_{\text{row-wise}} \left\{ \begin{bmatrix} 0 & 0 & 0 \\ -2 & -5 & 1 \\ 1 & -4 & -3 \end{bmatrix} \right\} = \begin{bmatrix} 0 \\ 1 \\ -3 \end{bmatrix}$$

The dilation of $B_{3\times1}$ and $C_{1\times3}$ will become

$$B_{3\times1} \oplus C_{1\times3} = \begin{bmatrix} 5 & 7 & 8 \\ \underline{6} & \underline{8} & 9 \\ 2 & \underline{4} & 5 \end{bmatrix},$$

and it is impossible to reconstruct the original structuring element because those underlined elements will be selected after the maximum operation.

2. Generate $S'_{N\times N}$ by comparing $S_{N\times N}$ and $B_{N\times1} \oplus C_{1\times N}$. That is, we set $S'_{(i,j)}$ to be $\times$ if the corresponding values in $S_{(i,j)}$ nd $[B_{N\times1} \oplus C_{1\times N}]_{(i,j)}$ are the same; otherwise, we set $S'_{(i,j)} = S_{(i,j)}$. Note that, (i, j) indicates the ith row and jth column coordinates. If all the elements in $S'_{N\times N}$ are $\times$, go to step 4.

3. Set $S_{N\times N} = S'_{N\times N}$, and go to step 1.

4. Decompose $B_{N\times1}$ and $C_{1\times N}$ into small SEs of sizes 2 and 3 using the aforementioned algorithm.

Example 9.19: An example of using the combined row-and-column decomposition is shown in Figure 9.24.

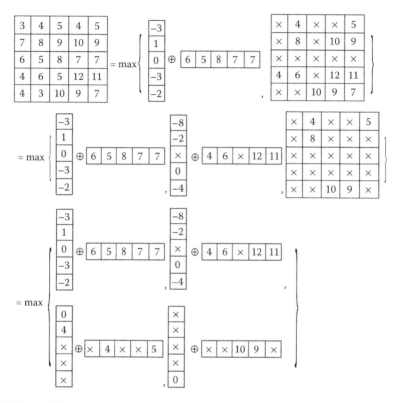

FIGURE 9.24 The combined row-and-column decomposition of a two-dimensional structuring element.

9.3.4 Complexity Analysis

In this section, we discuss the computational complexity of the structuring element decomposition. The worst case in decomposing the one-dimensional arbitrary grayscale structuring element is when the size is n, it is decomposed into $(n - 1)/2$ smaller structuring components of size 3 as shown in Figure 9.25.

Supposing that an image is performed several dilations by l SEs of size N, $S_N^1, S_N^2, \ldots, S_N^l$, for each pixel it needs $l \cdot N$ processing time and $l \cdot N$ comparisons (maximum operation) in a general sequential architecture as shown in Figure 9.26a. If all the SEs are decomposable without utilizing the maximum selection operation as shown in Figure 9.26b, it needs $3 \cdot [(N - 1)/2 + (l - 1)]$ processing time and $3 \cdot l \cdot (N - 1)/2$ comparisons in a pipeline architecture with l stages. However, in the worst case if all the SEs can only be decomposed into the format as shown in Figure 9.25, it needs $3 \cdot [(N - 1)/2 + (l - 1)]$ processing time and $3 \cdot l \cdot (N - 1)/2 + l \cdot N$ comparisons in a pipelined architecture with l stages. Table 9.4 shows the computational complexity of decomposing one-dimensional SEs in different conditions.

$$S_N = [\,a_1, a_2, a_3, \ldots, a_N\,]$$

$$= \max\{[a_1, a_2, a_3, \ldots, a_{N-2}] \oplus [0, \times, \times], [\times, a_{N-1}, a_N]_{N-3}\}$$

$$= \max \begin{cases} [a_1, a_2, a_3, \ldots, a_{N-2}] \oplus [0, \times, \times] \oplus [0, \times, \times], \\ [0, \times, \times] \oplus [\times, a_{N-3}, a_{N-2}]_{N-5}, [\times, a_{N-1}, a_N]_{N-3} \end{cases}$$

$$\underbrace{}_{\frac{N-3}{2}\text{ terms}}$$

$$= \max \left\{ \begin{array}{c} [\,a_1, a_2, a_3] \oplus [0, \times, \times] \oplus [0, \times, \times] \oplus \cdots \oplus [0, \times, \times] \\ [\times, a_4, a_5]_2 \oplus [0, \times, \times] \oplus \cdots \oplus [0, \times, \times] \\ \underbrace{}_{\frac{N-5}{2}\text{ terms}} \\ \vdots \\ [\times, a_{N-3}, a_{N-2}]_{N-5}, \oplus [0, \times, \times] \\ [\times, a_{N-1}, a_N]_{N-3} \end{array} \right\} \quad \frac{N-1}{2}\text{ terms}$$

$$= \max\{[a_1, a_2, a_3], [\times, a_4, a_5]_2, \ldots, [\times, a_{N-3}, a_{N-2}]_{N-5}, [\times, a_{N-1}, a_N]_{N-3}\}$$

FIGURE 9.25   The worst case of one-dimensional structuring element decomposition.

For the two-dimensional arbitrary grayscale structuring element decomposition, the computational complexity of row- or column-based algorithm is based on the one-dimensional structuring element decomposition. The worst case in the combined row-and-column algorithm is when a two-dimensional structuring element of size $n \times n$ is decomposed into n pairs of dilations of row and column vectors. Figure 9.27 shows a row-major example of decomposing a 5×5 structuring element. In the first iteration, the first row is selected for the decomposition. Note that the first combined pair of row and column can determine only some pixels as shown in Figure 9.27b. The symbol • indicates that the value cannot be determined in the current iteration and will be determined in the next iteration, and the symbol × indicates the don't care pixel in the current iteration. Figures 9.27c through f show the determined pixels of the SEs in the 2nd–5th pairs of the combined row and column decompositions, respectively. Sussner and Ritter [1997, 2000] indicated that the best case occurs when the rank of a matrix is 1, and the decomposition of two-dimensional SEs only needs one pair of combined row and

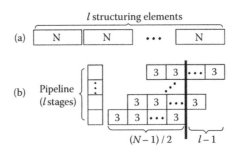

FIGURE 9.26   The sequential and pipelined architectures.

TABLE 9.4

The Computational Complexity of Decomposing One-Dimensional
SEs in Different Conditions

	Processing Time	Comparison
Sequential architecture	$O(l \cdot N)$	$O(l \cdot N)$
Pipelined and decomposable	$O(l + N)$	$O(l \cdot N)$
Pipelined and indecomposable	$O(l + N)$	$O(l \cdot N)$

(a)

$$\begin{bmatrix} a_1 & a_2 & a_3 & a_4 & a_5 \\ a_6 & a_7 & a_8 & a_9 & a_{10} \\ a_{11} & a_{12} & a_{13} & a_{14} & a_{15} \\ a_{16} & a_{17} & a_{18} & a_{19} & a_{20} \\ a_{21} & a_{22} & a_{23} & a_{24} & a_{25} \end{bmatrix}$$

(b)

$$\begin{bmatrix} a_1 & a_2 & a_3 & a_4 & a_5 \\ \bullet & a_7 & \bullet & \bullet & \bullet \\ \bullet & \bullet & \bullet & a_{14} & \bullet \\ \bullet & \bullet & a_{18} & \bullet & \bullet \\ \bullet & \bullet & \bullet & \bullet & a_{25} \end{bmatrix}$$

(c)

$$\begin{bmatrix} \times & \times & \times & \times & \times \\ a_6 & \times & a_8 & a_9 & a_{10} \\ \bullet & a_{12} & \bullet & \times & \bullet \\ a_{16} & \bullet & \times & \bullet & \bullet \\ \bullet & \bullet & \bullet & a_{24} & \times \end{bmatrix}$$

(d)

$$\begin{bmatrix} \times & \times & \times & \times & \times \\ \times & \times & \times & \times & \times \\ a_{11} & \times & a_{13} & \times & a_{15} \\ \times & \bullet & \times & a_{19} & \bullet \\ a_{21} & \bullet & \bullet & \times & \times \end{bmatrix}$$

(e)

$$\begin{bmatrix} \times & \times & \times & \times & \times \\ \times & \times & \times & \times & \times \\ \times & \times & \times & \times & \times \\ \times & a_{17} & \times & \times & a_{20} \\ \times & \bullet & a_{23} & \times & \times \end{bmatrix}$$

(f)

$$\begin{bmatrix} \times & \times & \times & \times & \times \\ \times & \times & \times & \times & \times \\ \times & \times & \times & \times & \times \\ \times & \times & \times & \times & \times \\ \times & a_{22} & \times & \times & \times \end{bmatrix}$$

FIGURE 9.27 An example of the two-dimensional combined row-and-column decomposition algorithm.

TABLE 9.5

The Computational Complexity of Decomposing Two-Dimensional SEs
in Different Conditions

	Processing Time	Comparison
$N \times N$ structuring element	$O(N^2)$	$O(N^2)$
Decomposed to dilation of row and column one-dimensional structuring element	$O(N)$	$O(N)$
Worst case in a pipelined architecture with $2N$ stages	$O(N)$	$O(N^2)$

column. The proposed algorithms provide a general decomposition strategy and do not need additional pre-processes, such as checking the type of SEs and the decomposition rules.

Given a dilation of an image with a structuring element of size $N \times N$, we need N^2 processing time and N^2 comparisons (maximum operation) without decomposition. If the structuring element can be decomposed into a dilation of row and column one-dimensional SEs, we need $2N$ processing time and $2N$ comparisons in a sequential architecture. However in the worst case, if it can be decomposed into N pairs of dilations of row and column SEs, then $6(N-1)$ processing time and $4N^2 - 3N$ comparisons are needed in a pipelined architecture with $2N$ stages. Table 9.5 shows the computational complexity of decomposing two-dimensional SEs in different conditions.

References

Anelli, G., Broggi, A., and Destri, G., "Decomposition of arbitrarily shaped binary morphological structuring elements using genetic algorithms," *IEEE Trans. Pattern Analysis and Machine Intelligence*, vol. 20, no. 2, pp. 217–224, Feb. 1998.

Camps, O. I., Kanungo, T., and Haralick, R. M., "Grayscale structuring element decomposition," *IEEE Trans. Image Processing*, vol. 5, no. 1, pp. 111–120, Jan. 1996.

Gader, P. D., "Separable decompositions and approximations of grayscale morphological templates," *CVGIP: Image Understanding*, vol. 53, no. 3, pp. 288–296, May 1991.

Haralick, R. M., Sternberg, S. R., and Zhuang, X., "Image analysis using mathematical morphology," *IEEE Trans. Pattern Analysis and Machine Intelligence*, vol. 9, no. 7, pp. 532–550, July 1987.

Hashimoto, R. F. and Barrera, J., "A note on Park and Chin's algorithm," *IEEE Trans. Pattern Analysis and Machine Intelligence*, vol. 24, no. 1, pp. 139–144, Jan. 2002.

Herrera, F., Lozano, M., and Verdegay, J. L., "Applying genetic algorithms in fuzzy optimization problems," *Fuzzy Systems and Artificial Intelligence*, vol. 3, no. 1, pp. 39–52, Jan. 1994.

Ho, S.-Y., Chen, H.-M., and Shu, L.-S., "Solving large knowledge base partitioning problems using the intelligent genetic algorithm," *Proc. Inter. Conf. Genetic and Evolutionary Computation*, pp. 1567–1572, 1999.

Holland, J. H., *Adaptation in Natural and Artificial Systems*, The University of Michigan Press, Ann Arbor, MI, 1975.

Holland, J. H., *Adaptation in Natural and Artificial System: An Introductory Analysis with Applications to Biology, Control, and Artificial Intelligence*, MIT Press, Cambridge, MA, 1992.

Park, H. and Chin, R. T., "Optimal decomposition of convex morphological structuring elements for 4-connected parallel array processors," *IEEE Trans. Pattern Analysis and Machine Intelligence*, vol. 16, no. 3, pp. 304–313, March 1994.

Park, H. and Chin, R. T., "Decomposition of arbitrarily shaped morphological structuring elements," *IEEE Trans. Pattern Analysis and Machine Intelligence*, vol. 17, no. 1, pp. 2–15, Jan. 1995.

Richardson, C. H. and Schafer, R. W., "A lower bound for structuring element decompositions," *IEEE Trans. Pattern Analysis and Machine Intelligence*, vol. 13, no. 4, pp. 365–369, April 1991.

Serra, J., *Image Analysis and Mathematical Morphology*, Academic Press, New York, 1982.

Shih, F. Y. and Mitchell, O. R., "Threshold decomposition of grayscale morphology into binary morphology," *IEEE Trans. Pattern Analysis and Machine Intelligence*, vol. 11, no. 1, pp. 31–42, Jan. 1989.

Shih, F. Y. and Mitchell, O. R., "Decomposition of grayscale morphological structuring elements," *Pattern Recognition*, vol. 24, no. 3, pp. 195–203, Mar. 1991.

Shih, F. Y. and Mitchell, O. R., "A mathematical morphology approach to Euclidean distance transformation," *IEEE Trans. Image Processing*, vol. 1, no. 2, pp. 197–204, April 1992.

Shih, F. Y. and Puttagunta, P., "Recursive soft morphological filters," *IEEE Trans. Image Processing*, vol. 4, no. 7, pp. 1027–1032, July 1995.

Shih, F. Y. and Wu, Hong, "Decomposition of geometric-shaped structuring elements using morphological transformations on binary image," *Pattern Recognition*, vol. 25, no. 10, pp. 1097–1106, 1992.

Sussner, P., Pardalos, P. M., and Ritter, G. X., "On integer programming approaches for morphological template decomposition problems in computer vision," *Journal of Combinatorial Optimization*, vol. 1, no. 2, pp. 165–178, 1997.

Sussner, P. and Ritter, G. X., "Decomposition of grayscale morphological templates using the rank method," *IEEE Trans. Pattern Analysis and Machine Intelligence*, vol. 19, no. 6, pp. 649–658, June 1997.

Sussner, P. and Ritter, G. X., "Rank-based decomposition of morphological templates," *IEEE Trans. Image Processing*, vol. 19, no. 6, pp. 1420–1430, June 2000.

Tang, K.-S., Man, K.-F., Liu, Z.-F., and Kwong, S., "Minimal fuzzy memberships and rules using hierarchical genetic algorithms," *IEEE Trans. Industrial Electronics*, vol. 45, no. 1, pp. 162–169, Jan. 1998.

Yang, H.-T. and Lee, S.-J., "Decomposition of morphological structuring elements with integer linear programming," *IEE Proc. Vision, Image and Signal Processing*, vol. 152, no. 2, pp. 148–154, April 2005.

Zhuang, X. and Haralick, R. M., "Morphological structuring element decomposition," *Computer Vision, Graphics, and Image Processing*, vol. 35, no. 3, pp. 370–382, Sep. 1986.

10

Architectures for Mathematical Morphology

General-purpose digital computers have traditionally been used for image processing, despite the fact that they are not architecturally suited for image data. The serial nature of conventional computers seems to restrict their usefulness as image processors. Morphological operations are the essence of the cellular logic machines, such as Golay logic processor [Golay 1969], Diff3 [Graham and Norgren 1980], PICAP [Kruse 1977], Leitz texture analysis system (TAS) [Klein and Serra 1977], CLIP processor arrays [Duff 1979], cytocomputers [Lougheed and McCubbrey 1980], and Delft image processor (DIP) [Gerritsen and Aardema 1981]. Most of them were optimized for local transformations and are referred to as *cellular transformations*. However, they have the weakness of input/output bottlenecks, low reliability, and difficult programming.

Handley [2003, 2005] presented a real-time, compact architecture for translation-invariant windowed nonlinear discrete operators represented in computational mathematical morphology. The architecture enables output values to be computed in a fixed number of operations and thus can be pipelined. Louverdis and Andreadis [2003] presented a hardware implementation of a fuzzy processor suitable for morphological color image processing applications. The proposed digital hardware structure is based on a sequence of pipeline stages, and the parallel processing is used to minimize computational time.

This chapter is organized as follows. Section 10.1 introduces the threshold decomposition architecture of grayscale morphology into binary morphology. Section 10.2 describes the implementation of morphological operations using programmable neural networks. Section 10.3 presents the multilayer perceptron (MLP) as morphological processing modules. Section

10.4 describes a systolic array architecture. Section 10.5 discusses the implementation on multicomputers.

10.1 Threshold Decomposition of Grayscale Morphology into Binary Morphology

Binary morphological operations of dilation and erosion have been successfully extended to grayscale image processing. But grayscale morphological operations are difficult to implement in real time. A superposition property called *threshold decomposition* and another property called *stacking* are introduced and have been found to successfully apply to grayscale morphological operations. These properties allow grayscale signals to be decomposed into multiple binary signals, which can be processed in parallel, and the binary results are combined to produce the final grayscale result.

In this section, we present the threshold decomposition architecture and the stacking property, which allows the implementation of the parallel architecture. Grayscale operations are decomposed into binary operations with the same dimensionality as the original operations. This decomposition allows grayscale morphological operations to be implemented using only logic gates in new VLSI architectures, which significantly improve speed as well as give new theoretical insight into the operations [Shih and Mitchell 1989].

10.1.1 Threshold Decomposition Algorithm for Grayscale Dilation

The traditional grayscale dilation operation has been implemented using one of the following two methods:

(a) Let $F, K \subseteq E^{N-1}$ and $f: F \to E$ and $k: K \to E$. The *grayscale dilation* of f by k, denoted by $f \oplus_g k$, is defined as

$$f \oplus_g k = T[U[f] \oplus_b U[k]],\qquad (10.1)$$

where $U[f]$ denotes the umbra of f and $T[A]$ denotes the top or top surface of A.

(b) Let $f(x, y): F \to E$ and $k(m, n): K \to E$. Then $(f \oplus_g k)(x, y): F \oplus_g K \to E$ can be computed by

$$(f \oplus_g k)(x, y) = \max\{f(x - m, y - n) + k(m, n)\},\qquad (10.2)$$

for all $(m, n) \in K$ and $(x - m, y - n) \in F$.

Equation 10.1 requires a three-dimensional binary dilation (for two-dimensional images) followed by conversion back to a two-dimensional image by the top surface operation. Equation 10.2 requires many grayscale additions and a local maximum operation for each pixel in an input image. What is presented here is a different approach of calculating grayscale dilation, which may provide significant improvement in speed if appropriate architectures are available. This method is based on threshold decomposition of the grayscale image and structuring element and results in calculations that require binary dilations of one less dimension than that of Equation 10.1. First we will derive the basis and validity of this algorithm, then present the algorithm itself, and finally consider its computational complexity.

To derive this algorithm, we first introduce some new notations and definitions. Then, through viewing functions as sets, we can produce the result of the threshold decomposition algorithm.

10.1.1.1 Notations and Definitions

Vector notation will be used for sequences, so that an input image f of length n will be written as $\vec{f}_n = (f_0, f_1, \dots, f_{n-1})$. The constant vector $\vec{k}_{c,n}$ has n terms in total, and each term has a value the constant k; that is, $\vec{k}_{c,n} = (k, k, \dots, k)$. If a sequence is a binary sequence, it will be marked as $\vec{f}_{b,n}$ to make this clear. When the value of n is clear from the context, it will be dropped.

As for relations between sequences, notation from developed stack filter theory will be used [Fitch, Coyle, and Gallager 1985; Wendt, Coyle, and Gallagher 1986]. Suppose that $\vec{x}_{b,n} = (x_0, x_1, \dots, x_{n-1})$ and $\vec{y}_{b,n} = (y_0, y_1, \dots, y_{n-1})$ are two binary sequences of length n. We say $\vec{x}_{b,n} = \vec{y}_{b,n}$ if and only if $x_i = y_i$ for all i. We say $\vec{x}_{b,n} \leq \vec{y}_{b,n}$ if $x_i = 1$ implies $y_i = 1$ for all i; if we also have $\vec{x}_{b,n} \neq \vec{y}_{b,n}$, we write $\vec{x}_{b,n} < \vec{y}_{b,n}$.

Definition 10.1: The *threshold decomposition* of an M-level sequence $\vec{f}$ is the set of M binary sequences, called *thresholded sequences*, $\vec{f0}_b, \vec{f1}_b, \dots, \vec{f(M-1)}_b$, whose elements are defined as

$$fi_b(j) = \begin{cases} 1 & \text{if } f(j) \geq i \\ 0 & \text{if } f(j) < i \end{cases} \tag{10.3}$$

Note that if there are some gray values, say $i, i+1$, which do not exist in the sequence, but gray value $i+2$ does exist, then we conclude that $\vec{fi}_b, \vec{f(i+1)}_b$, and $\vec{f(i+2)}_b$ are equal.

Definition 10.2: An ordered set of L binary sequences $\vec{f0}_b, \vec{f1}_b, \dots, \vec{f(L-1)}_b$, in which all the sequences have length n, is said to obey *the stacking property* if

$$\vec{f0}_b \geq \vec{f1}_b \geq \cdots \geq \vec{f(L-1)}_b \tag{10.4}$$

Definition 10.3: A binary operation Oper is said to have *the stacking property* if

$$\text{Oper}(\vec{x}) \geq \text{Oper}(\vec{y}) \text{ whenever } \vec{x} \geq \vec{y} \tag{10.5}$$

Recall that the binary dilation operation possesses the increasing property: If $A \subseteq B$ then $A \oplus_b D \subseteq B \oplus_b D$. Hence, binary dilation with any structuring element has the stacking property.

Definition 10.4: Let $F \subseteq E^{N-1}$ and $f: F \rightarrow E$. The *slice* of f, denoted by $S[f]$, $S[f] \subseteq F \times E$, is defined as

$$S[fi]_{(x,y)} = \begin{cases} 1 & \text{if } y = i \text{ and } f(x) \geq y \\ 0 & \text{elsewhere} \end{cases} \tag{10.6}$$

Example 10.1:

Note that for a discrete image with M levels $0, 1, \dots, M-1$, the umbra of the image can be decomposed into M slices:

$$U[f] = S[f0] \cup S[f1] \cup \cdots \cup S[f(M-1)]$$

$$= \bigcup_{i=0}^{M-1} S[fi] \tag{10.7}$$

Definition 10.5: Let $F \subseteq E^{N-1}$ and $f: F \rightarrow E$. The *slice complement* of f, denoted by $S'[f]$, $S'[f] \subseteq F \times E$, is defined as

$$S'[fi]_{(x,y)} = \begin{cases} 1 & \text{if } y = i \text{ and } f(x) \geq y \\ 0 & \text{if } y = i \text{ and } f(x) < y \\ 1 & \text{if } y \neq i \end{cases} \tag{10.8}$$

Example 10.2:

↱	0	1	2	3
0	2	1	0	2

f

	0	1	2	3
3	1	1	1	1
2	1	1	1	1
1	1	1	0	1
0	1	1	1	1

$S'[f1]$

Note that for a discrete image with M levels $0, 1, \ldots, M-1$, the umbra of the image can be decomposed into M slice complements:

$$U[f] = S'[f0] \cap S'[f1] \cap \cdots \cap S'[f(M-1)]$$

$$= \bigcap_{i=0}^{M-1} S'[fi] \tag{10.9}$$

10.1.1.2 Formulas' Derivation

Suppose z is the grayscale dilation of image f by structuring element k, $z = f \oplus_g k$. Then, from Equation 10.1, this is equivalent to the top surface of the binary dilation of the respective umbras of the image and structuring element.

$$f \oplus_g k = T\{U(f) \oplus_b U(k)\} \tag{10.10}$$

Decomposing the umbra of the image according to Equation 10.7:

$$= T\left\{ \left(\bigcup_{j=0}^{M-1} S[fj] \right) \%_b U[k] \right\} \tag{10.11}$$

If we assume the largest gray value in the image f is P, then

$$= T\left\{ \left(\bigcup_{j=0}^{P} S[fj] \right) \%_b U[k] \right\} \tag{10.12}$$

According to the property of binary dilation, dilation distributes itself over union:

$$= T\left\{ \bigcup_{j=0}^{P} [S[fj] \%_b U[k]] \right\} \tag{10.13}$$

Decomposing the structuring element k according to Equation 10.7,

$$= T \left\{ \bigcup_{j=0}^{P} \left[S[fj] \, \%_b \left(\bigcup_{i=0}^{N-1} S[ki] \right) \right] \right\} \tag{10.14}$$

Assuming the largest gray value in the structuring element k is Q,

$$= T \left\{ \bigcup_{j=0}^{P} \left[S[fj] \, \%_b \left(\bigcup_{i=0}^{Q} S[ki] \right) \right] \right\} \tag{10.15}$$

$$= T \left\{ \bigcup_{j=0}^{P} \bigcup_{i=0}^{Q} [S[fj] \, \%_b S[ki]] \right\} \tag{10.16}$$

The top surface of the union of the terms in brackets is the maximum at each sequence location over the top surfaces of every term:

$$= \max_{j=0}^{P} T \left\{ \bigcup_{j=0}^{Q} [S[fj] \, \%_b S[ki]] \right\} \tag{10.17}$$

Each slice above consists of only one nonzero row, so that the binary dilation can be performed in one less dimension. Due to the stacking property of the rows in the umbra and the binary dilation operator, the top surface operation can be equated to a summation of the reduced dimensionality stacked signals. There is a constant offset for each term since the image slice used in each term is at a different level, and due to the translation invariance property of dilation, this moves the top surface of the result by the same amount.

$$f \oplus_g k = \max \left\{ \left[\sum_{i=0}^{Q} \overrightarrow{f0_b} \oplus_b \overrightarrow{ki_b} \right] - \vec{1}, \sum_{i=0}^{Q} (\overrightarrow{f1_b} \oplus_b \overrightarrow{ki_b}), \left[\sum_{i=0}^{Q} (\overrightarrow{f2_b} \oplus_b \overrightarrow{ki_b}) \right] + \vec{1}, \dots \right\} \tag{10.18}$$

If the image and SEs are assumed to be nonnegative, then $\overrightarrow{f0_b} = \vec{1}$, and the first term is constant.

$$f \oplus_g k = \max \left\{ \vec{Q_c}, \sum_{i=0}^{Q} (\overrightarrow{f1_b} \oplus_b \overrightarrow{ki_b}), \left[\sum_{i=0}^{Q} (\overrightarrow{f2_b} \oplus_b \overrightarrow{ki_b}) \right] + \vec{1}, \dots \right\} \tag{10.19}$$

Consider the maximum over the first two terms in Equation 10.19. Since the result of each binary dilation has a maximum of one, the sequence

$\overrightarrow{Ki_b}$ obeys the stacking property, and there are $Q+1$ total terms in the summation, if the final term in the summation is a one at any location, the summation result will be $Q + 1$; otherwise, it will be less than or equal to Q.

$$\max\left\{\vec{Q}_c, \sum_{i=0}^{Q}(\overrightarrow{f1_b} \oplus_b \overrightarrow{ki_b})\right\} = \max\left\{\vec{Q}_c, \vec{Q}_c + (\overrightarrow{f1_b} \oplus_b \overrightarrow{kQ_b})\right\} \quad (10.20)$$

When this concept is extended to the third and later terms in Equation 10.19, an increasing number of final terms must be considered since the constant added to the summation is increasing.

$$f \oplus_g k = \max\left\{\vec{Q}_c, \vec{Q}_c + (\overrightarrow{f1_b} \oplus_b \overrightarrow{kQ_b}), \overrightarrow{(Q-1)_c} + \left[\sum_{i=Q-1}^{Q}(\overrightarrow{f2_b} \oplus_b \overrightarrow{ki_b})\right] + \vec{1}, \ldots\right\}$$

$$(10.21)$$

$$= \vec{Q}_c + \max\left\{\overrightarrow{f1_b} \oplus_b \overrightarrow{kQ_b}, \sum_{i=Q-1}^{Q}(\overrightarrow{f2_b} \oplus_b \overrightarrow{ki_b}), \sum_{i=Q-2}^{Q}(\overrightarrow{f3_b} \oplus_b \overrightarrow{ki_b}), \ldots\right\} \quad (10.22)$$

10.1.1.3 Algorithm Description

Let a discrete image have M levels $0, 1, \ldots, M-1$ and a discrete structuring element have N levels $0, 1, \ldots, N-1$. The image and structuring element can be one dimension or more dimensions. The algorithm is described below. A one-dimensional example is given in Figure 10.1 and a two-dimensional example is given in Figure 10.2.

1. Detect the highest gray value in both the image and the structuring element. Also determine what values between 0 and the highest gray value do not occur in each in order to construct two equality tables that will be used in step 3. Let the highest gray value of the image be P and the highest gray value of the structuring element be Q.

2. Decompose the M-level image into P binary images by thresholding it at each level between 1 and P. In Figure 10.1, this decomposition operation produces five binary SEs since the input structuring element has the highest value 5. Similarly, decompose the N-level structuring element into $Q+1$ binary SEs by thresholding it at each level between 0 and Q. In Figure 10.1, this decomposition operation produces five binary images since the input image has the highest value 4. Note that the sum over all thresholded sequences (excluding the zero thresholded sequence) is always equal to the original input sequence.

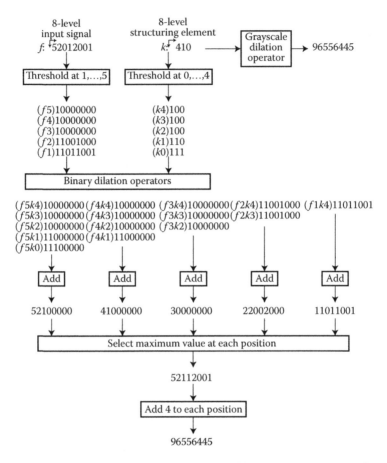

FIGURE 10.1 Grayscale morphological dilation the old way (top row) versus that using threshold decomposition. Both origins of the image and structuring element are on the left hand side. Note this is a one-dimensional example.

3. Pass these thresholded images and thresholded SEs sequence pairs through the binary dilation operators according to the following combinations: The first image and the Qth structuring element; the second image and the Qth structuring element, the second image and the $(Q - 1)$th structuring element; the third image and the Qth structuring element, the third image and the $(Q - 1)$th structuring element, the third image and the $(Q - 2)$th structuring element, and so forth (see Figures 10.1 and 10.2). Since each pair is processed independently, these operations can be implemented in parallel, as shown in Figure 10.3. The binary dilation is trivial to implement by using pipeline processors [Sternberg 1983]. Note that if there is a missing gray value as shown by the equality tables generated in step 1, then

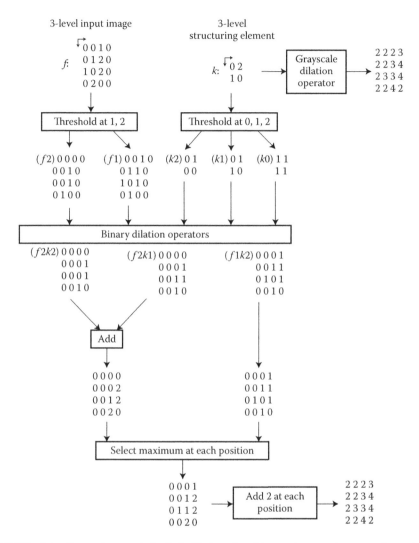

FIGURE 10.2 Grayscale morphological dilation the old way (top row) versus that using threshold decomposition. Both origins of the image and structuring element are at the upper-left corner. Note this is a two-dimensional example. The algorithm can be extended to any number of dimensions.

we skip passing the pair through the binary dilation operator and use the output due to the next existing gray value.

4. Add the stacking binary outputs for each thresholded image. We get P outputs in total with the highest gray value $Q + 1$ for the thresholded images. Note that if $P \geq Q$, then add 1 to the $(Q + 2)$th output at each position, add 2 to the $(Q + 3)$th output at each position, and so forth.

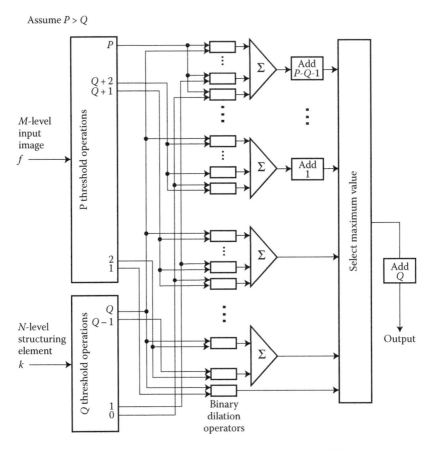

FIGURE 10.3   The threshold decomposition architecture for grayscale dilation.

5. Select the maximum value at each position from the P outputs, and
   then add Q to every position. The final result is exactly the same as
   the output of grayscale dilation operator, as shown in Figure 10.1.

The significance of the threshold decomposition is that it reduces the
analysis of the grayscale operation to multiple binary operations that can be
performed in parallel. It allows complex problems to be decomposed into
simpler problems.

10.1.1.4 Computational Complexity

An M-level $0, 1, \ldots, M - 1$ image does not necessarily have the highest value
$M - 1$. If the highest value of the image is P and the highest value of the structuring element is Q, then we can cut down the computation complexity some.
When we threshold the M-level image f, all binary images for thresholds
above P should be equal to $\vec{0}$. Hence, we skip those. Similarly for the N-level

structuring element, we skip all images for thresholds above Q. We analyze the computational complexity as follows.

Case 1: $P > Q$

$$f \oplus_g k = \vec{Q}_c + \max \left\{ \begin{array}{l} \overrightarrow{f1}_b \oplus_b \overrightarrow{kQ}_b, \sum_{i=Q-1}^{Q} (\overrightarrow{f2}_b \oplus_b \overrightarrow{ki}_b), \sum_{i=Q-2}^{Q} (\overrightarrow{f3}_b \oplus_b \overrightarrow{ki}_b), \ldots, \\[2ex] \sum_{i=0}^{Q} (\overrightarrow{f(Q+1)}_b \oplus_b \overrightarrow{ki}_b) \; \left[\sum_{i=0}^{Q} (\overrightarrow{f(Q+2)}_b \oplus_b \overrightarrow{ki}_b) + \vec{1} \right], \ldots, \\[2ex] \left[\sum_{i=0}^{Q} (\overrightarrow{fP}_b \oplus_b \overrightarrow{ki}_b) + \overrightarrow{(P-Q-1)}_c \right] \end{array} \right\}$$

(10.23)

The summation terms in Equation 10.23 have the stacking property since the ki_b terms have this property as well as the binary dilation operator. Therefore, each set of values to be added consists of a set of ones followed by zeros. Therefore, the sum can be found by locating the transition point. This can be done by a binary search operation such as is discussed in Wendt, Coyle, and Gallagher [1986].

This grayscale dilation requires 1 maximum operation, $P - Q$ additions, and

$$1 + 2 + \cdots + (Q+1) + (P-Q-1) \times (Q+1) = (Q+1) \times \left(P - \frac{Q}{2} \right)$$

binary dilations.

Case 2: $P \le Q$

$$f \oplus_g k = \vec{Q}_c + \max \left\{ \begin{array}{l} \overrightarrow{f1}_b \oplus_b \overrightarrow{kQ}_b, \sum_{i=Q-1}^{Q} (\overrightarrow{f2}_b \oplus_b \overrightarrow{ki}_b), \sum_{i=Q-2}^{Q} (\overrightarrow{f3}_b \oplus_b \overrightarrow{ki}_b), \ldots, \\[2ex] \sum_{i=Q-P+1}^{Q} (\overrightarrow{fP}_b \oplus_b \overrightarrow{ki}_b) \end{array} \right\}$$

(10.24)

Once again the summation can be performed by only a binary search. The other operations require one maximum, one addition, and

$$1 + 2 + \cdots + P = P \times \frac{P+1}{2}$$

binary dilations.

Note that if there are some gray values missing in either the image or the structuring element, then we skip passing the pairs through the binary dilation operators and reduce the total number of binary dilation operations. Hence, the above computational complexity is the worst-case complexity.

10.1.2 A Simple Logic Implementation of Grayscale Morphological Operations

In the previous section, we introduced a grayscale dilation algorithm that thresholds the grayscale image and the structuring element into binary images and SEs, then passes these through binary dilation operators, and sums up and takes the maximum value for each output pixel. We now present a faster improved grayscale dilation and erosion architecture that uses only logic gates (AND, OR, INVERSE) and can be implemented in VLSI.

10.1.2.1 Binary Morphological Operations

It is known that binary morphological operations may be done using simple logic gates.

1. *Binary Dilation Operator*: According to the definition of binary dilation operator,

$$A \%_b B = \bigcup_{b \in B} (A)_b$$

Suppose that A is a two-dimensional binary image and B is a two-dimensional binary structuring element. We may express binary dilation in this form:

$$(A \oplus_b B)_{(i,j)} = \text{OR}_{(m,n)}[B_{(m,n)} \cdot A_{(i-m,j-n)}] \tag{10.25}$$

Hence, we use AND, OR gates to produce the binary dilation operator.

2. *Binary Erosion Operator*: According to the definition of binary erosion operator,

$$A *_b B = \bigcap_{b \in B} (A)_{-b}$$

We may express two-dimensional erosion in this form:

$$(A \ominus_b B)_{(i,j)} = \underset{m,n}{\text{AND}} \{\text{OR}[\text{AND}(B_{(m,n)}, A_{(i+m, j+n)}), \bar{B}_{(m,n)}]\} \tag{10.26}$$

Equivalently, we may simplify

$$(A \ominus_b B)_{(i,j)} = \underset{m,n}{\text{AND}}[\text{OR}(A_{(i+m,j+n)}, \bar{B}_{(m,n)})] \tag{10.27}$$

Hence, we use AND, OR, INVERSE gates to produce the binary erosion operator.

10.1.2.2 Grayscale Morphological Operations

The grayscale morphological dilation operation requires the selection of a maximum over all output lines. In addition, the erosion operation using the morphological duality theorem requires several additional steps to compute. We will now present a method of eliminating these implementation drawbacks.

1. *Grayscale Dilation Operator*: We begin with the result of Equation 10.16:

$$f \oplus_g k = T \left\{ \bigcup_{j=0}^{P} \bigcup_{i=0}^{Q} \left[S[fj] \oplus_b S[ki] \right] \right\} \tag{10.28}$$

Each subgroup from left to right possesses the stacking property from bottom to top. Hence, the top surface of the union of those stacked subgroups is equated to a summation of the reduced dimensionality stacked signals. There is a constant offset "one" that we must count starting from zero level:

$$f \oplus_g k = \begin{cases} [\overrightarrow{f0_b} \oplus_b \overrightarrow{k0_b}] + [(\overrightarrow{f1_b} \oplus_b \overrightarrow{k0_b}) \cup (\overrightarrow{f0_b} \oplus_b \overrightarrow{k1_b})] + \cdots \\ + [(\overrightarrow{f(Q+1)_b} \oplus_b \overrightarrow{k0_b}) \cup (\overrightarrow{fQ_b} \oplus_b \overrightarrow{k1_b}) \cup \cdots \cup (\overrightarrow{f1_b} \oplus_b \overrightarrow{kQ_b})] + \cdots \\ + [(\overrightarrow{fP_b} \oplus_b \overrightarrow{k(Q-1)_b}) \cup (\overrightarrow{f(P-1)_b} \oplus_b \overrightarrow{kQ_b})] \cup [\overrightarrow{fP_b} \oplus_b \overrightarrow{kQ_b}] \end{cases} - \vec{1} \tag{10.29}$$

If the image and SEs are assumed to be non-negative, $\overrightarrow{f0_b} = \vec{1}$, then $\overrightarrow{f0_b} \oplus_b \overrightarrow{ki_b} = \vec{1}$.

$$f \oplus_g k = \begin{cases} \vec{1} + \vec{1} + \cdots + \vec{1} + \\ + [(\overrightarrow{f(Q+1)_b} \oplus_b \overrightarrow{k0_b}) + (\overrightarrow{fQ_b} \oplus_b \overrightarrow{k1_b}) + \cdots + (\overrightarrow{f1_b} \oplus_b \overrightarrow{kQ_b})] + \cdots \\ + [(\overrightarrow{fP_b} \oplus_b \overrightarrow{k(Q-1)_b}) + (\overrightarrow{f(P-1)_b} \oplus_b \overrightarrow{kQ_b})] + [\overrightarrow{fP_b} \oplus_b \overrightarrow{kQ_b}] \end{cases} - \vec{1} \tag{10.30}$$

$$f \oplus_g k = \begin{cases} \overrightarrow{(Q+1)_c} + \\ [(\overrightarrow{f(Q+1)_b} \oplus_b \overrightarrow{k0_b}) + (\overrightarrow{fQ_b} \oplus_b \overrightarrow{k1_b}) + \cdots + (\overrightarrow{f1_b} \oplus_b \overrightarrow{kQ_b})] + \cdots \\ + [(\overrightarrow{fP_b} \oplus_b \overrightarrow{k(Q-1)_b}) + (\overrightarrow{f(P-1)_b} \oplus_b \overrightarrow{kQ_b})] + [\overrightarrow{fP_b} \oplus_b \overrightarrow{kQ_b}] \end{cases} - \vec{1} \tag{10.31}$$

$$f \oplus_g k = \begin{cases} [(\overrightarrow{f(Q+1)_b} \oplus_b \overrightarrow{k0_b}) + (\overrightarrow{fQ_b} \oplus_b \overrightarrow{k1_b}) + \cdots + (\overrightarrow{f1_b} \oplus_b \overrightarrow{kQ_b})] + \cdots \\ + [(\overrightarrow{fP_b} \oplus_b \overrightarrow{k(Q-1)_b}) + (\overrightarrow{f(P-1)_b} \oplus_b \overrightarrow{kQ_b})] + [\overrightarrow{fP_b} \oplus_b \overrightarrow{kQ_b}] \end{cases} + \vec{Q_c} \tag{10.32}$$

Shown in Figure 10.4 is the architecture implied by Equation 10.32.

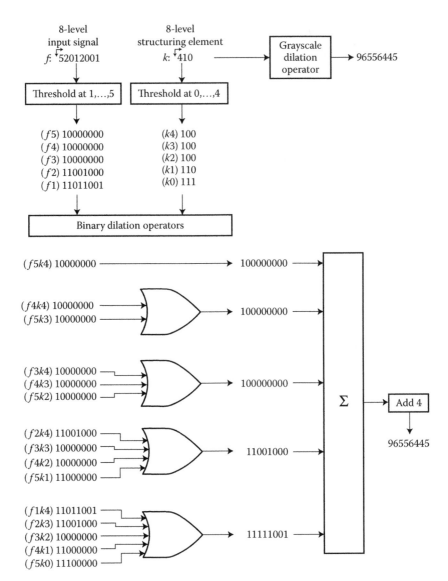

FIGURE 10.4 Grayscale morphological dilation the old way versus the new simple logic implementation.

2. *Grayscale Erosion Operator*: Although grayscale erosion can be implemented using the duality principle, we have found a direct implementation, which will now be presented.

Suppose z is the grayscale erosion of image f by structuring element k, $z = f \ominus_g k$. This is equivalent to the top surface of the binary erosion of the respective umbras of the image and structuring element.

$$f \ominus_g k = T\{\cup (f) \ominus_b \cup(k)\}$$

Decomposing the umbra of the image according to Equation 10.9 and decomposing the umbra of the structuring element according to Equation 10.7.

$$= T\left\{\left(\bigcap_{j=0}^{M-1} S'[fj]\right) \ominus_b \left(\bigcup_{i=0}^{N-1} S[ki]\right)\right\} \tag{10.33}$$

If we assume the largest gray value in the image f is P and in the structuring element is Q, then

$$= T\left\{\left(\bigcap_{j=0}^{P} S'[fj]\right) \ominus_b \left(\bigcup_{i=0}^{Q} S[ki]\right)\right\} \tag{10.34}$$

According to the distributive property of the binary erosion operation,

$$\begin{aligned}
f \ominus_g k = T\{&(S'[f0] \ominus_b S[kQ]) \cap [S'[f1] \ominus_b S[kQ]) \cap (S'[f0] \ominus_b S[k(Q-1)])] \cap \cdots \\
&\cap [(S'[fQ] \ominus_b S[KQ]) \cap (S'[f(Q-1)] \ominus_b S[k(Q-1)])] \cap \cdots \\
&\cap (S'[f0] \ominus_b S[k0])] \cap \cdots \\
&\cap [(S'[fP] \ominus_b S[kQ]) \cap (S'[f(P-1)] \ominus_b S[k(Q-1)])] \cap \cdots \\
&\cap (S'[f(P-Q)] \ominus_b S[k0]) \cap \cdots \cap (S'[fP] \ominus_b S[k0])]\}
\end{aligned} \tag{10.35}$$

Each subgroup from left to right possesses the stacking property from bottom to top. Hence, the top surface of the intersection of those stacked subgroups is equated to a summation of the reduced dimensionality stacked signals. There is a constant offset $\overrightarrow{(Q+1)}_c$ that we count starting from zero level.

$$\begin{aligned}
f \ominus_g k = \{&(\overrightarrow{f0}_b \ominus_b \overrightarrow{kQ}_b) + [(\overrightarrow{f1}_b \ominus_b \overrightarrow{kQ}_b) \cap (\overrightarrow{f0}_b \ominus_b \overrightarrow{k(Q-1)}_b)] + \cdots \\
&\cap [(\overrightarrow{fQ}_b \ominus_b \overrightarrow{kQ}_b) \cap \overrightarrow{f(Q-1)}_b \ominus_b \overrightarrow{k(Q-1)}_b \cap \cdots \cap (\overrightarrow{f0}_b \ominus_b \overrightarrow{k0}_b)] +) \cdots \\
&+ [(\overrightarrow{fP}_b \ominus_b \overrightarrow{kQ}_b) \cap \overrightarrow{f(P-1)}_b \ominus_b \overrightarrow{k(Q-1)}_b) \cap \cdots \\
&\cap \overrightarrow{f(P-Q)}_b \ominus_b \overrightarrow{k0}_b] + \cdots + (\overrightarrow{fP}_b \ominus_b \overrightarrow{k0}_b)\} - \overrightarrow{(Q+1)}_c
\end{aligned} \tag{10.36}$$

If the image and SEs are assumed to be nonnegative, $\vec{f0}_b = \vec{1}$, then $\vec{f0}_b \ominus_b \vec{ki}_b = \vec{1}$.

$$
\begin{aligned}
f \ominus_g k = \{\vec{1} + [(\vec{f1}_b \ominus_b \overline{k\vec{Q}_b})] + \cdots + [(\vec{fQ}_b \ominus_b \overline{k\vec{Q}_b}) \\
\cap (\overline{f(Q-1)}_b \ominus_b \overline{k(Q-1)}_b \cap \cdots \cap (\overline{f1}_b \ominus_b \overline{k1}_b)] + \cdots \\
+ [(\vec{fP}_b \ominus_b \overline{k\vec{Q}_b}) \cap (\overline{f(P-1)}_b \ominus_b \overline{k(Q-1)}_b) \cap \cdots \\
\cap (\overline{f(P-Q)}_b \ominus_b \overline{k0}_b)] + \vec{0} + \cdots + \vec{0}\} - \overline{(Q+1)}_c
\end{aligned}
\tag{10.37}
$$

$$
\begin{aligned}
f \ominus_g k = \{(\overline{f1}_b \ominus_b \overline{k\vec{Q}_b}) + \cdots + [(\vec{fQ}_b \ominus_b \overline{k\vec{Q}_b}) \\
\cap (\overline{f(Q-1)}_b \ominus_b \overline{k(Q-1)}_b) \cap \cdots \cap (\overline{f1}_b \ominus_b \overline{k1}_b)] + \cdots \\
+ [(\vec{fP}_b \ominus_b \overline{k\vec{Q}_b}) \cap (\overline{f(P-1)}_b \ominus_b \overline{k(Q-1)}_b) \cap \cdots \\
\cap (\overline{f(P-Q)}_b \ominus_b \overline{k0}_b)]\} - \vec{Q}_c
\end{aligned}
\tag{10.38}
$$

Shown in Figure 10.5 is the architecture implied by Equation 10.38.

10.1.2.3 Additional Simplification

An examination of the architectures in Figures 10.4 and 10.5 reveals that the final step (add or subtract the constant Q) is not really necessary. In fact, the number of binary lines summed in both cases is equal to the highest gray level in the image, so that without the constant addition or subtraction, the resulting grayscale morphological operation is guaranteed not to change the range of the input image. Such insights are easier to perceive from the threshold decomposition architecture.

If the image f and the structuring element k are not positive, for example, $f \geq -P'$ and $k \geq -Q'$, then the method works by doing:

$$
f \oplus_g k = [(f + P') \oplus_g (k + Q')] - (P' + Q')
$$

10.1.2.4 Implementation Complexity

Assume that an image has a highest gray value P and a structuring element has a highest gray value Q. According to Equation 10.32 and Equation 10.38, we analyze the implementation complexity as follows.

Case 1: $P > Q$: The grayscale dilation/erosion requires $P - Q$ additions and $(Q + 1) \times (P - (Q/2))$ binary dilations/erosions.

Case 2: $P \leq Q$: The grayscale dilation/erosion requires one addition and $P(P + 1)/2$ binary dilations/erosions.

Example 10.3: If both the image and structuring element have 16 gray levels and the size of the structuring element is 3×3, then the total number

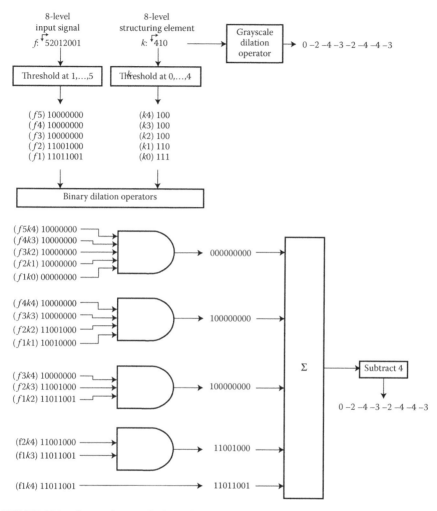

FIGURE 10.5 Grayscale morphological erosion the old way versus the new simple logic implementation. Both origins of the image and structuring element are on the left-hand side.

of gates we need is as follows. According to Equation 10.25, we need nine AND gates and one OR gate for each binary dilation operation. According to Equation 10.27, we need nine OR gates, nine INVERSE gates, and one AND gate for each binary erosion operation. The total number of gates is

1. For grayscale dilation,

$$15 \times \frac{16}{2} \times (9 + 1) + 14 = 1214 \text{ gates}$$

2. For grayscale erosion,

$$15 \times \frac{16}{2} \times (9 + 9 + 1) + 14 = 2294 \text{ gates}$$

TABLE 10.1

Approximate Number of Logic Gates for Grayscale Dilation

Approximate Number of Gates

Structuring Element Size	Bits of Gray Level		
	4	6	8
3 × 3	1214	20.2K	326.7K
4 × 4	2054	34.3K	555.1K
5 × 5	3134	52.5K	848.9K

TABLE 10.2

Approximate Number of Logic Gates for Grayscale Erosion

Approximate Number of Gates

Structuring Element Size	Bits of Gray Level		
	4	6	8
3 × 3	2294	38.4K	620.4K
4 × 4	3974	66.6K	1077.4K
5 × 5	6134	102.9K	1664.9K

Tables 10.1 and 10.2 describe the comparison of different gray levels and different sizes of a structuring element. Such a chip used in a pipeline processor could produce grayscale morphological output at real-time video rates. An example pipeline implementation is shown in Figure 10.6 for a 3×3 structuring element. The input image is shifted through a register (or delay line), so that a 3×3 window of data is available for calculation of one output point. Each of these nine values is thresholded simultaneously to produce nine threshold decomposition outputs ($I1–I9$). The 3×3 structuring element is also decomposed ($S1–S9$). Binary dilation is performed using logic gates to implement Equation 10.25. Note that the order of either the image or the structuring element must be reversed from normal raster scan to produce the reflection required by Equation 10.25.

It should be noted that if a particular structuring element will always be used, the architecture of Figure 10.6 can be significantly simplified by elimination of all structuring element thresholding devices, all AND gates, and many of the lines.

10.2 Implementing Morphological Operations Using Programmable Neural Networks

Neural networks have been studied for decades to achieve human-like performances. The neural network approach exhibits two major characteristics

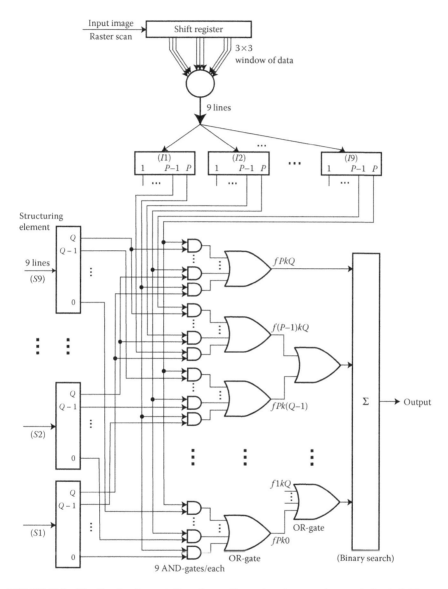

FIGURE 10.6 Pipeline implementation of two-dimensional grayscale morphological dilation for a 3×3 structuring element with gray levels ranging from 0 to Q and with image gray levels ranging from 0 to P. Each input image shift results in a new output point in real time.

in the field of problem solving. First, neural network architectures are parallel and connectionist. The connections between pairs of neurons play a primary role in the function performed by the whole system. The second characteristic is that the prescriptive computer programming routine is replaced by heuristic learning processes in establishing a system to solve a particular problem. It means that a neural network is provided with representative data, also called *training data*, and some learning process, which induces the data regularities. The modification of the transfer function at each neuron is derived empirically according to the training data set in order for maximizing the goal.

Two operating phases, learning and execution, are always encountered in neural networks. During the *learning phase*, the networks are programmed to take input from the training set, and adjust the connection weights to achieve the desired association or classification. During the *execution phase*, the neural networks are to process input data from outside world to retrieve corresponding outputs. These two phases are different in both the processing types and the time constraints involved.

Neural networks can be considered as signal processors, which are "value passing" architectures, conceptually similar to traditional analog computers. Each neuron may receive a number of inputs from the input layer or some other layers, and then produces a single output, which can be as an input to some other neurons, or a global network output. Many neural network models, such as Rosenblatt's and Minsky's *Perceptron* [Minsky and Papert 1969; Rosenblatt 1962] and Fukushima's *Cognitron* [Fukushima 1975], which use binary inputs and produce only binary outputs, can implement Boolean functions.

In this section, we are mainly concerned with the feed-forward layered neural network, which form a simple hierarchical architecture. In such a network, the execution of each layer is performed in a single step. The input neurons in the first layer can be viewed as entry points for sensor information from the outside world, called *sensoring neurons*, and the output neurons in the last layer are considered as decision points control, called *decision neurons*. The neurons in the intermediate layers, often called *hidden neurons*, play the role of interneurons. The network implicitly memorizes an evaluated mapping from the input vector and the output vector at the execution time, each time the network traversed. The mapping, represented as $\mathbf{X} \rightarrow \mathbf{Y}$, is just a transfer function that takes the n-dimensional input vectors $\mathbf{X} = (x_1, x_2, \ldots, x_n)$ and then outputs m-dimensional input vectors $\mathbf{Y} = (y_1, y_2, \ldots, y_m)$.

10.2.1 Programmable Logic Neural Networks

The concept of building neural network structure with dynamically programmable logic modules (DPLMs) is based on early efforts aimed at providing general design of logic circuits by using universal logic modules [Armstrong and Gecsei 1979]. The neural networks are constructed with node modules, which have only a limited number of inputs. Figure 10.7 shows a module of 2-bit input DPLM that consists of a commonly used 4-to-1 multiplexer and a 4-bit control register. In this circuit, the role of 4-bit input and 2-bit control is exchanged. We use the 2-bit control ($\mathbf{X}$) to select any bit of the 4-bit value in the register (C) as the output ($\mathbf{Y}$).

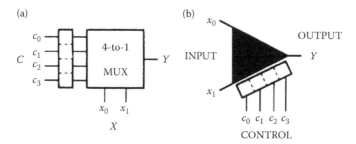

FIGURE 10.7 (a) A neuron cell consists of a 4-to-1 multiplexer and a 4-bit control register, (b) symbol for a 2-input DPLM with a 4-bit control register.

In Figure 10.7, $X = (x_0, x_1)$ denotes the input vector having four combinations: {00, 01, 10, 11}. The control register C determines the value of the single-bit output Y from the input vector. Therefore, by loading a 4-bit control word, one of the 16 (i.e., 2^4) possible Boolean functions for two input variables can be chosen. These Boolean functions and associated control codes (i.e., the truth table) are given in Table 10.3. A 2-input DPLM with all 16 possible Boolean functions is called a universal module. If the 16 possible functions are not fully utilized, the size of the control register could be reduced [Vidal 1988].

TABLE 10.3

Truth Tables of 16 Possible Boolean functions

Function	Control Code	Input $(x_0 x_1)$			
		00	01	10	11
FALSE	0000	0	0	0	0
AND	0001	0	0	0	1
LFT	0010	0	0	1	0
LFTX	0011	0	0	1	1
RIT	0100	0	1	0	0
RITX	0101	0	1	0	1
XOR	0110	0	1	1	0
OR	0111	1	1	1	1
NOR	1000	1	0	0	0
XNOR	1001	1	0	0	1
NRITX	1010	1	0	1	0
NRIT	1011	1	0	1	1
NLFTX	1100	1	1	0	0
NLFT	1101	1	1	0	1
NAND	1110	1	1	1	0
TRUE	1111	1	1	1	1

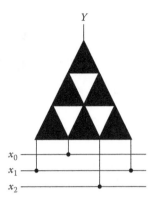

FIGURE 10.8 A 3-input primitive unit composed of six 2-input DPLMs.

Two-input modules are the simplest primitive units. Figure 10.8 illustrates a general-purpose 3-input DPLM composed of six 2-input modules. The desired function is selected by the corresponding control codes loaded to the six 2-input modules. For example, the AND codes are loaded to all six modules, acting as the AND function of the three inputs; this is called CONVERGE. By loading OR codes, similarly, it performs the OR function of the inputs, called DIVERGE. The accumulative combinations of primitive blocks can be used to achieve more complex architectures for various applications.

In this section, we are concerned with neighborhood processing processed especially for morphological operations. Hence, the structure of a 3-input DPLM in Figure 10.8 could be reduced into the structure in Figure 10.9. In the neighborhood operations, only three 2-input modules are needed in the pyramid structure. That is, the module in the upper level will receive inputs from the two in the lower level. Instead of loading the same control code to all modules, the new architecture can be loaded with different control codes

FIGURE 10.9 A 3-input primitive unit composed of three 2-input DPLMs.

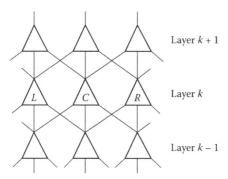

FIGURE 10.10 The structure of two-dimensional pyramid neural network composed of 3-input DPLMs.

in each level, so that more functions could be performed. A number of these functions are discussed in the next section.

10.2.2 Pyramid Neural Network Structure

To illustrate the process of pyramid model, we use a simple example constructed with 3-input DPLMs. Figure 10.10 shows the structure of the network. In the network, every node (C_{k+1}) in the ($k + 1$)th layer receives inputs from the three neighbors (nodes L_k, C_k, and R_k) in the kth layer, and all nodes in the same layer are assigned the same function. Hence, only one control register is required for each layer. Some functions are illustrated in Table 10.4.

It has to be noted that the CONVERGE and DIVERGE operations are actually the AND and the OR Boolean function, respectively. They are equivalent to the morphological, erosion, and dilation. The RIGHT-ON operation is used to detect the right neighbor and the LEFT-ON operation the left neighbor. Similarly, the CENTER-ON operation detects the center pixel.

TABLE 10.4

Boolean Functions Performed by 3-Input DPLMs

		Control	
Function Names	**Equations**	**Upper**	**Lower**
CONVERGE	$C_{k+1} \leftarrow L_k C_k R_k$	AND	AND
DIVERGE	$C_{k+1} \leftarrow L_k + C_k + R_k$	OR	OR
NONE	$C_{k+1} \leftarrow \bar{L}_k \bar{C}_k \bar{R}_k$	AND	NOR
CENTER-ON	$C_{k+1} \leftarrow C_k$	RITX	LFTX
		LFTX	RITX
RIGHT-ON	$C_{k+1} \leftarrow R_k$	RITX	RITX
LEFT-ON	$C_{k+1} \leftarrow L_k$	LFTX	LFTX

10.2.3 Binary Morphological Operations by Logic Modules

To implement morphological operations, a new neural architecture by the use of the pyramid model is constructed. The connections among the four 3-input DPLMs are shown in Figure 10.11 for a 9-input three-dimensional pyramid module. The layers k and $k + 1$ are called *regular layer* and the layer $k + \frac{1}{2}$ is called *intermediate layer*.

In the structure of 9-input module, each neuron $U_{k+\frac{1}{2}}(i, j)$ in the intermediate layer $k + \frac{1}{2}$ receives three inputs of its column-major neighbors ($U_k(i, j - 1)$, $U_k(i, j)$, $U_k(i, j + 1)$) in the preceding layer k. The neuron $U_{k+1}(i,j)$ then collects the three outputs of its three row-major neighbors ($U_{k+\frac{1}{2}}(i - 1,j)$, $U_{k+\frac{1}{2}}(i,j)$, $U_{k+\frac{1}{2}}(i + 1, j)$) in the preceding intermediate layer $k + \frac{1}{2}$. In other words, the module will extract one-dimensional local features from each column, and then combine the three adjacent 3×1 features into 3×3 two-dimensional local feature. The overall structure of the network is shown in Figure 10.12. There is no limitation on the number of layers in a neural network; that is, one could concatenate as many layers as desired and then assign different Boolean functions to the nodes for specific applications.

Since only one operation is applied on the whole original image at a certain time, loading different control codes to DPLMs on the same major dimension is not required. Only two control registers (one for the upper level and the other for the lower level) are connected for each column in any intermediate layer (layer $k + \frac{1}{2}$) as the same as for each row in any regular layer (layer k). That is to apply a certain function with respect to a column or a row. Some examples of morphological operations performed by this neural network for extracting the local features are shown in Table 10.5.

As we can see in Table 10.5, for all dilations, the upper level of the modules should be loaded with the DIVERGE operation, and for all erosions, the CONVERGE operation is employed. The lower level is loaded with different

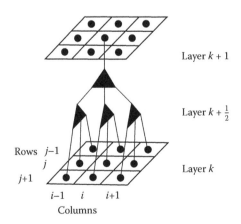

FIGURE 10.11   The 9-input pyramid module composed of four 3-input DPLMs.

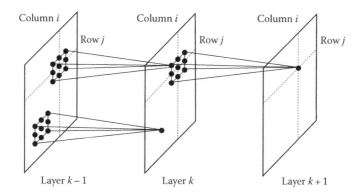

FIGURE 10.12 The overall structure of neural network formed with 3×3 input DPLMs for image morphological processing.

TABLE 10.5

Morphological Operations Using Pyramid Networks

Structuring Elements	Morphological Operations	Intermediate Layers	Regular Layers
111 111 111	DILATION	DIVERGE	DIVERGE
111	EROSION	CONVERGE	CONVERGE
000 111	DILATION	CENTER-ON	DIVERGE
000	EROSION	CENTER-ON	CONVERGE
010 010	DILATION	NONE, CONVERGE, NONE	DIVERGE
010	EROSION	NONE, CONVERGE, NONE	CONVERGE
010 111	DILATION	CENTER-ON, CONVERGE, CENTER-ON	DIVERGE
010	EROSION	CENTER-ON, CONVERGE, CENTER-ON	CONVERGE
100 010	DILATION	LEFT-ON, CENTER-ON, RIGHT-ON	DIVERGE
001	EROSION	LEFT-ON, CENTER-ON, RIGHT-ON	CONVERGE
001 010	DILATION	RIGHT-ON, CENTER-ON, LEFT-ON	DIVERGE
100	EROSION	RIGHT-ON, CENTER-ON, LEFT-ON	CONVERGE

operations for different SEs. Suppose that an erosion is applied on a binary image using the following 3×3 structuring element:

1	1	1
1	1	1
1	1	1

in the layer k and pass the result to the layer $k + 1$. One may simply load CONVERGE to all 3-input DPLMs in both intermediate layer $k + \frac{1}{2}$ and regular layer $k + 1$. The neurons in layer $k + \frac{1}{2}$ will output a "1" if the three neighbors in column are all ones. The neurons in layer $k + 1$ will output a "1" if the three neighboring intermediate neurons in row have all ones; that is, the nine (3×3) neighboring pixels are all ones. Similarly, for dilation with the same structuring element, we may simply load DIVERGE to all neurons.

If the structuring element is changed to:

0	0	0
1	1	1
0	0	0

the erosion requires applying CENTER-ON to neurons in the layer $k + \frac{1}{2}$ and CONVERGE to neurons in layer $k + 1$, and the dilation requires applying CENTER-ON to neurons in the layer $k + \frac{1}{2}$ and DIVERGE to neurons in the layer $k + 1$. Another example using a different structuring element:

1	0	0
0	1	0
0	0	1

is illustrated as follows. The LEFT-ON may be loaded to the first columns, CENTER-ON to the second columns, and RIGHT-ON to the third columns of each pyramid module in the layer $k + \frac{1}{2}$. At the same time, CONVERGE is loaded in the layer $k + 1$. Therefore, any occurrence of line segments with $135°$ is extracted in the layer $k + 1$. Similarly, DIVERGE is also applied to the layer $k + 1$ to dilate any occurrence of "1" in the layer k to a diagonal line segment with the length 3.

A more complicated neural network could be constructed by extending DPLMs and their connections. By assigning various combinations of Boolean functions, a neural network may act in quite different behaviors. Thus, DPLM networks can be viewed as a class of "digital perceptron" which can support discrete forms of distributed parallel problem solving. They offer a new conceptual framework as well as a promising technology for developing artificial neural networks.

10.2.4 Grayscale Morphological Operations by Tree Models

Dilation (erosion) is the local maximum (minimum) of an image adding (subtracting) a structuring element. A neural architecture can be used to extract the maximum or minimum. Two comparator subnets, which use threshold logic nodes to pick up the maximum or minimum of two analog inputs, are shown in Figure 10.13 [Martin 1970]. These nets are suited for the output selected from one of the input samples.

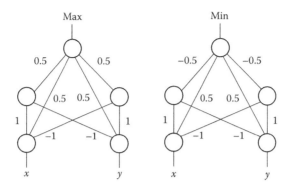

FIGURE 10.13 Two comparator subnets which can select the maximum and minimum of two analog inputs x and y at the same time.

The function of the above bi-comparator subnets can be easily verified. Since all the cells use threshold logic nonlinearity, the output values should be defined as follows:

$$\begin{aligned}
\max &= \frac{1}{2}x + \frac{1}{2}y + \frac{1}{2}f_t(x-y) + \frac{1}{2}f_t(y-x) \\
&= \frac{1}{2}x + \frac{1}{2}y + \frac{1}{2}|x-y| = \begin{cases} x & \text{if } x \geq y \\ y & \text{if } x < y \end{cases}
\end{aligned}$$
(10.39)

$$\begin{aligned}
\min &= \frac{1}{2}x + \frac{1}{2}y - \frac{1}{2}f_t(x-y) - \frac{1}{2}f_t(y-x) \\
&= \frac{1}{2}x + \frac{1}{2}y - \frac{1}{2}|x-y| = \begin{cases} x & \text{if } x \leq y \\ y & \text{if } x > y \end{cases}
\end{aligned} \quad ,$$
(10.40)

where f_t is the threshold logic nonlinearity

$$f_t(\alpha) = \begin{cases} \alpha & \text{if } x \geq y \\ 0 & \text{if } x < y \end{cases}$$
(10.41)

Comparator subnets can be layered into roughly $\log_2 M$ levels of a binary tree structure to pick up the maximum or minimum within a 3×3 mask and that is illustrated in Figure 10.14. Detailed implementation of morphological operations using programmable neural networks can be referred to in Shih and Moh [1992].

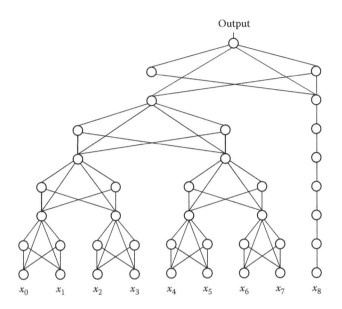

FIGURE 10.14   A binary tree structure to pick up the maximum or minimum within a 3×3 mask.

10.2.5 Improvement by Tri-Comparators

The binary tree structure discussed requires eight layers of processing elements. This design will consume much processing time. A better design is to extract the maximum (minimum) of three inputs and then the maximum (minimum) of three selected maxima (minima). Figure 10.15 shows the

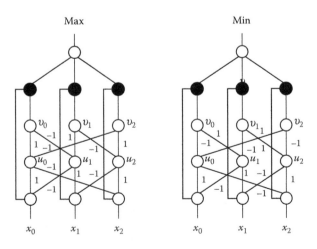

FIGURE 10.15   Tri-comparator subnets.

tri-comparator subnet for extracting the maximum or minimum of three analog inputs. The black circles indicate the Enable/Disable function and the blank circles denote the simple mathematical function. Let x_i denote the analog inputs, u_i the output neurons in the first hidden layer, and v_i the output in the second hidden layer. Suppose that $f_{h=\theta}$ denotes the hard limiter function with threshold value θ. We have the following relationship:

$$v_0 = f_{h=0}(u_0 - u_1)$$
$$= f_{h=0}(f_{h=0}(x_0 - x_2) - f_{h=0}(x_1 - x_0)) \tag{10.42}$$

$$v_1 = f_{h=0}(u_1 - u_2)$$
$$= f_{h=0}(f_{h=0}(x_1 - x_0) - f_{h=0}(x_2 - x_1)) \tag{10.43}$$

$$v_2 = f_{h=0}(u_2 - u_0)$$
$$= f_{h=0}(f_{h=0}(x_2 - x_1) - f_{h=0}(x_0 - x_2)), \tag{10.44}$$

where

$$f_{h=\theta}(\alpha) = \begin{cases} 1 & \text{if } \alpha \geq \theta \\ -1 & \text{if } \alpha < \theta \end{cases}$$

It can be easily proven that for $i, j = 0, 1, 2$ and $i \neq j$,

$$v_I = \begin{cases} 1 & \text{if } x_i \geq x_j \\ -1 & \text{if } x_i < x_j \end{cases} \tag{10.45}$$

A feed-forward net, which determines the maximum or minimum of nine inputs can be constructed by feeding the outputs of three tri-comparators at the lower layer forward to the higher layers. This forms a triple tree structure shown in Figure 10.16.

10.2.6 Another Improvement

The triple tree model containing at least six layers of processing elements still takes much time to extract the maximum (minimum) of inputs. Another improved method has been developed to reduce the number of layers and to speed up the neural network.

To extract the maximum (or minimum) of n inputs, we need $C(n, 2) = n(n-1)/2$ nodes (say v-nodes) in the first hidden layer to indicate whether any input x_i is greater than or equal to another input x_j. Also, we need n nodes (say u-nodes) in the second hidden layer to indicate if the corresponding input x_i is the maximum (or minimum) of the n inputs. For nine inputs, we need $C(9, 2) = 36$

Output

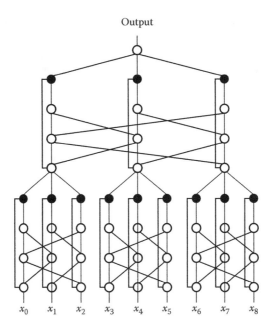

FIGURE 10.16 A feed-forward net that determines which of nine inputs is maximum or minimum using a triple tree structure of comparator subnets shown in Figure 10.15.

v-nodes in the first hidden layer as shown in Figure 10.17. The output function of each v-node is defined as follows:

$$v_{ij} = f_{h=0}((1)x + (-1)x) \quad \text{for } i < j \tag{10.46}$$

These units are used to determine which of two inputs is larger.

In the second hidden layer, we need nine u-nodes to sum up all the "larger/smaller" information from first hidden layer together to determine whether the particular input is the maximum or minimum. Let $u_i^{\max}$ denote the u-nodes for extracting the maximum and $u_i^{\min}$ for the minimum. Suppose x_i is the maximum of nine inputs, then for all $k \neq i$, the $x_i - x_k$ should be greater than (or equal to) zero. Thus, the following conditions should be satisfied:

$$f_{h=\theta}(x_i - x_k) = \begin{cases} 1 & \text{for all } k \geq i \\ -1 & \text{for all } k < i \end{cases} \tag{10.47}$$

Also, for the minimum value x_i,

$$f_{h=\theta}(x_i - x_k) = \begin{cases} 1 & \text{for all } k \leq i \\ -1 & \text{for all } k > i \end{cases} \tag{10.48}$$

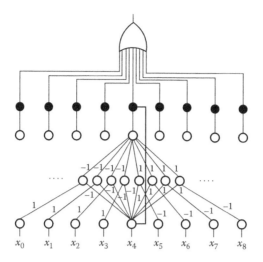

FIGURE 10.17 A new design of neural net for extracting the maximum or minimum of nine inputs.

The above conditions can be trivially verified since the maximum value must be greater than or equal to any other inputs, and the minimum value must be less than or equal to any other inputs. Therefore, the output function of node $u_i^{\max}$ for extracting the maximum value is defined by

$$u_i^{\max} = f_{h=7.5}\left(\sum_{k\geq i} v_{ik} + (-1)\sum_{k<i} v_{ki}\right), \tag{10.49}$$

where

$$f_{h=7.5}(x) = \begin{cases} 1 & \text{if } x \geq 7.5 \\ -1 & \text{if } x < 7.5 \end{cases}$$

And the output function of node $u_i^{\min}$ unit for indicating the minimum value is defined by

$$u_i^{\min} = f_{h=7.5}\left((-1)\sum_{k\geq i} v_{ik} + \sum_{k<i} v_{ki}\right), \tag{10.50}$$

Again, we may use transistors at output circuit of nodes in the second layer. The value 1 of $u_i^{\max}$ (or $u_i^{\min}$) will "enable" the input x to be conducted to the output of the whole circuit. In this design, we need no more than three layers of processing elements. Because of the parallelism of processing elements, this circuit can execute much faster than the past models. But, the trade-off is the large number of connections and processing elements in the neural network.

10.2.7 Application of Morphological Operations on Neocognitron

One of the most famous neural network models for visual pattern recognition is *"neocognitron,"* which is proposed by Fukushima [1983, 1988]. It is a multilayered neural network consisting of a cascade of layers of neuron-like cells and can be trained to achieve character recognition.

The hierarchical structure of the network is illustrated in Figure 10.18. There are forward connections between cells in adjoining layers. The input layer, represented by U_0, consists of a two-dimensional array of sensing neurons. Each of the succeeding stages has a layer of "S-cells" followed by a layer of "C-cells." Thus, layers of S-cells and C-cells are arranged alternatively in the whole architecture. We use U_{S_l} and U_{C_l} to represent the layers of S-cells and C-cells of the *l*th stage, respectively.

According to the activation of the whole architecture, we may say that S-cells are feature-extracting cells. Connection weights converging to feature-extracting S-cells are variable and are reinforced during the learning phase. Generally speaking, in the lower stages, local features, such as a particular orientation, are extracted, and in the higher stages, more global features, such

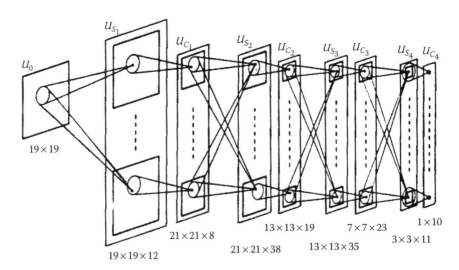

FIGURE 10.18 Hierarchical neural network architecture of neocognitron. The numerals at the bottom of the figure show the total numbers of S- and C-cells in individual layers of the network, which are used for the handwritten numeral recognition system. (Courtesy of Fukushima 1988.)

as a longer line segment or a part of input pattern, are extracted. The use of C-cells allows positional errors in the features of the stimulus, which defines the acceptable tolerance range. Connection weights from S-cells to C-cells are fixed and unchangeable. Each C-cell receives signals from a group of S-cells, which extract the same feature from slightly different positions. The C-cell is activated if at least one of these S-cells is active (same as Boolean OR operation). Hence, the C-cell's response is less sensitive to translation of the input patterns.

In the first stage, the function of S-cells can be considered as erosion operations using the twelve different local features as SEs. And the function of C-cells is the dilation operation with some other SEs, which are orthogonal to the related local features.

It is the reason why U_C layers may respond to the same pattern with small translation. It can be verified by the following simple example shown in Figure 10.19.

To simplify the problem, suppose the input patterns are 5×5 binary images and will be presented in the input layer U_0. In the first stage, there are two 5×5 U_{S_1}-cell planes: one for extracting a vertical line segment and the other for horizontal line segment, and two 7×7 U_{C_1} planes where each of them corresponds to either U_{S_1}-cell planes. Each U_{S_1}-cell receives the output signals from the eight neighboring pixels in the input layer. The U_{C_1} cells can be active if any U_{S_1} cell in the neighborhood is activated. In the example, since there exists a vertical line segment going through the three pixels $U_0(1, 2)$, $U_0(2, 2)$, and $U_0(3, 2)$, the $U_0(2, 2)$ is activated and $U_{C_1}(2, 1)$, $U_{C_1}(2, 2)$, and $U_{C_1}(2, 3)$ are also activated.

In the second stage, we use four 3×3 U_{S_2} cell-planes to extract corners of four different orientations with 5×5 SEs. According to the overall architecture, the connections are from U_{C_1} to U_{S_2}. It is obvious that the 3×3 pyramid DPLM, shown in Figure 10.11, is not able to handle such operations. A more complicated structure, which is not going to be discussed in detail, should be employed in this stage.

The same situation happens in the third stage. There are four single U_{C_2} cells used to indicate the orientation of the input pattern. Each of these neurons has connections from the four cell planes of the U_{C_2} layer, which results in sophisticated connections. After training (i.e. loading appropriate control codes), this simplified neural network can be used to recognize the orientation of any U-like pattern of different sizes.

10.3 MLP as Processing Modules

MLPs are feed-forward nets with one or more layers, called *hidden layers*, of neural cells, called *hidden neurons*, between the input and output layers.

Initial stage U_0

	1	2	3	4	5
1	0	1	0	1	0
2	0	1	0	1	0
3	0	1	0	1	0
4	0	1	1	1	0
5	0	0	0	0	0

Erosion
010
010
010

000
111
000

First stage U_{s_1}

	1	2	3	4	5
1	0	0	0	0	0
2	0	1	0	1	0
3	0	1	0	1	0
4	0	0	0	0	0
5	0	0	0	0	0

	1	2	3	4	5
1	0	0	0	0	0
2	0	0	0	0	0
3	0	0	0	0	0
4	0	0	1	0	0
5	0	0	0	0	0

Dilation
000
111
000

010
010
010

First stage U_{c_1}

	0	1	2	3	4	5	6
0	0	0	0	0	0	0	0
1	0	0	0	0	0	0	0
2	0	1	1	1	1	1	0
3	0	1	1	1	1	1	0
4	0	0	0	0	0	0	0
5	0	0	0	0	0	0	0
6	0	0	0	0	0	0	0

	0	1	2	3	4	5	6
0	0	0	0	0	0	0	0
1	0	0	0	0	0	0	0
2	0	0	0	0	0	0	0
3	0	0	0	1	0	0	0
4	0	0	0	1	0	0	0
5	0	0	0	1	0	0	0
6	0	0	0	0	0	0	0

Second stage U_{s_2}

	2	3	4
2	0	0	0
3	0	0	0
4	0	0	0

	2	3	4
2	0	0	0
3	0	0	0
4	1	0	0

	2	3	4
2	0	0	0
3	0	0	0
4	0	0	1

	2	3	4
2	0	0	0
3	0	0	0
4	0	0	0

Second stage U_{c_2}

	2	3	4
2	0	0	0
3	0	0	0
4	0	0	0

	2	3	4
2	0	0	0
3	1	1	0
4	1	1	0

	2	3	4
2	0	0	0
3	0	1	1
4	0	1	1

	2	3	4
2	0	0	0
3	0	0	0
4	0	0	0

Third stage U_{s_3}

1	⊔
0	⊏
0	⊓
0	⊐

FIGURE 10.19 A simple example illustrating the activation of neocognitron.

The learning process of MLPs is conducted with the error back-propagation learning algorithm derived from the *generalized delta rule* [Rumelhart, Hinton, and Williams 1986]. According to Lippmann [1987], no more than three layers of neurons are required to form arbitrarily complex decision regions in the hyperspace spanned by the input patterns. By using the sigmoidal non-linearities and the decision rule in selecting the largest output, the decision regions are typically bounded by smooth curves instead of line segments. If training data are sufficiently provided, an MLP can be trained to discriminate input patterns successfully. A discussion on limits of the number of hidden neurons in MLPs can be found in Huang and Huang [1991].

Most of image processing operations, such as smoothing, enhancement, edge detection, noise removal, and morphological operations, require checking the values of neighboring pixels. Two basic morphological operations, dilation (similar to "expansion") and erosion (similar to "shrink"), are often combined in sequences for image filtering and feature extraction. The size of neighborhood may vary with applications. Eight-neighbor along with the center pixel is often used. The neighborhood processing can be implemented by look up tables to associate the relationship between input and output values. The input–output association can be realized by neural network models with nine input neurons (assuming that the 8-neighbor area is used) and one output neuron with one or more layers of hidden neurons in between. By iteratively providing input vectors to compute output from the look up table and comparing with the desired output, the error will be used to adjust the weights of connections. Therefore, MLPs can gradually and adaptively learn the input–output association. If the training MLP converges with a set of weights, it can be used to perform the same transformation on any image.

Figure 10.20 shows the MLP modules designed as the general-purpose image processing modules. The more hidden layers are used and the more complicated discriminant regions it forms in the domain space spanned with input vectors. There is a tradeoff between the complexity of MLP connections and the converge time of the MLP. Since the general purpose MLP module is concerned, reasonably short training time to adapting the MLP and big capacity of connection weights are expected.

The error back-propagation training algorithm was given in Rumelhart, Hinton, and Williams [1986]. A sigmoidal logistic nonlinearity

$$f(\alpha) = \frac{1}{1 + e^{-(\alpha - \theta)}} \tag{10.51}$$

is used as the output function of neurons, where θ is a bias (or a threshold). Let $x_i^{(n)}$ denote the output of neuron i in layer n. Also, let w_{ij} denote the connection weight from neuron i in a lower layer to neuron j in the immediately higher layer. The activation values are calculated by

$$x_j^{(n+1)} = f\left(\sum_i w_{ij} x_i^{(n)} - \theta_j\right). \tag{10.52}$$

The major part of the algorithm is to adjust the connection weights according to the difference between the actual output of Equation 10.52 and the desired output provided in the training patterns. The Equation for weight adjustment is

$$w_{ij}(t + 1) = w_{ij}(t) + \eta \delta_j x_i^{(n)}, \tag{10.53}$$

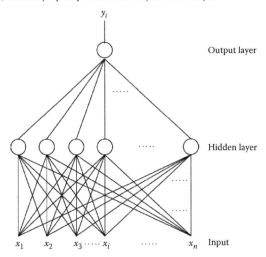

(a) A multilayer perceptron includes only one hidden layer

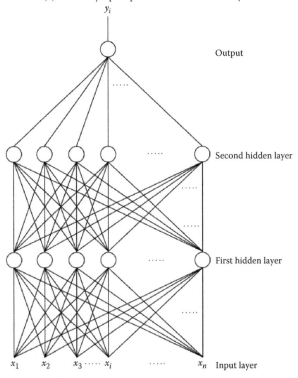

(b) A three-layer perceptron includes two hidden layers

FIGURE 10.20 MLP modules for the general purpose image processing.

0	1	0
1	1	1
0	1	0

FIGURE 10.21 A MSE.

where η is the gain term (or called *learning rate*), δ_j is the error term for neuron j, and x_i is either the output of neuron i or an input when $n = 1$. The error term δ_j is

$$\delta_j = \begin{cases} x_j^{(n+1)}(1 - x_j^{(n+1)})(d_j - x_j^{(n+1)}) & \text{if neuron } j \text{ is in the output layer} \\ x_j^{(n+1)}(1 - x_j^{(n+1)})\sum_k \delta_k w_{jk} & \text{if neuron } j \text{ is in a hidden layer} \end{cases}, \tag{10.54}$$

where k is with respect to all the neurons in the layer where neuron j is located.

The training patterns can be generated by all the variations in a local window of the specified low-level image operations. For example, in a 3×3 neighboring area, there are up to $2^9 = 512$ different training patterns, with a vector of nine binary values and one desired output. For instance, a morphological erosion with the structuring element shown in Figure 10.21 will respond output 1 at the central pixel for the 16 input patterns in Figure 10.22 and output 0 for other input patterns.

Figure 10.23 shows the overall architecture consisting of MLP modules for low-level image operations. There is one MLP module corresponding to each pixel in the input image. Every MLP module applies the operation, which has been trained to the pixel and its $m \times m$ neighboring pixels, where m is the window size of neighborhood. The window size can be changed depending upon the desired operations for training.

Since the same operation is simultaneously applied to all the modules for all the pixels in the input image, the set of connection weights are considered

0	1	0		1	1	0		0	1	0		0	1	0		0	1	1		1	1	0		0	1	0		0	1	1
1	1	1		1	1	1		1	1	1		1	1	1		1	1	1		1	1	1		1	1	1		1	1	1
0	1	0		0	1	0		1	1	0		0	1	1		0	1	0		1	1	0		1	1	1		0	1	1

1	1	1		1	1	0		0	1	1		1	1	0		0	1	1		1	1	1		1	1	1		1	1	1
1	1	1		1	1	1		1	1	1		1	1	1		1	1	1		1	1	1		1	1	1		1	1	1
0	1	0		0	1	1		1	1	0		1	1	1		1	1	1		0	1	1		1	1	0		1	1	1

FIGURE 10.22 Patterns having desired output 1 with erosion by the structuring element in Figure 10.21.

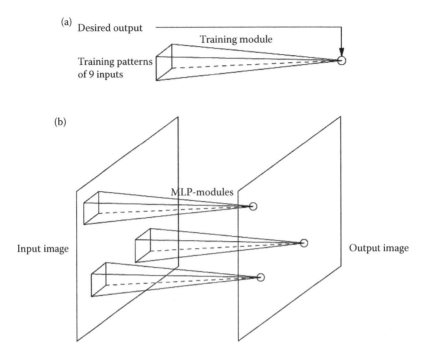

FIGURE 10.23 Overall system architecture with MLP modules for low-level image operations. (a) The pyramid is used for training tasks and storage of the trained weights. (b) Each pyramid represents an MLP module and receives m^2 inputs from its $m \times m$ neighboring pixels.

the same for all MLP modules. That is, all the MLP modules in the same stage can share the same set of trained connection weights. This will significantly reduce the required number of local memory for connections weights in MLP module. In Figure 10.23a, the pyramid called *the training module*, is used for training task and provides shared memory for connection weights. Thus, no local memory is required for individual MLP modules in between the input and output images to hold the connection weights. The training phase is first turned up at the training module. After the weight vector in the training module converges, the connection weights are frozen, shared, and retrieved by all the MLP modules to determine the activation value of each pixel after the operation is applied.

With the introduction of the error back-propagation learning algorithm, the MLP modules are capable of adapting themselves to the expected operations. It is also more flexible than those designed for dilation and erosion operations only [Krishnapuram and Chen 1993].

During the experiments with different image operations, the training sets for some operations, such as dilation, erosion, and noise removal, have a small number of patterns with the desired output 1 or 0. For example, an erosion with the structuring element in Figure 10.21 has only 16 patterns

1	1	1
1	1	1
1	1	1

FIGURE 10.24   The structuring element with all 1s.

with output 1, and an erosion with the structuring element shown in Figure 10.24 expects an output 1 only if all nine inputs are 1s.

By checking the training patterns, the values of some m^2 inputs do not even affect the output. Without considering a certain value of the central pixel, the training pattern set can be reduced. The same effect can be achieved by connecting the output neuron directly with the central input neuron and the MLP module operation is bypassed when the central pixel equals 0 or 1. For instance, applying an erosion to the central pixel with value 0 will never output 1. Also, applying a dilation to the central pixel with value 1 will always output 1. Such bypassing connections can be specified with the desired output as weights shared in all the MLP modules. The bypassing connections are intended to perform an exclusive OR between the central pixel and the unchanged input, which is defined with respect to operations. The operation of an MLP is disabling if the exclusive OR gets output 1. It means that the central pixel has the same value as specified in the shared module. Figure 10.25 illustrates the bypassing connection and expresses how it affects the operation of an MLP module. The bypassing control is to enable/disable the activation of neurons in the modules and is determined by

$$E = x_c \textbf{ XOR } U, \tag{10.55}$$

where E means the enable/disable control to neurons, x_c is the value of the central pixel, and U is the expected input. For instance, $U = 1$ is set for dilation and $U = 0$ for erosion. The dashed lines with arrows in Figure 10.25 show the enable/disable controls to neurons, while the dotted lines indicate the bypassing connection for passing the value of the central pixel directly to the output. The output is defined by

$$\text{Output} = y \cdot E + x_c \cdot \overline{E}, \tag{10.56}$$

where y is the output of the MLP module.

The proposed architecture can be used as a basis to organize multistage or recurrent networks. One may stack up with more than one stage as shown in Figure 10.26. By training the stages with a different set of patterns, one can provide and apply a sequence of image operations to the input image. The outputs of the last stage may be connected to the inputs of the first stage to

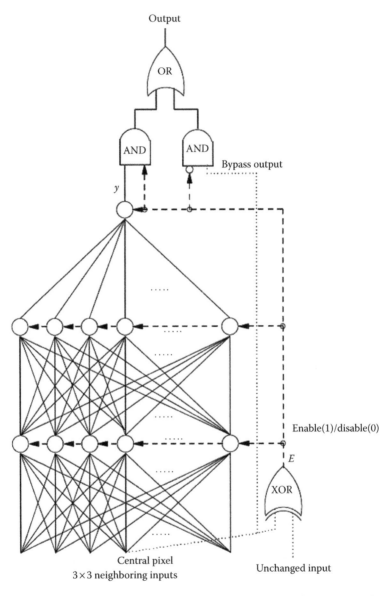

FIGURE 10.25 MLP module with a bypass control to enable/disable the activation of neurons in the modules. The dashed lines with arrows show the enable/disable controls to neurons, while the dotted lines indicate the bypassed connection for passing the value of the center pixel to the output.

form a recurrent architecture so that the same operations in sequence can be applied to the image for more than one iteration. The detailed training procedure and the application of the MLPs to thinning can be referred to in Moh and Shih [1995].

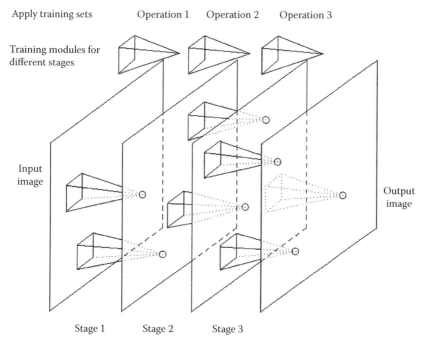

FIGURE 10.26 Stacked MLP-based architecture for multiple image operations. Only some MLP-modules are shown to simplify the figure. The separated pyramids at the top of the figure are the training modules for different stages. There may be more stages stacked on the architecture.

10.4 A Systolic Array Architecture

A systolic array architecture for the recursive morphological operations is presented in this section. The design will be introduced step-by-step in three subsections—the basic cell design, the systolic array for processing only one scan of the input image, and the complete array for recursive morphology which processes the input image in two scans.

10.4.1 Basic Array Design

Consider that the recursive morphological erosion is applied over row i of an $n \times n$ image f. The scanning order will affect the output image in recursive morphology. Starting with a corner pixel, the scanning sequences in a two-dimensional digital image is in general divided into the following:

1. *LT* denotes left-to-right and top-to-bottom. (Note: this is a usual television raster scan.)

2. *RB* denotes right-to-left and bottom-to-top.
3. *LB* denotes left-to-right and bottom-to-top.
4. *RT* denotes right-to-left and top-to-bottom.

The structuring element k_{LT} in the following equation sweeps through f in the *LT* direction.

$$
k_{LT} = \begin{bmatrix} A_1 & A_2 & A_3 \\ A_4 & A_5 & \times \\ \times & \times & \times \end{bmatrix}, \quad
k_{RB} = \begin{bmatrix} \times & \times & \times \\ \times & A_5 & A_6 \\ A_7 & A_8 & A_9 \end{bmatrix},
$$

$$
k_{LB} = \begin{bmatrix} \times & \times & \times \\ A_4 & A_5 & \times \\ A_7 & A_8 & A_9 \end{bmatrix}, \quad
k_{RT} = \begin{bmatrix} A_1 & A_2 & A_3 \\ \times & A_5 & A_6 \\ \times & \times & \times \end{bmatrix},
$$

(10.57)

where "×" denotes don't care. Each pixel, $h_{i,j}$, where $1 \le j \le n$ of row i is generated as follows:

$$ h_{i,j} = \min\{h_{i-1,j-1} - A_1, h_{i-1,j} - A_2, h_{i-1,j+1} - A_3, h_{i,j-1} - A_4, f_{i,j} - A_5\}. \quad (10.58) $$

To update $h_{i,j}$, we need to access the updated pixels of row $i-1$ (i.e., $h_{i-1,j-1}$, $h_{i-1,j}$, $h_{i-1,j+1}$), $h_{i,j-1}$, and the original pixel $f_{i,j}$. A systolic array for processing one row of the image is shown in Figure 10.27a.

There are six cells in the systolic array design. Figure 10.27b, illustrates the operation performed in each cell. Input pixels in rows $i-1$ and i are fed

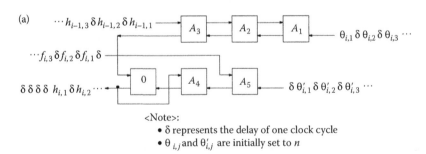

FIGURE 10.27 (a) Array for processing one row of the image, (b) the operation performed in each cell.

into the array from the left and flow to the right. Two streams of output data (θ and θ'), flowing from right to left, are used to collect the partial results. These two streams are combined at the leftmost cell to give the minimum (i.e., the updated pixel value). It is possible to use two independent streams to compare the minimum because the minimization operation is associative.

Figure 10.28 shows the operations of the array at four consecutive clock cycles, starting when the evaluation of pixel $h_{i,2}$ begins (time T_2). It can be

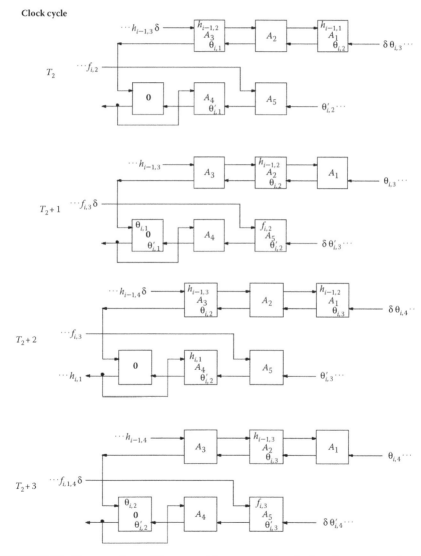

FIGURE 10.28 The operations of the array at four consecutive clock cycles, starting when the evaluation of pixel $h_{i,2}$ begins (time T_2).

seen that four clock cycles are needed to evaluate one pixel. However, due to the pipeline operations, we can update one pixel in every two-clock cycles. Thus, to process one row of n pixels, we need $4 + 2(n - 1)$ cycles in total. One problem with this design is that during processing half of the cells are idle. Utilization of the array is only 50%.

10.4.2 A Systolic Array for Processing One Scan of the Whole Image

Based on the aforementioned design for processing one row of the image, we now propose a systolic array for processing one of the two scans of the whole image. The basic idea is to concatenate the basic arrays (as shown in Figure 10.27), so that the outputs from one array become the inputs to the next array (see Figure 10.29). For an $n \times n$ image, n basic arrays are needed to form the larger array for processing the whole image. The delay, which a pixel experiences in each basic array, is four cycles. Thus, after the first pixel of the first row ($\theta_{1,1}$) enters into the top basic array in Figure 10.29, it takes $4n$ cycles

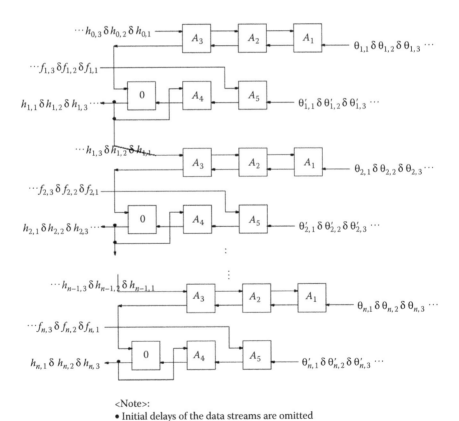

<Note>:
• Initial delays of the data streams are omitted
• Assume that $h_{0,j} = 0$, where $1 \le j \le n$

FIGURE 10.29 The systolic array for processing the whole image.

to obtain the first updated pixel of row $n(h_{n,1})$ from the bottom array. Thereafter, one new pixel of row n is generated for every two cycles. Total execution time of this systolic array for one scan of the whole image is $4n + 2(n - 1)$.

Further improvement of this 2D array is possible. Note that the total propagation delay from the top basic array to the bottom one is $4n$ cycles, whereas the last pixel of row $1(h_{1,n})$ will be generated at $4 + 2(n - 1)$. The number of basic arrays can be reduced by half if a feedback path from the output of $\lceil n/2 \rceil$th basic array is provided to the input of the first array. Thus, the first basic array will process pixels of row $\lceil n/2 \rceil + 1$ after it finishes the first row of pixels, and the second array processes rows 2 and $\lceil n/2 \rceil + 2$, and so on. In this way, the amount of hardware used is cut into half, and the total execution time remains the same.

10.4.3 The Complete Systolic Array for Processing Two Scans of the Whole Image

A straightforward way of processing two scans of the whole image is to use the systolic array in Figure 10.29 twice. In the first time, the first scan of *LT* is performed, and then, the second scan of *RB* is processed. Thus, the process of two scans of an image takes twice as much time as 1 scan of the image.

In fact, it is possible to process the two scans in parallel using the same amount of hardware. Note that the operations performed in the two scans are almost the same, and there is no restriction on the order of *LT* and *RB* scans. Thus, when performing the *LT* scan, the second scan of *RB* can be processed during the idle cycles. The systolic array design is shown in Figure 10.30.

Here the structure of a cell is somewhat more complex if the structuring element k is nonsymmetric. Specifically, the cells must use two different SEs alternatively between clock cycles to process two intermixed input streams. Also, care must be taken while feeding the cells so that only the most up-to-date values of neighboring pixels are fed in, especially when the two scans cross each other in the middle of the image.

The total execution time of the systolic array for processing two scans is the same as that in Figure 10.29 for processing only one scan. The same number of cells and interconnections are used, except that the structure of a cell is more complicated for nonsymmetric k—we need to add in each cell the storage for one more structuring element and some extra logic for alternating the values between clock cycles.

10.5 Implementation on Multicomputers

In this section, the implementations of the recursive morphological operations on multicomputers are discussed. The characteristics of multicomputers

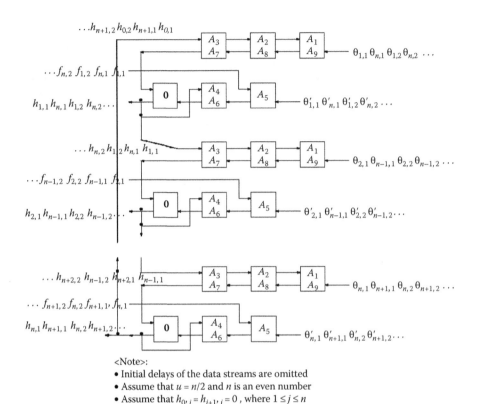

<Note>:
- Initial delays of the data streams are omitted
- Assume that $u = n/2$ and n is an even number
- Assume that $h_{0,j} = h_{j+1,j} = 0$, where $1 \leq j \leq n$

FIGURE 10.30 The array for processing two scans of the whole image.

essential to our discussions are first introduced and then the implementation of a two-scan algorithm is discussed. The implementation of a blocked two-scan algorithm is presented, and the performance results using the recursive morphology are compared with the standard MM.

10.5.1 Characteristics of Multicomputers

Multicomputers [Athas and Seitz 1988; Hwang 1987] refer to a class of Multiple Instruction stream, Multiple Data stream (MIMD) computers that consist of multiple loosely coupled processors with distributed local memories. The communication between processors is accomplished via message passing, which incurs non-negligible overhead. With the current technology, the communication overhead in most multicomputers is at least an order of magnitude higher than the computation.

To reduce the communication overhead (especially the startup delay), it is desirable to pack more data into one message. That means the amount of computation between communications (i.e., the granularity) must be increased to generate the message data. By controlling the granularity, a proper balance between the computation and communication can be achieved.

10.5.2 Implementation of the Two-Scan Algorithm

The recursive morphological operations are essentially sequential. For example, in Equation 10.58, the pixel $h_{i,j}$ cannot be updated unless the pixels $h_{i-1,j-1}$, $h_{i-1,j}$, $h_{i-1,j+1}$, and $h_{i,j-1}$ have been all updated. The only parallelism that can be exploited is through pipelining along the rows and columns of the image, as in the systolic array design. Thus, on multicomputers we must partition the problem in such a way that pipelining can take place among the processors and the granularity can be controlled.

The partitioning scheme shown in Figure 10.31, which adopts from the concept of pipelined data parallel algorithm [King and Ni 1988], will allow the granularity to vary. The updating of any pixel in a processor, say u in p_2 in Figure 10.31, will never depend on the pixels allocated to processors to its right, that is, p_3. In this way, the computation can proceed as a wave front from top-to-bottom and left-to-right. Each processor will handle two stripes of pixels in both LT and RB scans. Since there is no mutual dependence between the partitions, the granularity can be controlled. For example, in

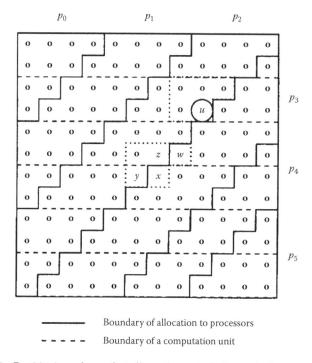

Boundary of allocation to processors

- - - - - Boundary of a computation unit

FIGURE 10.31 Partitioning scheme that allows the control of granularity.

Figure 10.31 two rows of pixels are processed together and the two ending pixels are sent as a single message to the next processor.

10.5.3 Implementation of the Blocked Two-Scan Algorithm

The blocked two-scan algorithm is a modified version of the two-scan algorithm, which is more suitable for parallel execution. The algorithm uses a rectangular partitioning scheme. Each processor handles the updating of one rectangular block. In Figure 10.32 solid boxes represent blocks of pixels and the dashed box indicates the extra storage needed for the boundary pixels of the neighboring processors, which are needed for proper updating of local boundary pixels.

Processors will execute the following operations:

Repeat

Exchange boundary pixels with their neighbors;

Perform a two-scan operation on the local pixels;

until the whole image converges.

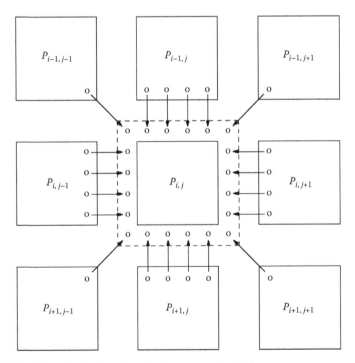

FIGURE 10.32 Boundary information that a processor has to keep.

Since blocks of the image are updated independently, processors might not have complete information from other processors during the initial two-scans. Thus, several iterations of two scans might be required before the image converges. This is in contrast to the original two-scan algorithm, which arrives at the root in just two scans (one LT scan and one RB scan) regardless of the image. From our experiments it seems that the number of iterations performed in blocked two-scan algorithm depends more on the number of partitions and less on the images and their sizes.

10.5.4 Performance of Two-Scan Algorithms

The algorithms have been implemented using an Intel iPSC to solve the distance transform problem. The iPSC is a multicomputer with a static interprocessor connection of a hypercube. For the purpose of comparison, the algorithm using the standard morphology is also implemented. The algorithm is very similar to that of blocked two-scan algorithm, except that a 3×3 mask is used to scan the local pixels once in each iteration. The execution times measured in the simulator for 64×64 and 32×32 images are plotted in Figure 10.33. The images used contain symmetric objects. Images containing other kinds of objects exhibit similar behaviors. The time unit used is that of a VAX clock because the simulator was run on a VAX 8530. The time plotted is chosen as the minimum of all possible configurations for a given number of processors.

From Figure 10.34, it can be seen that the blocked two-scan algorithm performs the best and the two-scan algorithm ranks next. Both of them improve dramatically over the standard morphology algorithm. For the blocked two-scan algorithm, the improvement is almost seven times. For some images, the improvement can go as high as 20 times.

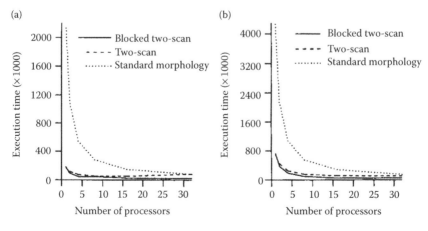

FIGURE 10.33 Execution time of the two-scan algorithm on an iPSC simulator. (a) 32×32 image, (b) 64×64 image.

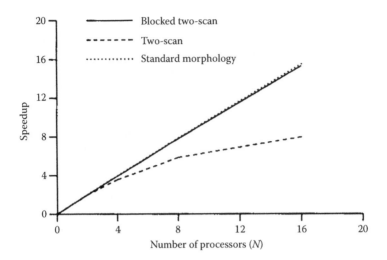

FIGURE 10.34 Speedup of the DT algorithms.

The speedup of a parallel algorithm is defined as

$$\text{speedup} = \frac{\text{execution time using one processor}}{\text{execution time using } p \text{ processors}} \quad (10.59)$$

Figure 10.34 depicts the speedup of the three algorithms. Again the blocked two-scan algorithm is almost linear with respect to the number of processors used. The standard morphology algorithm also performs well in this category. This is because both algorithms partition the given image into blocks and update blocks independently and concurrently. Therefore, there is a very high degree of parallelism. On the other hand, the speedup of the two-scan algorithm drops as the number of processors increases. This indicates (and confirms) that the two-scan algorithm is sequential in nature and wastes precious time in synchronization. There is not enough parallelism for large number of processors to exploit.

References

Armstrong, W. W. and Gecsei, J., "Adaptation algorithms for binary tree networks," *IEEE Trans. Syst., Man, Cybern.*, vol. 9, no. 5, pp. 276–285, May 1979.

Athas, W. L. and Seitz, C. L., "Multicomputers: Message-passing concurrent computers," *IEEE Trans. Computers*, vol. 21, no. 8, pp. 9–24, Aug. 1988.

Duff, M. J. B., "Parallel processors for digital image processing," in P. Stucki (Ed), *Advances in Digital Image Processing*, Plenum Press, London, pp. 265–279, 1979.

Fitch, J. P., Coyle, E. J., and Gallager, N. C., "Threshold decomposition of multidimensional ranked order operations," *IEEE Trans. Circuits and Systems*, vol. 32, no. 5, pp. 445–450, May 1985.

Fukushima, K., "Cognitron: a self-organizing multilayered neural network," *Biological Cybernetics*, vol. 20, no. 3–4, pp. 121–136, Sep. 1975.

Fukushima, K., "Neocognitron: a neural network model for a mechanism of visual pattern recognition," *IEEE Trans. Syst., Man, Cybern.*, vol. 13, no. 5, pp. 826–834, Sep. 1983.

Fukushima, K., "Neocognitron: a hierarchical neural network capable of visual pattern recognition," *Neural Network*, vol. 1, no. 2, pp. 119–130, April 1988.

Gerritsen, F. A. and Aardema, L. G., "Design and use of DIP-1: A fast flexible and dynamically microprogrammable image processor," *Pattern Recognition*, vol. 14, no. 1–6, pp. 319–330, 1981.

Golay, M. J. E., "Hexagonal parallel pattern transformations," *IEEE Trans. Computers*, vol. 18, no. 8, pp. 733–740, Aug. 1969.

Graham, D. and Norgren, P. E., "The Diff3 analyzer: A parallel/serial Golay image processor," in Onoe, Preston, and Rosenfeld (Eds), *Real Time Medical Image Processing*, Plenum, London, pp. 163–182, 1980.

Handley, J. C., "Bit vector architecture for computational mathematical morphology," *IEEE Trans. Image Processing*, vol. 12, no. 2, pp. 153–158, Feb. 2003.

Handley, J. C., "Minimal-memory bit-vector architecture for computational mathematical morphology using subspace projections," *IEEE Trans. Image Processing*, vol. 14, no. 8, pp. 1088–1095, Aug. 2005.

Huang, S. C. and Huang, Y. F., "Bounds on the number of hidden neurons in multilayer perceptrons," *IEEE Trans. Neural Networks*, vol. 2, no. 1, pp. 47–55, Jan. 1991.

Hwang, K., "Advanced parallel processing with supercomputer architectures," *Proc. IEEE*, vol. 75, no. 10, pp. 1348–1379, Oct. 1987.

King, C. T. and Ni, L. M., "Large-grain pipelining on hypercube multiprocessors," *Proc. Third ACM Conf. Hypercube Concurrent Computers and Applications*, pp. 1583–1591, 1988.

Klein, J. C. and Serra, J., "The texture analyzer," *Journal of Microscopy*, vol. 95, pp. 349–356, 1977.

Krishnapuram, R. and Chen, L. F., "Implementation of parallel thinning algorithms using recurrent neural networks," *IEEE Trans. Neural Networks*, vol. 4, no.1, pp. 142–147, Jan. 1993.

Kruse, B., "Design and implementation of a picture processor," Science and Technology Dissertations, no. 13, University of Linkoeping, Linkoeping, Sweden, 1977.

Lippmann, R. P., "An introduction to computing with neural nets," *IEEE ASSP Magazine*, vol. 42, no. 4, pp. 4–22, Apr. 1987.

Lougheed, R. M. and McCubbrey, D. L. "The cytocomputer: A practical pipelined image processor," *Proc. IEEE Annual Symp. Computer Architecture*, La Baule, France, pp. 271–278, 1980.

Louverdis, G. and Andreadis, I., "Design and implementation of a fuzzy hardware structure for morphological color image processing," *IEEE Trans. Circuits and Systems for Video Technology*, vol. 13, no. 3, pp. 277–288, March 2003.

Martin, T., "Acoustic recognition of a limited vocabulary in continuous speech," Ph.D. Thesis, Dept. Electrical Engineering Univ. Pennsylvania, 1970.

Minsky, M. L. and Papert, S. A., *Perceptrons*, MIT Press, Cambridge, MA, 1969.

Moh, J. and Shih, F. Y., "A general purpose model for image operations based on multilayer perceptrons," *Pattern Recognition*, vol. 28, no. 7, pp. 1083–1090, July 1995.

Rosenblatt, F., *Principles of Neurodynamics*, New York, Spartan, 1962.

Rumelhart, D. E., Hinton, G. E., and Williams, R. J., "Learning representations by back-propagating errors," *Nature*, vol. 323, pp. 533–536, Oct. 1986.

Shih, F. Y. and Mitchell, O. R., "Threshold decomposition of grayscale morphology into binary morphology," *IEEE Trans. Pattern Analysis and Machine Intelligence*, vol. 11, no. 1, pp. 31–42, Jan. 1989.

Shih, F. Y. and Moh, J., "Implementing morphological operations using programmable neural networks," *Pattern Recognition*, vol. 25, no. 1, pp. 89–99, Jan. 1992.

Sternberg, S. R., "Biomedical image processing," *Computer*, vol. 16, no. 1, pp. 22–34, Jan. 1983.

Vidal, J. J., "Implementing neural nets with programmable logic," *IEEE Trans. ASSP*, vol. 36, no. 7, pp. 1180–1190, July 1988.

Wendt, P. D., Coyle, E. J., and Gallagher, N. C., "Stack filters," *IEEE Trans. Acoust., Speech, Signal Processing*, vol. 34, no. 6, pp. 898–911, Aug. 1986.

11

General Sweep Mathematical Morphology

Traditional MM uses a designed structuring element that does not allow geometric variations during the operation to probe an image. General sweep MM provides a new class of morphological operations that allow one to select varying shapes and orientations of SEs while processing an image. Such a class holds syntactic characteristics similar to algebraic morphology as well as sweep geometric modeling. Traditional morphology is simply a subclass of general sweep morphology.

Sweep morphological dilation/erosion provides a natural representation of sweep motion in the manufacturing process, and sweep opening/closing delivers variant degrees of smoothing in image filtering. The theoretical framework for representation, computation, and analysis of sweep morphology is presented here. Its applications to deformations, image enhancement, edge linking, shortest path planning, and geometric modeling are also discussed. We present a framework for geometric modeling and representation by sweep MM. A set of grammatical rules that govern the generation of objects belonging to the same group is defined. In the screening process Earley's parser serves to determine whether a pattern is a part of the language. Sweep MM is demonstrated as an intuitive and efficient tool for geometric modeling and representation.

The organization of this chapter is as follows. Section 11.1 is an introduction to general sweep MM, and Section 11.2 its theoretical development. Section 11.3 describes the blending of sweep surfaces with deformations. Sections 11.4 and 11.5, respectively, introduce the applications to image enhancement and edge linking. Section 11.6 discusses geometric modeling and sweep MM. Section 11.7 introduces a formal language and representation scheme. Sections 11.8 and 11.9, respectively, present the grammar and parsing algorithm of the language.

11.1 Introduction

General sweeps of solids are useful in modeling the region swept out by a machine-tool cutting head or a robot following a path. General sweeps of two-dimensional cross-sections are known as *generalized cylinders* in computer vision, and are usually modeled as parameterized two-dimensional cross-sections swept at right angles along an arbitrary curve. Being the simplest of general sweeps, generalized cylinders are somewhat easy to compute. However, general sweeps of solids are difficult to compute since the trajectory and object shape may make the sweep object self-intersect [Blackmore, Leu, and Shih 1994; Foley et al. 1995].

MM involves geometric analysis of shapes and textures in images. Appropriately used, mathematical morphological operations tend to simplify image data presenting their essential shape characteristics and eliminating irrelevancies [Haralick, Sternberg, and Zhuang 1987; Serra 1982; Shih and Mitchell 1989, 1992]. Traditional morphological operations perform vector additions or subtractions by translating the structuring element to the object pixel. They are far from being capable of modeling the swept volumes of SEs moving with complex, simultaneous translation, scaling, and rotation in Euclidean space.

Pixel neighborhoods can be adaptively adjusted along the operations. Debayle and Pinoli [2006a,b] presented the theoretical aspects and practical applications of general adaptive neighborhood image processing. Bouaynaya and Schonfeld [2008] presented theoretical foundations of spatially variant (SV) MM for binary and gray-level images. The size of SEs may be modulated by the perspective function varying linearly with the vertical position in the image. For instance, when analyzing traffic control camera images, vehicles at the bottom of the image are closer and appear larger than those higher in the image. Roerdink [2000] presented polar morphology, constrained perspective morphology, spherical morphology, translation–rotation morphology, projective morphology, and differential morphology; all are brought together in the general framework of group morphology. Cuisenaire [2006] defined morphological operators with SEs whose size can vary over the image without any constraint, and showed that when the SEs are balls of a metric, locally adaptable erosion and dilation can be efficiently implemented as a variant of DT algorithms.

In this section, we present an approach that adopts sweep morphological operations to study the properties of swept volumes. We present a theoretical framework for representation, computation, and analysis of a new class of general sweep MM and its practical applications.

Geometric modeling is the foundation for CAD/CAM integration. The goal of automated manufacturing inspection and robotic assembly is to generate a complete process automatically. The representation must not only possess nominal geometric shapes, but also reason the geometric inaccuracies (or tolerances) into the locations and shapes of solid objects. Boundary representation (B-Rep) and Constructed Solid Geometry (CSG) representation are popularly used as internal database [Requicha 1980; Requicha and Voelcker 1982; Rossignac 2002] for geometric modeling. B-Rep consists of two kinds of information—topological information and geometric information including vertex coordinates, surface equations, and connectivity between faces, edges, and vertices. There are several advantages in B-Rep: large domain, unambiguity, uniqueness, and explicit representation of faces, edges, and vertices. There are also several disadvantages: verbose data structure, difficulty in creating the representation, difficulty in checking validity, and variational information unavailability.

CSG representation works by constructing a complex part by hierarchically combining simple primitives using Boolean set operations [Mott-Smith and Baer 1972]. There are several advantages of using CSG representation: large domain, unambiguity, easy-to-check validity, and it is easily created. There are also several disadvantages: non-uniqueness, difficulty in editing graphically, input data redundancy, and variational information unavailability.

The framework for geometric modeling and representation is sweep MM. The *sweep* operation provides a natural design tool to generate a volume by sweeping a primitive object along a space curve trajectory. The simplest sweep is *linear extrusion* defined by a two-dimensional area swept along a linear path normal to the plane of the area to create a volume [Chen, Wu, and Hung 1999]. Another sweep is *rotational sweep* defined by rotating a two-dimensional object about an axis [Shih, Gaddipati, and Blackmore 1994].

A generalized sweeping method for CSG modeling was developed by Shiroma, Kakazu, and Okino [1991] to generate swept morphological operations that tend to simplify data representing image volume. It is shown that complex solid shapes can be generated by a blending surface to join two disconnected solids, fillet volumes for rounding corners, and swept volumes formed by the movement of Numeric Control (NC) tools. Ragothama and Shapiro [1998] have presented a B-Rep method for deformation in parametric solid modeling.

11.2 Theoretical Development of General Sweep MM

Traditional morphological dilation and erosion perform vector additions or subtractions by translating a structuring element along an object. Obviously, these operations have the limitation of orientation-dependence and can represent a sweep motion that involves translation. However, by including not only translation but also rotation and scaling, the entire theoretical framework and practical applications become extremely useful. Sweep morphological dilation and erosion describe a motion of a structuring element that sweeps along the boundary of an object or an arbitrary curve by geometric transformations. The rotation angles and scaling factors are defined with respect to the boundary or the curve.

11.2.1 Computation of Traditional Morphology

Because rotation and scaling are inherently defined on each pixel of the curve, traditional morphological operations of an object by a structuring element need to be converted to sweep morphological operations. We assume that the sets considered are connected and bounded.

Definition 11.1: A set S is said to be *connected* if each pair of points $p, q \in S$ can be joined by a path that consists of pixels entirely located in S.

Definition 11.2: Given a set S, a *boundary* ∂S is defined as the set of points all of whose neighborhoods intersect both S and its complement S^c.

Definition 11.3: If a set S is connected and has no holes, it is called *simply connected*; if it is connected but has holes, it is called *multiply connected*.

Definition 11.4: Given a set S, the *outer boundary* $\partial_+ S$ of the set is defined as the closed loop of points in S that contains every other closed loop consisting of points of the set S; the *inner boundary* $\partial_- S$ is defined as the closed loop of points in S that does not contain any other closed loop in S.

Proposition 11.1: If a set S is simply connected, then ∂S is its boundary; if it is multiply connected, then

$$\partial S = \partial_+ S \cup \partial_- S. \tag{11.1}$$

Definition 11.5: The *positive filling* of a set S is denoted as $[S]_+$ and is defined as the set of all points that are inside the outer boundary of S; the *negative filling* is denoted as $[S]_-$ and is defined as the set of all points that are outside the inner boundary.

Note that if S is simply connected, then $[S]_-$ is a universal set. We can then find out whether S is simply or multiply connected, $S = [S]_+ \cap [S]_-$.

Proposition 11.2: Let A and B be simply connected sets. The dilation of A by B equals the positive filling of $\partial A \oplus B$; that is, $A \oplus B = [\partial A \oplus B]_+$. The significance is that if A and B are simply connected sets, we can compute the dilation of the boundary ∂A by the set B. This leads to a substantial reduction of computation.

Proposition 11.3: If A and B are simply connected sets, the dilation of A by B equals the positive filling of the dilation of their boundaries. That is

$$A \oplus B = [\partial A \oplus \partial B]_+ \tag{11.2}$$

This proposition further reduces the computation required for the dilation. Namely, the dilation of sets A by B can be computed by the dilation of the boundary of A by the boundary of B.

Proposition 11.4: If A is multiply connected and B is simply connected, then

$$A \oplus B = [\partial_+ A \oplus \partial B]_+ \cap [\partial_- A \oplus \partial B]_- \tag{11.3}$$

Since A and B possess the commutative property with respect to dilation, the following proposition can be easily obtained.

Proposition 11.5: If A is simply connected and B is multiply connected, then

$$A \oplus B = [\partial A \oplus \partial_+ B]_+ \cap [\partial A \oplus \partial_- B]_- \tag{11.4}$$

11.2.2 General Sweep MM

Sweep morphology can be represented as a 4-tuple, $\Psi(B, A, S, \Theta)$, where B is a structuring element set indicating a primitive object; A is either a curve path or a closed object whose boundary represents the sweep trajectory with a parameter t along which the structuring element B is swept; $S(t)$ is a vector consisting of scaling factors; and $\Theta(t)$ is a vector consisting of the rotation angles. Note that both scaling factors and rotation angles are defined with respect to the sweep trajectory.

Definition 11.6: Let ∂A denote the boundary of set A. If A is a simply connected object, the *sweep morphological dilation* of A by B in Euclidean space, denoted by $A \boxplus B$, is defined as

$$A \boxplus B = \{c \mid c = a + \hat{b} \ \text{ for some } a \in A \text{ and } \hat{b} \in [S(t) \times \Theta(t) \times B]\} \tag{11.5}$$

This is equivalent to dilation on the boundary of A (i.e., ∂A) and taking the positive filling as

$$A \boxplus B = \left\{ \bigcup_{0 \le t \le 1} \bigcup_{b \in B} [\partial A(t) + b \times S(t) \times \Theta(t)] \right\}_+ \tag{11.6}$$

If A is a curve path, that is, $\partial A = A$, then sweep morphological dilation of A by B is defined as

$$A \boxplus B = \bigcup_{0 \le t \le 1} \bigcup_{b \in B} [A(t) + b \times S(t) \times \Theta(t)] \tag{11.7}$$

Note that if B does not involve rotations (or B is rotation-invariant like a circle) and scaling, then the sweep dilation is equivalent to the traditional morphological dilation.

Example 11.1: Figure 11.1a shows a curve and Figure 11.1b shows an elliptical structuring element. The rotation angle θ is defined as $\theta(t) = \tan^{-1}(dy/dt)/(dx/dt)$ along the curve with parameter t in the range of $[0, 1]$. Traditional morphological dilation is shown in Figure 11.1c and sweep dilation using defined rotation is shown in Figure 11.1d.

A geometric transformation of the structuring element specifies the new coordinates of each point as functions of the old coordinates. Note that the new coordinates are not necessarily integers after a transformation is applied to a digital image. To make the results of the transformation into a digital image, they must be resampled or interpolated. Since we are transforming a two-valued (black-and-white) image, zero-order interpolation is adopted.

Sweep morphological erosion, unlike dilation, is defined with the restriction on a closed object only and its boundary represents the sweep trajectory.

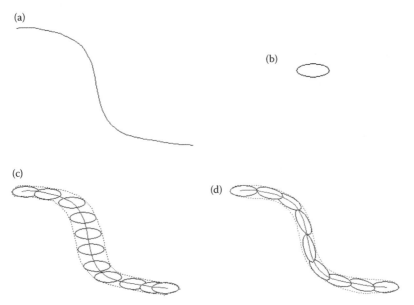

(a)

(b)

(c)

(d)

FIGURE 11.1 (a) An open curve path, (b) a structuring element, (c) the result of a traditional morphological dilation, and (d) the result of a sweep morphological dilation.

Definition 11.7: Let ∂A be the boundary of an object A and B be a structuring element. The *sweep morphological erosion* of A by B in Euclidean space, denoted by $A \boxminus B$, is defined as

$$A \boxminus B = \{c \mid c + \hat{b} \in A \text{ and for every } \hat{b} \in [S(t) \times \Theta(t) \times B]\}. \qquad (11.8)$$

An example of a sweep erosion by an elliptical structuring element whose semi-major axis is tangential to the boundary is shown in Figure 11.2. Like in traditional morphology, the general sweep morphological opening can be defined as a general sweep erosion of A by B followed by a general sweep dilation where A must be a closed object. Then sweep morphological closing can be defined in the opposite sequence; that is, a general sweep dilation of A by B followed by a general sweep erosion, where A can be either a closed object or a curve path.

The propositions of traditional morphological operations can be extended to sweep morphological operations.

Proposition 11.6: If the structuring element B is simply connected, sweep dilation of A by B is equal to the positive filling of sweep dilation by the boundary of B; that is, $A \boxplus B = [A \boxplus \partial B]_+$.

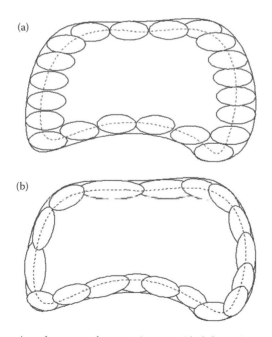

FIGURE 11.2 Sweeping of a square along a trajectory with deformation to a circle.

Extending this proposition to multiply connected objects, we get the following three cases.

Case 6(a): If A is multiply connected, that is, $\partial A = \partial_+ A \cup \partial_- A$, then

$$A \boxplus B = [\partial A_+ \boxplus B]_+ \cap [\partial A_- \boxplus B]_- \qquad (11.9)$$

Case 6(b): If B is multiply connected, that is, $\partial B = \partial_+ B \cup \partial_- B$, then

$$A \boxplus B = [A \boxplus \partial B_+]_+ \cap [A \boxplus \partial B_-]_- \qquad (11.10)$$

Case 6(c): Finally, if both A and B are multiply connected; that is, $\partial A = \partial_+ A \cup \partial_- A$ and $\partial B = \partial_+ B \cup \partial_- B$, then

$$A \boxplus B = [(\partial_+ A \cup \partial_- A) \boxplus \partial_+ B]_+ \cap [(\partial_+ A \cup \partial_- A) \boxplus \partial_- B]_- \qquad (11.11)$$

This leads to a substantial reduction of computation. We can make an analogous development for sweep erosion.

Proposition 11.7: If A and B are simply connected sets, then

$$A \boxminus B = A \boxminus \partial B. \qquad (11.12)$$

With the aforementioned propositions and considering the boundary of the structuring element, we can further reduce the computation of sweep morphological operations.

11.2.3 Properties of Sweep Morphological Operations

Property 11.1: Non-commutative. Because of the rotational factor in the operation, commutativity does not hold. That is

$$A \boxplus B \neq B \boxplus A \qquad (11.13)$$

Property 11.2: Non-associative. Because rotational and scaling factors are dependent on the characteristics of the boundary of the object, associativity does not hold. Hence,

$$A \boxplus (B \boxplus C) \neq (A \boxplus B) \boxplus C. \qquad (11.14)$$

But the associativity of regular dilation and sweep dilation holds; that is, $A \oplus (B \boxplus C) = (A \oplus B) \boxplus C$. As the structuring element is rotated based on the boundary properties of B and after $A \oplus B$, remain the boundary properties will be similar to that of B.

Property 11.3: Translational invariance:

$$A_x \boxplus B = (A \boxplus B)_x \qquad (11.15)$$

Proof:

$$A_x \boxplus B = \left[\bigcup_{0 \le t \le 1} \bigcup_{b \in B} [\partial A + x + b \times S(t) \times \theta(t)] \right]_+$$

$$= \left[\bigcup_{0 \le t \le 1} \bigcup_{b \in B} [\partial A + b \times S(t) \times \theta(t)] + x \right]_+$$

$$= \left[\bigcup_{0 \le t \le 1} \bigcup_{b \in B} [\partial A + b \times S(t) \times \theta(t)] \right]_+ + x = (A \boxplus B)_x \qquad \square$$

Sweep erosion can be derived similarly.

Property 11.4: Increasing property will not hold in general. If the boundary is smooth; that is, derivative exists everywhere, then increasing property will hold.

Property 11.5: Distributivity

(a) *Distributivity over union.* Dilation is distributive over the union of SEs. That is, dilation of A with a union of two SEs B and C is the same as the union of dilation of A with B and dilation of A with C.

$$A \boxplus (B \cup C) = (A \boxplus B) \cup (A \boxplus C) \qquad (11.16)$$

Proof:

$$A \boxplus (B \cup C) = \left[\bigcup_{0 \le t \le 1} \bigcup_{b \in B \cup C} [\partial A + b \times S(t) \times \theta(t)] \right]_+$$

$$= \left[\bigcup_{0 \le t \le 1} \left[\bigcup_{b \in B} [\partial A + b \times S(t) \times \theta(t)] \right. \right.$$

$$\left. \left. \cup \bigcup_{b \in C} [\partial A + b \times S(t) \times \theta(t)] \right] \right]_+$$

$$= \left[\left[\bigcup_{0 \le t \le 1} \bigcup_{b \in B} [\partial A + b \times S(t) \times \theta(t)] \right] \right.$$

$$\left. \cup \left[\bigcup_{0 \le t \le 1} \bigcup_{b \in C} [\partial A + b \times S(t) \times \theta(t)] \right] \right]_+$$

$$= \left[\bigcup_{0 \le t \le 1} \bigcup_{b \in B} [\partial A + b \times S(t) \times \theta(t)] \right]_+$$

$$\cup \left[\bigcup_{0 \le t \le 1} \bigcup_{b \in C} [\partial A + b \times S(t) \times \theta(t)] \right]_+$$

$$= (A \boxplus B) \cup (A \boxplus C) \qquad \square$$

(b) Dilation is not distributive over the union of sets. That is, dilation of $(A \cup C)$ with a structuring element B is not same as the union of dilation of A with B and dilation of C with B.

$$(A \cup C) \boxplus B \ne (A \boxplus B) \cup (C \boxplus B) \qquad (11.17)$$

(c) Erosion is anti-distributive over the union of SEs. That is, erosion of A with a union of two SEs B and C is the same as the intersection of erosion of A with B and erosion of A with C.

(d) Distributivity over intersection.

$$A \boxplus (B \cap C) \subseteq (A \boxplus B) \cap (A \boxplus C) \tag{11.18}$$

Proof:

$$\partial A \boxplus (B \cap C) = \left[\bigcup_{0 \le t \le 1} \left[\bigcup_{b \in B \cap C} [\partial A + b \times S(t) \times \theta(t)] \right] \right]$$

$$\Rightarrow \partial A \boxplus (B \cap C) \subseteq \left[\bigcup_{0 \le t \le 1} \left[\bigcup_{b \in B} [\partial A + b \times S(t) \times \theta(t)] \right] \right]$$

and also $\partial A \boxplus (B \cap C) \subseteq [\bigcup_{0 \le t \le 1} [\bigcup_{b \in C} [\partial A + b \times S(t) \times \theta(t)]]]$. Therefore, we have $\partial A \boxplus (B \cap C) \subseteq (\partial A \boxplus B) \cap (\partial A \boxplus C)$, which implies $[\partial A \boxplus (B \cap C)]_+ \subseteq [(\partial A \boxplus B)]_+ \cap [(\partial A \boxplus C)]_+$. That is, $A \boxplus (B \cap C) \subseteq (A \boxplus B) \cap (A \boxplus C)$. □

11.3 Blending of Sweep Surfaces with Deformations

By using the general sweep MM, a smooth sculptured surface can be described as a trajectory of a cross-section curve swept along a profile curve, where the trajectory of the former is the structuring element B and the profile curve is the open or closed curve C. It is very easy to describe the sculptured surface by specifying the two-dimensional cross-sections, and that the resulting surface be aesthetically appealing. The designer can envision the surface as a blended trajectory of cross-section curves swept along a profile curve.

Let ∂B denote the boundary of a structuring element B. A swept surface S_w (∂B, C) is produced by moving ∂B along a given trajectory curve C. The plane of B must be perpendicular to C at any instance. The contour curve is represented as a B-spline curve, and ∂B is represented as the polygon net of the actual curve. This polygon net is swept along the trajectory to get the intermediate polygon nets, and later they are interpolated by a B-spline surface. The curve can be deformed by twisting or scaling uniformly or by applying the deformations to selected points of ∂B. The curve can also be deformed by varying the weights at each of the points. When a uniform variation is desired, it can be applied to all the points and otherwise to some selected points. These deformations are applied to ∂B before it is moved along the trajectory C.

Let ∂B denote a planar polygon with n points and each point $\partial B_i = (x_i, y_i, z_i, h_i)$, where $i = 1, 2, \dots, n$. Let C denote any three-dimensional curve with m points and each point $C_j = (x_j, y_j, z_j)$, where $j = 1, 2, \dots, m$. The scaling factor, weight, and twisting factor for point j of C are denoted as sx_j, sy_j, sz_j, w_j, and θ_j, respectively. The deformation matrix is obtained as $[S_d] = [S_{sw}][R_\theta]$, where

$$[S_{sw}] = \begin{bmatrix} sx_j & 0 & 0 & 0 \\ 0 & sy_j & 0 & 0 \\ 0 & 0 & sz_j & 0 \\ 0 & 0 & 0 & w_j \end{bmatrix} \tag{11.19}$$

and

$$[R_\theta] = \begin{bmatrix} \cos\theta_j & \sin\theta_j & 0 & 0 \\ -\sin\theta_j & \cos\theta_j & 0 & 0 \\ 0 & 0 & 1 & 0 \\ 0 & 0 & 0 & 1 \end{bmatrix}$$

(11.20)

The deformed ∂B must be rotated in three-dimensional space with respect to the tangent vector at each point of the trajectory curve C. To calculate the tangent vector, we add two points, C_0 and C_{m+1}, to C, where $C_0 = C_1$ and $C_{m+1} = C_m$. The rotation matrix R_x about x-axis is given by

$$[R_x] = \begin{bmatrix} 1 & 0 & 0 & 0 \\ 0 & \cos\alpha_j & \sin\alpha_j & 0 \\ 0 & -\sin\alpha_j & \cos\alpha_j & 0 \\ 0 & 0 & 0 & 1 \end{bmatrix}$$

(11.21)

where

$$\cos\alpha_j = \frac{C_{y_{j-1}} - C_{y_{j+1}}}{h_x}, \quad \sin\alpha_j = \frac{C_{z_{j+1}} - C_{z_{j-1}}}{h_x}, \quad \text{and} \quad h_x = \sqrt{(C_{y_{j-1}} - C_{y_{j+1}})^2 + (C_{z_{j+1}} - C_{z_{j-1}})^2}.$$

The rotation matrices about the y- and z-axes can similarly be derived. Finally, ∂B must be translated to each point of C and the translation matrix C_{xyz} is given by

$$[C_{xyz}] = \begin{bmatrix} 1 & 0 & 0 & 0 \\ 0 & 1 & 0 & 0 \\ 0 & 0 & 1 & 0 \\ C_{x_j} - C_{x_1} & C_{y_j} - C_{y_1} & C_{z_j} - C_{z_1} & 1 \end{bmatrix}$$

(11.22)

The polygon net of the sweep surface will be obtained by

$$[B_{i,j}] = [\partial B_i] [S_d] [Sw_C],$$

(11.23)

where $[Sw_C] = [R_x][R_y][R_z][C_{xyz}]$. The B-spline surface can be obtained from the polygon net by finding the B-spline curve at each point of C. To get the whole swept surface, the B-spline curves at each point of the trajectory C have to be calculated. This computation can be reduced by selecting a few polygon nets and calculating the B-spline surface.

Example 11.2: Sweeping of a circle along a trajectory with deformation to a square.
Here the deformation is only the variation of the weights. The circle is represented as a rational B-spline curve. The polygon net is a square with nine

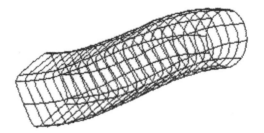

FIGURE 11.3 Structuring element assignment using general sweep morphology.

points with the first and last being the same, and the weights of the corner vary from 5 to $\sqrt{2}/2$ as it is being swept along the trajectory C, which is given in the parametric form as $x = 10s$ and $y = \cos(\pi s) - 1$. The sweep transformation is given by

$$[Sw_T] = \begin{bmatrix} \cos\psi & \sin\psi & 0 & 0 \\ -\sin\psi & \cos\psi & 0 & 0 \\ 0 & 0 & 1 & 0 \\ 10s & \cos\pi s & 0 & 1 \end{bmatrix}$$

where $\psi = \tan^{-1}(-\pi \sin(\pi s)/10)$. Figure 11.3 shows the sweeping of a square along a trajectory with deformation to a circle.

11.4 Image Enhancement

Because they are adaptive to local properties of an image, general sweep morphological operations can provide variant degrees of smoothing for noise removal while preserving an object's features. Some research on statistical analysis of traditional morphological operations has been done previously. Stevenson and Arce [1987] developed the output distribution function of opening with flat SEs by threshold decomposition. Morales and Acharya [1993] presented general solutions for the statistical analysis of morphological openings with compact, convex, and homothetic SEs.

Traditional opening can remove noise as well as object features whose sizes are smaller than the size of the structuring element. With general sweep morphological opening, features of similar shape and that are larger than the structuring element will be preserved while removing noise. In general, the highly varying parts of the image are assigned with smaller SEs, whereas the slowly varying parts with larger ones. The SEs can be assigned based on the contour gradient variation. An example is illustrated in Figure 11.4.

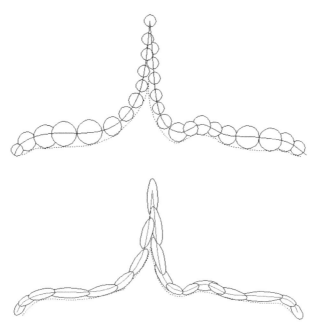

FIGURE 11.4 The elliptic structuring element.

Step edge is an important feature in an image. Assume a noisy step edge is defined as

$$f(x) = \begin{cases} N & \text{if } x < 0 \\ h + N_x & \text{if } x \geq 0 \end{cases}, \tag{11.24}$$

where h is the strength of the edge, and N is independent identically distributed (i.i.d.) Gaussian random noise with mean value 0 and variance 1. For image filtering with a general sweep morphological opening, we essentially adopt a smaller sized structuring element for a important feature points and larger one for other locations. Therefore, the noise in an image is removed while the features are preserved. For instance, in a one-dimensional image, this can be easily achieved by computing the gradient by $f(x) - f(x - 1)$ and setting those points accordingly, in which the gradient values are larger than the predefined threshold, with smaller SEs.

In Chen, Wu, and Hung [1999], the results of noisy step edge filtering by both traditional morphological opening and so-called *space-varying* opening (involving both scaling and translation in our general sweep morphology model) were shown and compared by computing the mean and variance of output signals. The mean value of the output distribution follows the main shape of the filtering result well and this gives evidence of the shape preserving ability of the proposed operation. Additionally, the variance of output distribution coincides with noise variance, and this shows the corresponding noise removal ability. It is observed that general sweep opening possesses approximately the same noise removing ability as the traditional one. Moreover, it can be observed that the relative edge strength with respect to the variation between the transition interval, say [−2, 2], for general sweep opening is larger than that of the traditional one. This explains why the edge is degraded in traditional morphology but is enhanced in general sweep. Although a step-edge model was tested successfully, other complicated cases need further elaboration. The statistical analysis for providing a quantitative approach to general sweep morphological operations will be further investigated.

Chen, Wu, and Hung [1996] presented image filtering using adaptive signal processing, which is nothing but sweep morphology with only scaling and translation. The method uses space-varying SEs by assigning different filtering scales to features. To adaptively assign structuring elements, they developed the progressive umbra-filling (PUF) procedure. This is an iterative process. Experimental results have shown that this approach can successfully eliminate noise without oversmoothening the important features of a signal.

11.5 Edge Linking

Edge is a local property of a pixel and its immediate neighborhood, and edge detector is a local process that locates sharp changes in the intensity function. An ideal edge has a step like cross-section, as gray levels change abruptly across the border. In practice, edges in digital images are generally slightly blurred due to the effects of sampling and noise.

There are many edge detection algorithms and the basic idea underlying most edge detection techniques is the computation of a local derivative operator [Gonzalez and Woods 2002]. Some algorithms like the Laplacian of Gaussian (LoG) filter produce closed edges; however, false edges are generated when blur and noise appear in an image. Some algorithms like the Sobel operator produce noisy boundaries that do not actually lie on the borders, and there are broken gaps where border pixels should reside. That is because of noise and breaks present in the boundary from nonuniform illumination and other effects that introduce spurious intensity discontinuities. Thus, edge detection algorithms are typically followed by linking and other boundary detection procedures designed to assemble edge pixels into meaningful boundaries.

Edge linking by the tree search technique was proposed by Martelli [1976] to link the edge sequentially along the boundary between pixels. The cost of each boundary element is defined by the step size between the pixels on either side. The larger the intensity difference, the larger the step size, which is assigned a lower cost. The path of boundary elements with the lowest cost is linked up as an edge. The cost function was redefined by Cooper et al. [1980], who extended the edge through a path having the maximal local likelihood. Similar efforts were made by Eichel et al. [1988] and by Farag and Delp [1995].

Basically, the tree search method is time-consuming and requires the suitable assignment of root points. Another method locates all of the endpoints of the broken edges and uses a relaxation method to pair them up, so that line direction is maintained, lines are not allowed to cross, and closer points are matched first. However, this suffers from problems if unmatched endpoints or noise is present.

A simple approach to edge linking is a morphological dilation of points by some arbitrarily selected radius of circles followed by the OR operator of the boundary image with the resulting dilated circles, and the result is finally skeletonized [Russ 1992]. This method, however, has the problem that some of the points may be too far apart for the circles to touch, while conversely the circles may obscure details by touching several existing lines. To overcome this, sweep MM is used, as it allows variation of the structuring element according to local properties of the input pixels.

Let B denote an elliptic structuring element shown in Figure 11.5, where p and q denote, respectively, the semi-major and semi-minor axes. That is

$$\partial B \equiv \{[x, y]^T \mid x^2/p^2 + y^2/q^2 = 1\} \tag{11.25}$$

An edge-linking algorithm was proposed by Shih and Cheng [2004] based on sweep dilation, thinning, and pruning. This is a three-step process, and is explained below.

Step 1: *Sweep Dilation.* Broken line segments can be linked by using sweep morphology provided the structuring element has been suitably adjusted. Using the input signal plotted in Figure 11.6a, sweep morphological dilation is illustrated in Figure 11.6b. Extending the line segments in the direction of local slope performs the linking. The basic shape of the structuring element is an ellipse, where the major axis is always aligned with the tangent of the signal.

An elliptical structuring element is used to reduce noisy edge points and small insignificant branches. The width of the ellipse is

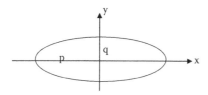

FIGURE 11.5 (a) Input signal and (b) sweep dilation with elliptical structuring elements.

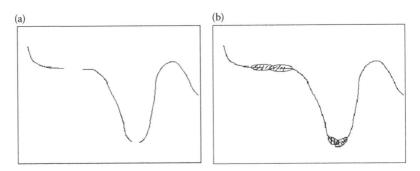

FIGURE 11.6   (a) An input signal, (b) sweep dilation.

selected so as to accomplish this purpose. The major axis of the ellipse should be adapted to the local curvature of the input signal to prevent overstretching at a high curvature point. At high curvature points, a short major axis is selected, and vice versa.

Step 2: *Thinning.* After sweep dilation by directional ellipses, the edge segments are extended in the direction of the local slope. Because tolerance (or the minor axis of the ellipse) is added, the edge segments grow a little fat. To suppress this effect, morphological thinning is adopted.

A thinning algorithm using MM was proposed by Jang and Chin [1990]. The skeletons generated by their algorithm are connected, are one pixel width, and closely follow the medial axes. The algorithm is an iterative process based on a hit/miss operation. Four SEs are constructed to remove boundary pixels from four directions, and another four are constructed to remove the extra pixels at skeleton junctions. There are four passes in each iteration. Three of the eight predefined SE templates are applied simultaneously in each pass. The iterative process is performed until the result converges. As the thinning algorithm will not shorten the skeletal legs it is applied to the sweep dilated edges.

Step 3: *Pruning.* The dilated edge segments after thinning may still produce a small number of short skeletal branches. These short branches should be pruned. In a skeleton, any pixel that has three or more neighbors is called a root. Starting from each neighbor of the root pixel, the skeleton is traced outward. Those paths whose lengths are shorter than a given threshold k are treated as branches and are pruned.

Figure 11.7a shows the edge of an industrial part and Figure 11.7b shows its randomly discontinuous edge. Figure 11.7c shows the result of using the sweep morphological edge-linking algorithm. Figure 11.8a shows the edge with added uniform noise and Figure 11.8b shows the edge after removing noise. Figure 11.8c shows the result of using the sweep morphological edge-linking algorithm. Note that for display clarity, the intensities of the

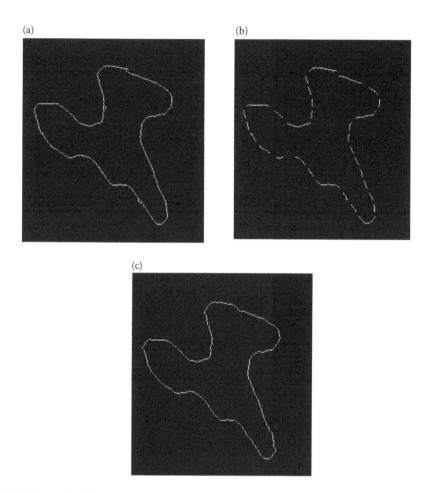

FIGURE 11.7 (a) The edge of an industrial part, (b) its randomly discontinuous edge, and (c) using the sweep morphological edge-linking algorithm.

images in Figures 11.7 and 11.8 are inverted; that is, the white edge now becomes black.

Figure 11.9a shows a face image with the originally detected broken edge. Figure 11.9b shows the face image with the edge linked by the sweep morphological edge-linking algorithm.

11.6 Geometric Modeling and Sweep MM

Dilation can be represented in matrix form as follows. Let $A(t)$ be represented by the matrix $[a_x(t), a_y(t), a_z(t)]$, where $0 \le t \le 1$. For every t, let the scaling factors

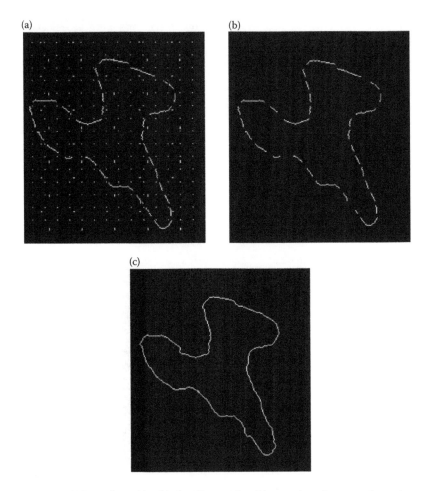

FIGURE 11.8 (a) Part edge with added uniform noise, (b) part edge after removing noise, and (c) using the sweep morphological edge-linking algorithm.

be $s_x(t)$, $s_y(t)$, $s_z(t)$ and the rotation factors be $\theta_x(t)$, $\theta_y(t)$, $\theta_z(t)$. By using the homogeneous coordinates, the scaling transformation matrix can be represented as

$$
[S(t)] = \begin{bmatrix} s_x(t) & 0 & 0 & 0 \\ 0 & s_y(t) & 0 & 0 \\ 0 & 0 & s_z(t) & 0 \\ 0 & 0 & 0 & 1 \end{bmatrix}
\tag{11.26}
$$

(a) 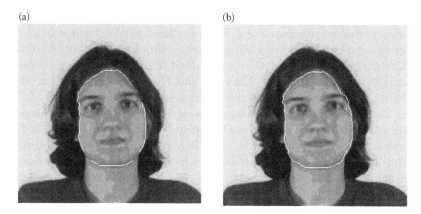 (b)

FIGURE 11.9 (a) Face image with the originally detected broken edge, (b) face image with the edge linked by the sweep morphological edge-linking algorithm.

The rotation matrix about x-axis is represented as

$$[R_x(t)] = \begin{bmatrix} 1 & 0 & 0 & 0 \\ 0 & \cos\theta_x(t) & \sin\theta_x(t) & 0 \\ 0 & -\sin\theta_x(t) & \cos\theta_x(t) & 0 \\ 0 & 0 & 0 & 1 \end{bmatrix} \tag{11.27}$$

where $\cos\theta_x(t) = [a_y(t-1) - a_y(t+1)]/h_x$, $\sin\theta_x(t) = [a_z(t+1) - a_z(t-1)]/h_x$, and $h_x = \sqrt{(a_y(t-1) - a_y(t+1))^2 + (a_z(t+1) - a_z(t-1))^2}$. The rotation matrices about the y- and z-axes can be similarly derived. Finally, the structuring element is translated by using

$$[A(t)] = \begin{bmatrix} 1 & 0 & 0 & 0 \\ 0 & 1 & 0 & 0 \\ 0 & 0 & 1 & 0 \\ a_x(t) & a_y(t) & a_z(t) & 1 \end{bmatrix}. \tag{11.28}$$

Therefore, sweep dilation is equivalent to the concatenated transformation matrices:

$$(A \boxplus B)(t) = [B][S(t)][R_x(t)][R_y(t)][R_z(t)][A(t)] \quad \text{where } 0 \le t \le 1. \tag{11.29}$$

Schemes based on sweep representation are useful in creating solid models of two-and-a-half-dimensional objects that include both solids of uniform thickness in a given direction and axis-symmetric solids. Computer representation of the swept volume of a planar surface has been used as a primary modeling scheme in solid modeling systems [Brooks 1981; Shiroma, Okino, and Kakazu 1982]. Representation of the swept volume of a three-dimensional object [Wang and Wang 1986; Voelcker and Hunt 1981; Pennington, Bloor, and Balila 1983], however, has received limited attention.

Leu, Park, and Wang [1986] presented a method for representing the swept volumes of translating objects using B-Rep and ray in–out classification. Their method is restricted to translation only. Representing the swept volumes of moving objects under a general motion is a more complex problem. A number of researchers have examined the problem of computing swept volumes, including Korein [1985] for rotating polyhedra, Kaul [1993] using Minkowski sums for translation, Wang and Wang [1986] using envelop theory, and Martin and Stephenson [1990] using envelop theory and computer algebraic techniques.

Because of the geometric nature of morphological operators and their nonlinear properties, some modeling problems will become simple and intuitive. This framework can be used for modeling not only swept surface and volumes but also for tolerance modeling in manufacturing.

11.6.1 Tolerance Expression

Tolerances constrain an object's features to lie within regions of space called *tolerance zones*. Tolerance zones in Rossignac and Requicha [1985] were constructed by expanding a nominal feature to obtain the region bounded by the outer closed curve, shrinking a nominal feature to obtain the region bounded by the inner curve, and then subtracting the two resulting regions. This procedure is equivalent to the morphological dilation of the offset inner contour with the tolerance-radius disc structuring element. Figure 11.10a shows an annular tolerance zone that corresponds to a circular hole, and Figure 11.10b shows the tolerance zone for an elongated slot. Both can be constructed by dilating the nominal contour with a tolerance-radius disc structuring element as shown in Figure 11.11. The tolerance zone for testing

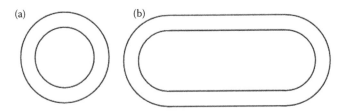

FIGURE 11.10. Tolerance zones (a) an annular tolerance zone that corresponds to a circular hole, (b) a tolerance zone for an elongated slot.

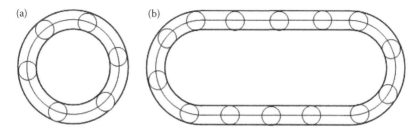

FIGURE 11.11 An example of adding tolerance by a morphological dilation.

the size of a round hole is an annular region lying between two circles with the specified maximal and minimal diameters; the zone corresponding to a form constraint for the hole is also an annulus defined by two concentric circles whose diameters must differ by a specified amount but are otherwise arbitrary.

Sweep MM supports the conventional limit (±) tolerance on "dimensions" that appear in engineering drawings. The positive deviation is equivalent to the dilated result and the negative deviation is equivalent to the eroded result. The industrial parts adding tolerance information can be expressed using dilation with a circle.

11.6.2 Sweep Surface Modeling

A simple sweep surface is generated by a profile sweeping along a spine with or without deformation. This is the sweep mathematical dilation of two curves. Let $P(u)$ be the profile curve, $B(w)$ be the spine, and $S(u, w)$ be the sweep surface. The sweep surface can be expressed as

$$S(u, w) = P(u) \boxplus B(w). \tag{11.30}$$

A sweep surface with initial and final profiles $P_1(u)$ and $P_2(u)$ at relative locations O_1 and O_2 respectively, and with the sweeping rule $R(w)$ is shown in Figure 11.12 and can be expressed as

$$S(u, w) = \{[1-R(w)][P_1(u) \boxplus (B(w) - O_1)]\} + \{R(w)[P_2(u) \boxplus (B(w) - O_2)]\}. \tag{11.31}$$

11.7 Formal Language and Sweep Morphology

Our representation framework is formulated as follows. Let E^N denote the set of points in the N-dimensional Euclidean space and $p = (x_1, x_2, \ldots, x_N)$

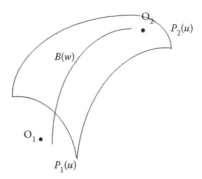

FIGURE 11.12   Modeling of sweep surface.

represent a point in E^N. Thus, any object is a subset in E^N. The formal model for geometric modeling is a context-free grammar, G, consisting of a four-tuple [Fu 1982; Gosh 1988; Shih 1991]:

$$G = (V_N, V_T, P, S),$$

where V_N is a set of nonterminal symbols such as complicated shapes; V_T is a set of terminal symbols that contain two sets: one has decomposed primitive shapes, such as lines and circles, and the other has shape operators; P is a finite set of rewriting rules or productions denoted by $A \rightarrow \beta$, where $A \in V_N$ and β is a string over $V_N \cup V_T$; S is the start symbol that represents the solid object. The operators used include sweep morphological dilation, set union, and set subtraction. Note that such a production allows the non-terminal A to be replaced by the string β independent of the context in which A appears.

A context-free grammar is defined as a derived form: $A \rightarrow \beta$, where A is a single nonterminal and β is a nonempty string of terminals and nonterminals. The languages generated by the context-free grammars are called context-free languages. Object representation can be viewed as a task of converting a solid shape into a sentence in the language, whereas object classification is the task of "parsing" a sentence.

The criteria for primitive selection are influenced by the nature of data, the specific application in the question, and the technology available for implementing the system. The following serve as general guidelines for primitive selection.

1. The primitives should be the basic shape elements that can provide a compact but adequate description of the object shape in terms of the specified structural relations (e.g., the concatenation relation).

2. The primitives should be easily extractable by the existing nonsyntactic (e.g., decision-theoretic) methods, because they are considered to be simple and compact shapes and their structural information is not important.

11.7.1 Two-Dimensional Attributes

Commonly used two-dimensional attributes are rectangles, parallelograms, triangles, rhombus, circles, and trapezoids. They can be represented easily by using sweep morphological operators. The expressions are not unique, and preference depends on the simplest combination and the least computational complexity. The common method is to decompose the attributes into smaller components and apply morphological dilations to grow these components. Let a and b represent unit vectors in the x- and y-axis, respectively. The unit vector could represent 1 m, 0.1 m, 0.01 m, and so on as needed. Note that when sweep dilation is not associated with rotation and scaling, it is equivalent to traditional dilation.

(a) *Rectangle*: It is represented as a unit x-axis vector a swept along a unit y-axis vector b, that is, $b \boxplus a$ with no rotation or scaling.

(b) *Parallelogram*: Let k denote a vector sum of a and b that are defined in (a). It is represented as $k \boxplus a$ with no rotation or scaling.

(c) *Circle*: Using sweep rotation, a circle can be represented as a unit vector a swept about a point p through 2π degrees, that is, $p \boxplus a$.

(d) *Trapezoid*: $b \boxplus a$ with a linear scaling factor to change a magnitude of a into c as it is swept along b as shown in Figure 11.13a. Let $0 \le t \le 1$. The scaling factor along the trajectory b is $S(t) = (c/a)t + (1 - t)$.

(e) *Triangle*: $b \boxplus a$, similar to trapezoid but with a linear scaling factor that changes a magnitude of a into zero as it is swept along b,

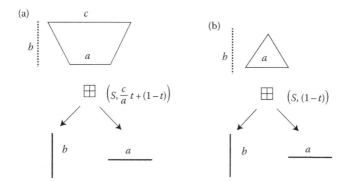

FIGURE 11.13   The decomposition of two-dimensional attributes.

as shown in Figure 11.13b. Note that the shape of triangles (e.g., an equilateral or right triangle) is determined by the fixing location of the reference point (i.e., the origin) of the primitive line *a*.

11.7.2 Three-Dimensional Attributes

Three-dimensional attributes can be applied by a similar method. Let *a*, *b*, *c* denote unit vectors in the *x*-, *y*-, and *z*-axis, respectively. Their formal expressions are presented below.

(a) *Parallelepiped*: It is represented as a unit vector *a* swept along a unit vector *b* to obtain a rectangle, and then it is swept along a unit vector *c* to obtain the parallelepiped; that is, *c* ⊞ (*b* ⊞ *a*).

(b) *Cylinder*: It is represented as a unit vector *a* swept about a point *p* through 2π degrees to obtain a circle, and then it is swept along a unit vector *c* to obtain the cylinder; that is, *c* ⊞ (*p* ⊞ *a*).

(c) *Parallelepiped with a corner truncated by a sphere*: A unit vector *a* is swept along a unit vector *b* to obtain a rectangle. A vector *r* is swept about a point *p* through 2π degrees to obtain a circle, and then it is subtracted from the rectangle. The result is swept along a unit vector *c*; that is, *c* ⊞ [(*b* ⊞ *a*) − (*p* ⊞ *r*)], as shown in Figure 11.14.

(d) *Sweep dilation of a square along a trajectory with deformation to a circle*: The square is represented as a rational B-spline curve. The polygon net is specified by a square with nine points with the first and the last being the same, and the weights of the corner vary from 5 to $\sqrt{2}/2$ as it is swept along the trajectory *C* that is defined in the parametric form as $x = 10s$ and $y = \cos(\pi s) - 1$. The sweep transformation is given by

$$[Sw_T] = \begin{bmatrix} \cos\psi & \sin\psi & 0 & 0 \\ -\sin\psi & \cos\psi & 0 & 0 \\ 0 & 0 & 1 & 0 \\ 10s & \cos\pi s & 0 & 1 \end{bmatrix}, \qquad (11.32)$$

where $\psi = \tan^{-1}(-\pi \sin(\pi s)/10)$. The formal expression is *C* ⊞ *B*. The sweeping of a square along a trajectory with deformation to a circle is shown in Figure 11.15.

(e) *Parallelepiped with a cylindrical hole*: A unit vector *a* is swept along a unit vector *b* to obtain a rectangle. A vector *r* is swept about a point *p* through 2π degrees to obtain a circle, and it is subtracted from the rectangle. The result is swept along a unit vector *c*; that is, *c* ⊞ [(*b* ⊞ *a*) − (*p* ⊞ *r*)], as shown in Figure 11.16.

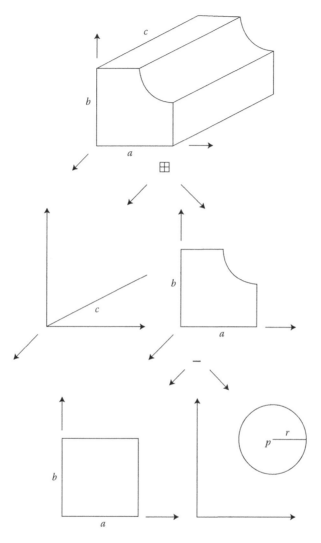

FIGURE 11.14 Sweep dilation of a rectangle with a corner truncated by a circle.

(f) *U-shape block*: A unit vector *a* is swept along a unit vector *b* to obtain a rectangle. A vector *r* is swept about a point *p* through π degrees to obtain a half circle, and it is dilated along the rectangle to obtain a rectangle with two rounded corners, which is then subtracted from another rectangle to obtain a U-shaped two-dimensional object. The result is swept along a unit vector *c* to obtain the final U-shaped object; that is, $c \boxplus [(b' \boxplus a') - [(b \boxplus a) \boxplus (p \boxplus r)]]$, as shown in Figure 11.17.

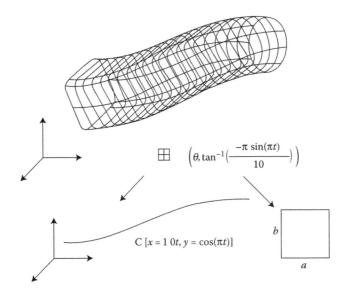

FIGURE 11.15   Sweeping of a square along a trajectory with deformation to a circle.

Note that the proposed sweep MM model can be applied to the NC machining process. For example, the ball-end milling cutter can be viewed as the structuring element and it can be moved along a predefined path to cut a work piece. During the movement, the cutter can be rotated to be perpendicular to the sweep path. If the swept volume is subtracted from the work piece, the remaining part can be obtained.

11.8 Grammars

In this section, we describe grammars in two-dimensional and three-dimensional attributes. We have experimented on many geometric objects. The results show that our model works successfully.

11.8.1 Two-Dimensional Attributes

All the primitive two-dimensional objects can be represented by the following grammar using the sweep MM model:

$$G = (V_N, V_T, P, S),$$

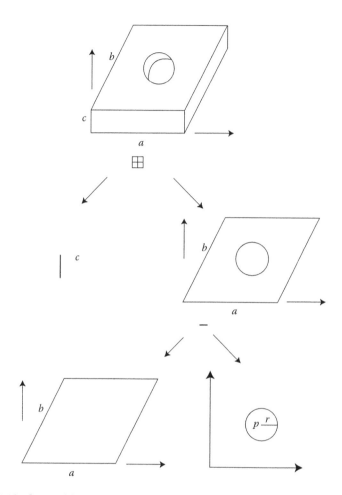

FIGURE 11.16 Sweep dilation of rectangle with a circular hole.

where,

$$V_N = \{S, A, B, K\}, \quad V_T = \{a, b, k, p, \boxplus\},$$
$$P: S \rightarrow B \boxplus A \mid B \boxplus K \mid p \boxplus A,$$
$$A \rightarrow aA \mid a,$$
$$B \rightarrow bB \mid b,$$
$$K \rightarrow kK \mid k$$

The sweep dilation $\boxplus$ can be $\oplus$ $(S = 0, \ \theta = 0)$, $\oplus$ $(S = (c/a)t + (1 - t), \ \theta = 0)$, or $\oplus$ $(S = 0, \ \theta = 2\pi)$. Note that the repetition of a unit vector in a generated string is the usual way of representing it grammatically. We can shorten the

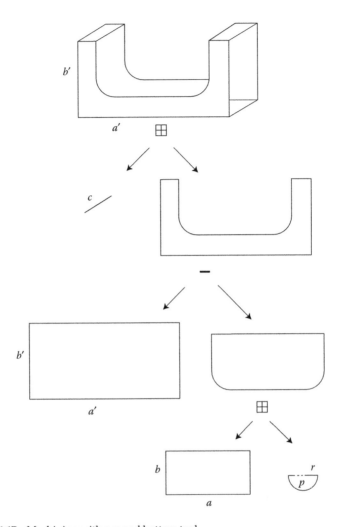

FIGURE 11.17 Machining with a round bottom tool.

length of a string by adopting a repetition symbol *. For example, *5a denotes *aaaaa.*

(a) A *rectangle* can be represented by the string *bb* ⊞ *aa,* with *a*s and *b*s repeated any number of times depending on the required size.

(b) *A parallelogram* can be represented by the string *kk* ⊞ *aaa,* with *a*s and *k*s repeated any number of times depending on the required size.

(c) A *circle* can be represented by the string *p* ⊞ *aaa,* with *a*s repeated any number of times depending on the required size and with ⊞ as ⊕ ($S = 0, \theta = 2\pi$).

(d) A *trapezoid* can be represented by the string $bb \boxplus aa$, with as and bs repeated any number of times depending on the required size and with $\boxplus$ as $\oplus$ $(S = (c/a)t + (1 - t), \theta = 0)$.

(e) A *triangle* can be represented by the string $bb \boxplus aa$, with as and bs repeated any number of times depending on the required size and with $\boxplus$ as $\oplus$ $(S = (1 - t), \theta = 0)$.

11.8.2 Three-Dimensional Attributes

All the primitive three-dimensional objects can be categorized into the following grammar:

$$G = (V_N, V_T, P, S),$$

where

$$V_N = \{S, A, B, C\}, \quad V_T = \{a, b, c, p, (,), \boxplus\},$$
$$P: S \rightarrow C \boxplus (B \boxplus A) \mid C \boxplus (p \boxplus A),$$
$$A \rightarrow aA \mid a,$$
$$B \rightarrow bB \mid b,$$
$$C \rightarrow cC \mid c$$

The sweep dilation $\boxplus$ can be either $\oplus$ $(S = 0, \theta = 0)$ or $\oplus$ $(S = 0, \theta = 2\pi)$.

(a) A *parallelepiped* can be represented by the string $ccc \boxplus (bb \boxplus aaa)$, with as, bs, and cs repeated any number of times depending on the required size.

(b) A *cylinder* can be represented by the string $cccc \boxplus (p \boxplus aaa)$, with as and cs repeated any number of times depending on the required size, and with the first dilation operator $\boxplus$ as $\oplus$ $(S = 0, \theta = 2\pi)$ and the second dilation as the traditional dilation.

(c) Consider the grammar

$$G = (V_N, V_T, P, S),$$

where

$$V_N = \{A, B, C, N\}, \quad V_T = \{a, b, c, p, \boxplus, -, (,)\},$$
$$P: S \rightarrow C \boxplus N,$$
$$N \rightarrow \text{rectangle} - \text{circle},$$
$$C \rightarrow cC \mid c$$

The grammar for creating rectangles and circles was given in the previous section.

(c.1) The *sweep dilation of a rectangle with a corner truncated by a circle* can be represented by the string $cc \boxplus ((bb \boxplus aaa) - (p \boxplus aa))$, with as, bs, and cs repeated any number of times depending on the required size and with the second dilation operator $\boxplus$ as $\oplus$ ($S = 0$, $\theta = 2\pi$).

(c.2) The *sweep dilation of a rectangle with a circular hole* can be represented by the string $cc \boxplus ((bb \boxplus aaa) - (p \boxplus a))$, with as, bs, and cs repeated any number of times depending on the required size and with the second dilation operator $\boxplus$ as $\oplus$ ($S = 0$, $\theta = 2\pi$). The difference from the previous one is that the circle lies completely within the rectangle, and hence we obtain a hole instead of a truncated corner.

(d) The grammar for the U-shaped a block can be represented as follows:

$$G = (V_N, V_T, P, S),$$

where

$$V_N = \{A, B, C, N, M, \text{half\_circle}\}, \quad V_T = \{a, b, c, p, \boxplus, -\},$$

$$P: S \rightarrow C \boxplus N,$$

$$N \rightarrow \text{rectangle} - M,$$

$$C \rightarrow cC \mid c,$$

$$M \rightarrow \text{rectangle} \boxplus \text{half\_circle},$$

$$\text{half\_circle} \rightarrow p \boxplus A$$

The *U-shaped block* can be represented by the string $ccc \boxplus (bbb \boxplus aaaa - ((bb \boxplus aa) \boxplus (p \boxplus a)))$, with as, bs, and cs repeated any number of times depending on the required size and with the fourth dilation operator $\boxplus$ as $\oplus$ ($S = 0$, $\theta = \pi$).

11.9 Parsing Algorithms

Given a grammar G and an object representation as a string, the string can be parsed to find out whether it belongs to the given grammar. There are various parsing algorithms, among which Earley's parsing algorithm for context-free grammars is very popular. Let V^* denote the set of all sentences composed of elements from V. The algorithm is described as follows:

Input: Context-free Grammar $G = (V_N, V_T, P, S)$, and an input string $w = a_1 a_2 \ldots a_n$ in V_T^*.

Output: The parse lists $I_0, I_1, \ldots, I_n$.

Method: First construct I_0 as follows:

1. If $S \to \alpha$ is a production in P, add $[S \to .\alpha, 0]$ to I_0.

Now, perform steps 2 and 3 until no new item can be added to I_0.

2. If $[B \to \gamma., 0]$ is on I_0, add $[A \to \alpha B.\beta, 0]$ for all $[A \to \alpha.B\beta, 0]$ on I_0.

3. Suppose that $[A \to \alpha.B\beta, 0]$ is an item in I_0. Add to I_0 for all productions in P of the form $B \to \gamma$, the item $[B \to .\gamma, 0]$ (provided this item is not already in I_0).

Now, we construct I_j by having constructed $I_0, I_1, \ldots, I_{j-1}$.

4. For each $[B \to \alpha.a\beta, i]$ in I_{j-1} such that $a = a_j$, add $[B \to \alpha a.\beta, i]$ to I_j.

Now, perform steps 5 and 6 until no new items can be added.

5. Let $[A \to \gamma., i]$ be an item in I_j. Examine I_i for items of the form $[B \to \alpha.A\beta, k]$. For each one found, add $[B \to \alpha A.\beta, k]$ to I_j.

6. Let $[A \to \alpha.B\beta, i]$ be an item in I_j. For all $B \to \gamma$ in P, add $[B \to .\gamma, j]$ to I_j.

The algorithm then is to construct I_j for $0 < j \leq n$. Some examples of the parser are shown below.

Example 11.3: Let a rectangle be represented by the string $b \boxplus aa$. The given grammar is

$$G = (V_N, V_T, P, S),$$

where

$$V_N = \{S, A, B\}, \quad V_T = \{a, b, \boxplus\}$$

$$P: S \to B \boxplus A,$$

$$A \to aA \mid a,$$

$$B \to bB \mid b$$

The parsing lists obtained are as follows:

I_0	I_1	I_2	I_3	I_4
$[S \rightarrow .B \boxplus A, 0]$	$[B \rightarrow b.B, 0]$	$[S \rightarrow B \boxplus .A, 0]$	$[A \rightarrow a.A, 2]$	$[A \rightarrow a.A, 3]$
$[B \rightarrow .bB, 0]$	$[B \rightarrow b., 0]$	$[A \rightarrow .aA, 2]$	$[A \rightarrow a., 2]$	$[A \rightarrow a., 3]$
$[B \rightarrow .b, 0]$	$[S \rightarrow B. \boxplus A, 0]$	$[A \rightarrow .a, 2]$	$[S \rightarrow B \boxplus A., 0]$	$[A \rightarrow aA., 2]$
	$[B \rightarrow .bB, 1]$		$[A \rightarrow .aA, 3]$	$[A \rightarrow .aA, 4]$
	$[B \rightarrow .b, 1]$		$[A \rightarrow .a, 3]$	$[A \rightarrow .a, 4]$
				$[S \rightarrow B \boxplus A., 0]$

Since $[S \rightarrow A \boxplus B., 0]$ is on the last list, the input belongs to the language $L(G)$ generated by G.

Example 11.4: Consider the input string $b \boxplus ba$. The given grammar is given by

$$G = (V_N, V_T, P, S),$$

where

$$V_N = \{S, A, B\}, \quad V_T = \{a, b, \boxplus\},$$
$$P: S \rightarrow B \boxplus A,$$
$$A \rightarrow aA \mid a,$$
$$B \rightarrow bB \mid b$$

The parsing lists obtained are as follows:

I_0	I_1	I_2	I_3
$[S \rightarrow .B \boxplus A, 0]$	$[B \rightarrow b.B, 0]$	$[S \rightarrow B \boxplus .A, 0]$	*Nil*
$[B \rightarrow .bB, 0]$	$[B \rightarrow b., 0]$	$[A \rightarrow .aA, 2]$	
$[B \rightarrow .b, 0]$	$[S \rightarrow B. \boxplus A, 0]$	$[A \rightarrow .a, 2]$	
	$[B \rightarrow .bB, 1]$		
	$[B \rightarrow .b, 1]$		

Since there is no production starting with S on the last list, the input does not belong to the language $L(G)$ generated by G.

A question arises as how we can construct a grammar that will generate a language to describe any kind of solid object. Ideally, it would be nice to have a grammatical inference machine that would infer a grammar from a set of given strings describing the objects under study. Unfortunately, such a machine is not available except for some very special cases. In most cases, the designer constructs the grammar based on the *a priori* knowledge available and past experience. In general, the increased descriptive power of a language is paid for in terms of increased complexity of the analysis system. The trade-off between the descriptive power and analysis efficiency of a grammar for a given application is almost completely defined by the designer.

As a final note in this chapter, Bouaynaya, Charif-Chefchaouni, and Schonfeld [2006] used the theory of SV MM to extend and analyze two important image-processing applications: morphological image restoration and skeleton representation of binary images. In most applications of MM, the structuring element remains constant in shape and size as the image is probed. Hence, the focus of MM has been mostly devoted to translation-invariant operators. An extension of the theory to SV operators has emerged due to the requirements of some applications, such as traffic spatial measurements [Beucher, Blosseville, and Lenoir 1987] and range imagery [Verly and Delanoy 1993]. A fundamental result in lattice morphology, which provides the representation of a large class of nonlinear and non-necessarily translation-invariant operators in terms of lattice erosions and dilations has been presented in Banon and Barrera [1993]. A separate approach to the representation of SV MM, which preserves the geometrical structuring element, in the Euclidean space was introduced by Serra [1988]. A comprehensive development of the theory of SV MM in the Euclidean space has been presented in Bouaynaya, Charif-Chefchaouni, and Schonfeld [2004].

References

Banon, G. and Barrera, J., "Decomposition of mappings between complete lattices by mathematical morphology, Part I: General lattices," *Signal Processing*, vol. 30, no. 3, pp. 299–327, Feb. 1993.

Beucher, S., Blosseville, J. M., and Lenoir, F., "Traffic spatial measurements using video image processing," in *Proc. SPIE. Intelligent Robots and Computer Vision*, vol. 848, pp. 648–655, Nov. 1987.

Blackmore, D., Leu, M. C., and Shih, F. Y., "Analysis and modeling of deformed swept volumes," *Computer Aided Design*, vol. 26, no. 4, pp. 315–326, Apr. 1994.

Bouaynaya, N. and Schonfeld, D., "Theoretical foundations of spatially variant mathematical morphology—Part II: Gray-level images," *IEEE Trans. Pattern Analysis and Machine Intelligence*, vol. 30, no. 5, pp. 837–850, May 2008.

Bouaynaya, N., Charif-Chefchaouni, M., and Schonfeld, D., "Spatially variant mathematical morphology: A geometry-based theory," *Tech. Rep. MCL-TR-2005-01*, Multimedia Communications Lab., Univ. Illinois at Chicago, 2004.

Bouaynaya, N., Charif-Chefchaouni, M., and Schonfeld, D., "Spatially variant morphological restoration and skeleton representation," *IEEE Trans. Image Processing*, vol. 15, no. 11, pp. 3579–3591, Nov. 2006.

Bouaynaya, N. and Schonfeld, D., "Theoretical foundations of spatially variant mathematical morphology—Part II: Gray-level images," *IEEE Trans. Pattern Analysis and Machine Intelligence*, vol. 30, no. 5, pp. 837–850, May 2008.

Brooks, R. A., "Symbolic reasoning among three-dimensional models and two-dimensional images," *Artificial Intelligence*, vol. 17, no. 1–3, pp. 285–348, Aug. 1981.

Chen, C. S., Wu, J. L., and Hung, Y. P., "Statistical analysis of space-varying morphological openings with flat structuring elements," *IEEE Trans. Signal Processing*, vol. 44, no. 4, pp. 1010–1014, Apr. 1996.

Chen, C. S., Wu, J. L., and Hung, Y. P., "Theoretical aspects of vertically invariant gray-level morphological operators and their application on adaptive signal and image filtering," *IEEE Trans. Signal Processing*, vol. 47, no. 4, pp. 1049–1060, Apr. 1999.

Cooper, D., Elliott, H., Cohen, F., and Symosek, P., "Stochastic boundary estimation and object recognition," *Computer Graphics Image Process*, vol. 12, no. 4, pp. 326–356, Apr. 1980.

Cuisenaire, O., "Locally adaptable mathematical morphology using distance transformations," *Pattern Recognition*, vol. 39, no. 3, pp. 405–416, March 2006.

Debayle, J. and Pinoli, J. C., "General adaptive neighborhood image processing—Part I: Introduction and theoretical aspects," *Journal of Math. Imaging and Vision*, vol. 25, no. 2, pp. 245–266, Sep. 2006a.

Debayle, J. and Pinoli, J. C., "General adaptive neighborhood image processing—Part II: Practical application examples," *Journal of Math. Imaging and Vision*, vol. 25, no. 2, pp. 267–284, Sep. 2006b.

Eichel, P. H., Delp, E. J., Koral, K., and Buda, A. J., "A method for a fully automatic definition of coronary arterial edges from cineangiograms," *IEEE Trans. Medical Imaging*, vol. 7, no. 4, pp. 313–320, Dec. 1988.

Farag, A. A. and Delp, E. J., "Edge linking by sequential search," *Pattern Recognition*, vol. 28, no. 5, pp. 611–633, May 1995.

Foley, J., van Dam, A., Feiner, S., and Hughes, J., *Computer Graphics: Principles and Practice*, 2nd edition, MA: Addison-Wesley, 1995.

Fu, K. S., *Syntactic Pattern Recognition and Applications*, NJ: Prentice Hall, 1982.

Gonzalez, R. C. and Woods, R. E., *Digital Image Processing*, New York: Addison-Wesley, 2002.

Gosh, P. K., "A mathematical model for shape description using Minkowski operators," *Compt. Vision, Graphics, and Image Processing*, vol. 44, no. 3, pp. 239–269, Dec. 1988.

Haralick, R. M., Sternberg, S. K., and Zhuang, X., "Image analysis using mathematical morphology," *IEEE Trans. Pattern Analysis and Machine Intelligence*, vol. 9, no. 4, pp. 532–550, July 1987.

Jang, B. K. and Chin, R. T., "Analysis of thinning algorithms using mathematical morphology," *IEEE Trans. Pattern Analysis and Machine Intelligence*, vol. 12, no. 6, pp. 541–551, June 1990.

Kaul, A., *Computing Minkowski Sums*, Ph.D. thesis, Department of Mechanical Engineering, Columbia University, 1993.

Korein, J., *A Geometric Investigating Reach*, MA: MIT Press, 1985.

Leu, M. C., Park, S. H., and Wang, K. K., "Geometric representation of translational swept volumes and its applications," *ASME Journal of Engineering for Industry*, vol. 108, no. 2, pp. 113–119, 1986.

Martelli, A., "An application of heuristic search methods to edge and contour detection," *Communication ACM*, vol. 19, no. 2, pp. 73–83, Feb. 1976.

Martin, R. R. and Stephenson, P. C., "Sweeping of three dimensional objects," *Computer Aided Design*, vol. 22, no. 4, pp. 223–234, May 1990.

Morales, A. and Acharya, R., "Statistical analysis of morphological openings," *IEEE Trans. Signal Processing*, vol. 41, no. 10, pp. 3052–3056, Oct. 1993.

Mott-Smith, J. C. and Baer, T., "Area and volume coding of pictures," in *Picture Bandwidth Compression*, edited by T. S. Huang and O. J. Tretiak, New York: Gordon and Beach, 1972.

Pennington, A., Bloor, M. S., and Balila, M., "Geometric modeling: A contribution toward intelligent robots," *Proc. 13th Inter. Symposium on Industrial Robots*, pp. 35–54, 1983.

Ragothama, S. and Shapiro, V., "Boundary representation deformation in parametric solid modeling," *ACM Transactions on Graphics*, vol. 17, no. 4, pp. 259–286, Oct. 1998.

Requicha, A. A. G., "Representations for rigid solids: Theory, methods, and systems," *ACM Computing Surveys*, vol. 12, no. 4, pp. 437–464, Dec. 1980.

Requicha, A. A. G. and Voelcker, H. B., "Solid modeling: A historical summary and contemporary assessment," *IEEE Computer Graphics and Applications*, vol. 2, no. 2, pp. 9–24, Mar. 1982.

Roerdink, J. B. T. M., "Group morphology," *Pattern Recognition*, vol. 33, no. 6, pp. 877–895, June 2000.

Rossignac, J. and Requicha, A. A. G., "Offsetting operations in solid modeling," Production Automation Project, University of Rochester, NY Tech, Memo 53, 1985.

Rossignac, J., "CSG-Brep duality and compression," *Proc. ACM Symposium on Solid Modeling and Applications*, Saarbrucken, Germany, pp. 59–66, 2002.

Russ, J. C., *The Image Processing Handbook*, CRC Press, 1992.

Serra, J., *Image Analysis and Mathematical Morphology*, London: Academic Press, 1982.

Serra, J., *Image Analysis and Mathematical Morphology*, New York: Academic Press, vol. 2, 1988.

Shih, F. Y., "Object representation and recognition using mathematical morphology model," *Journal of Systems Integration*, vol. 1, no. 2, pp. 235–256, Aug. 1991.

Shih, F. Y. and Mitchell, O. R., "Threshold decomposition of gray-scale morphology into binary morphology," *IEEE Trans. on Pattern Analysis and machine Intelligence*, vol. 11, no. 1, pp. 31–42, Jan. 1989.

Shih, F. Y. and Mitchell, O. R., "A mathematical morphology approach to Euclidean distance transformation," *IEEE Trans. on Image Processing*, vol. 1, no. 2, pp. 197–204, Apr. 1992.

Shih, F. Y. and Cheng, S., "Adaptive mathematical morphology for edge linking," *Information Sciences*, vol. 167, no. 1–4, pp. 9–21, Dec. 2004.

Shih, F. Y., Gaddipati, V., and Blackmore, D., "Error analysis of surface fitting for swept volumes," *Proc. Japan-USA Symp. Flexible Automation*, Kobe, Japan, pp. 733–737, July 1994.

Shiroma, Y., Okino, N., and Kakazu, Y., "Research on three-dimensional geometric modeling by sweep primitives," *Proc. of CAD*, Brighton, United Kingdom, pp. 671–680, 1982.

Shiroma, Y., Kakazu, Y., and Okino, N., "A generalized sweeping method for CSG modeling," *Proc. ACM Symposium on Solid Modeling Foundations and CAD/CAM Applications*, Austin, Texas, pp. 149–157, 1991.

Stevenson, R. L. and Arce, G. R., "Morphological filters: statistics and further syntactic properties," *IEEE Trans. Circuits and Systems*, vol. 34, no. 11, pp. 1292–1305, Nov. 1987.

Verly, J. G. and Delanoy, R. L., "Adaptive mathematical morphology for range imagery," *IEEE Trans. Image Process.*, vol. 2, no. 2, pp. 272–275, Apr. 1993.

Voelcker, H. B. and Hunt, W. A., "The role of solid modeling in machining process modeling and NC verification," *SAE Tech. Paper #810195*, 1981.

Wang, W. P. and Wang, K. K., "Geometric modeling for swept volume of moving solids," *IEEE Computer Graphics and Applications*, vol. 6, no. 12, pp. 8–17, Dec. 1986.

12

Morphological Approach to Shortest Path Planning

Advances in robotics, artificial intelligence, and computer vision have stimulated considerable interest in the problems of robot motion and shortest path planning [Latombe 1991]. The path-planning problem is concerned with finding paths connecting different locations in an environment (e.g., a network, a graph, or a geometric space). Depending on specific applications, the desired paths often need to satisfy some constraints (e.g., obstacle-avoiding) and optimize certain criteria (e.g., variant distance metrics and cost functions).

This chapter presents an MM-based approach to finding the shortest path of an arbitrarily shaped moving object with rotations amid obstacles of arbitrary shapes. By utilizing rotational morphology along with the continuous turning-angle constraint, the representation framework of the motion-planning problem becomes simple and practical. Two models of an object's movement with different driving and turning systems are used to solve the path-planning problem and their strategies are presented. Simulation results show that the proposed algorithm works successfully under various conditions.

This chapter is organized as follows. Section 12.1 introduces the topic and Section 12.2 describes the relationships between shortest-path-finding and MM. Section 12.3 presents rotational MM. Section 12.4 presents the shortest-path-finding algorithm. Section 12.5 provides experimental results and discussions. Section 12.6 describes dynamic rotational mathematical morphology (DRMM). Section 12.7 states the rule of distance functions for shortest path planning.

12.1 Introduction

Modern manufacturing and other high technology fields of robotics and artificial intelligence have stimulated considerable interest in motion planning or the shortest-path-finding problems [Latombe 1991]. The main objective is to plan a collision-free path (and preferably an optimal path) for a moving object in a workspace populated with obstacles. Given an initial position and orientation for a robot and a target position and orientation, a path is formed from start to end positions, which is collision-free and satisfies some criteria such as minimal cost and continuous turning angles.

There are several theoretical approaches in optimal path planning, and each has its own particular advantages and drawbacks [Hwang and Ahuja 1992]. In general, two classes are used. One is a via configuration space [Lozano-Perez 1987; Pei, Lai, and Shih 1998], wherein the geometry of the object that is to be moved through a space filled with obstacles is essentially set and theoretically added to the obstacle to create a new space. The space created by removing the enlarged obstacles is called *the configuration space*. In the context of the configuration space, the path-planning problem is reduced to that of finding optimal paths of a reference point in the object from an initial to an end position. This approach is used in the traditional MM method when path planning for nonrotating objects [Ghosh 1990; Lin and Chang 1993].

The second approach uses computational geometry to search the free space directly without transforming the workspace into a configuration space [Brooks 1983; Kambhampati and Davis 1994]. This method allows robot movement with rotations, so that it is close to real-world applications. However, it becomes very complicated if obstacles gather closely. To make those algorithms tractable, some limitations, such as polyhedral assumption (i.e., all obstacles are assumed to be polyhedral in shape), must be added to simplify problems.

The goal is to use a novel morphological approach to solve the motion planning problem. Computational complexity can thereby be reduced, compared to computational geometry methods, while allowing for rotations of moving objects. To take the rotation movement into consideration in modeling the moving object, a new technique, named *rotational MM*, is proposed to process both the orientation and geometric information, thus overcoming the disadvantages of the traditional morphological approach, which is limited to nonrotating moving objects. All the properties of traditional morphology [Serra 1982; Shih and Mitchell 1989] are inherited in rotational morphology.

The shape decomposition property and parallel processing architecture can be used to speed up the computations.

12.2 Relationships between Shortest Path Finding and MM

To solve the shortest-path-finding problem by using MM, the relationships between shortest-path-finding and MM are first explored. Assume that an object of arbitrary shape moves from a starting point to an ending point in a finite region containing arbitrarily shaped obstacles. To relate this model to MM, the finite region consists of a free space set with values equal to one and an obstacle set with values equal to zero, the moving object is modeled as a structuring element. Thus, the shortest-path-finding problem is equivalent to applying morphological erosion to a free space, followed by a distance transform (DT) on the region with the grown obstacles excluded, and then tracing back the resultant distance map from the target point to its neighbors with the minimum distance until the starting point is reached.

The drawback of using traditional MM to solve the path-planning problem is the fixed direction of the structuring element used throughout the processes. A fixed directional movement results in a poor path, much longer than the optimal path of real-world applications. To model the actual moving actions of a car-like robot by incorporating rotations, a rotational morphology is proposed such that the structuring element can be rotated simultaneously during translations.

Figure 12.1a shows the result of a shortest-path-finding algorithm [Lin and Chang 1993] for a nonrotating object. By incorporating rotations into the motion of a moving object, Figure 12.1b gives a more realistic solution to the shortest-path-finding algorithm.

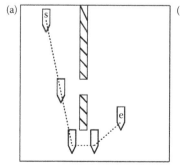
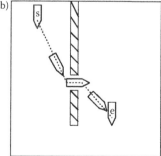

FIGURE 12.1 Results of a path-finding algorithm for (a) nonrotating and (b) rotating vehicle models, where "s" and "e" denote the starting and ending points, respectively, and the slashed regions denote the obstacles set. Note that the resultant path in (a) is longer than the path in (b).

12.3 Rotational MM

12.3.1 Definitions

Traditional morphological dilation and erosion perform vector additions or subtractions by translating a structuring element along an object. Since the structuring element is fixed in its orientation during processes, these operations cannot deal with the complex variation of directional features in an image. Moreover, their applications to the sweep motion involving both translation and rotation are limited. We propose rotational MM for an enhanced theoretical framework with a wider range of practical applications.

Assume that the entire 360° angles are equally divided into N parts (or directions). Let B_i denote the rotation of a structuring element B by the degree of $360i/N$, where $i = 0, 1, \ldots, N - 1$, and let $\mathbf{y}$ denote two-dimensional coordinates. Let A and B be a binary image and a binary structuring element, respectively. The rotational dilation of A by B is defined as

$$(A \, \tilde{\oplus} \, B)(\mathbf{y}) = [A \oplus B_{N-1}, A \oplus B_{N-2}, \ldots, A \oplus B_1, A \oplus B_0](\mathbf{y})$$

$$= [P_{N-1}, P_{N-2}, \ldots, P_1, P_0](\mathbf{y}), \qquad (12.1)$$

$$= \mathbf{P}(\mathbf{y})$$

where $\mathbf{P}$ is a N-bit codeword matrix.

Similarly, the *rotational erosion* of A by B is defined as

$$(A \, \tilde{\ominus} \, B)(\mathbf{y}) = [A \ominus B_{N-1}, A \ominus B_{N-2}, \ldots, A \ominus B_1, A \ominus B_0](\mathbf{y}) \qquad (12.2)$$

Figure 12.2 shows an example of rotational erosion. Compared to conventional MM, rotational morphology provides not only geometric shape but also orientation information. In other words, the results of rotational morphological operations indicate which direction of the structuring element is included or excluded in a local object region. Since the rotational morphology is equivalent to executing N binary morphology modules as shown in Figure 12.3, it can be computed in parallel and its computational complexity remains the same as in binary morphology.

Note that $N = 8$ is often used in digital image processing applications. Therefore, $\mathbf{P} = [P_7, P_6, \ldots, P_1, P_0]_2$. That is, $\mathbf{P}$ is an 8-bit codeword with base 2. If all the bits are summed up, it ranges from 0 to 8. After normalization, it ranges from 0 to 1, and the value possesses a fuzzy interpretation. Let us consider a fuzzy rotational erosion. Value 0 indicates that the structuring element cannot be completely contained in the object region in any direction. Value 1 indicates that the structuring element is entirely located within the object region for all orientations. Therefore, the value can be interpreted as the degree of the structuring element's orientations that can be fitted into the

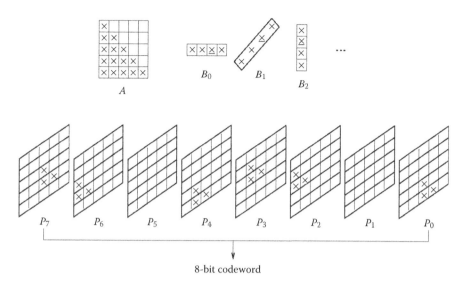

FIGURE 12.2 An example of a rotational erosion operation of *A* by *B*.

object region. Readers should note that if the structuring element is symmetric with respect to all directions such as in a circle, the fuzzy rotational morphology is equivalent to binary morphology.

The all-directional dilation, based on the intersection of all bits of the codeword, is defined as

$$(A \stackrel{\vee}{\oplus} B)(\mathbf{y}) = \bigwedge_{i=0}^{N-1} P_i(\mathbf{y}) \tag{12.3}$$

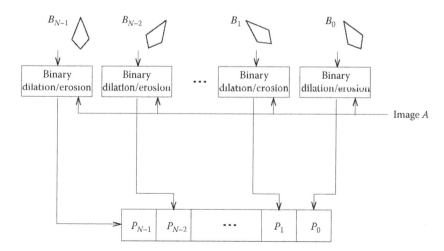

FIGURE 12.3 The parallel processing structure of the rotational morphology.

While the all-directional erosion is defined as

$$(A \check{\ominus} B)(\mathbf{y}) = \bigvee_{i=0}^{N-1} P_i(\mathbf{y}) \tag{12.4}$$

Since the all-directional dilation represents the growth of obstacles, positions with all bits equal to 1 represent absolute collision-free space where no obstacles grow. Note that because the all-directional erosion is used to indicate collision-free positions in path finding such that the moving vehicle can be contained in the free space for at least one orientation, it is defined as a union of the codeword bits.

12.3.2 Properties

All the properties of binary morphology are also true in rotational morphology since it is composed of N binary morphological operations. Some properties used in our shortest-path-finding algorithm are described below.

Property 12.1: Rotational morphology is translation invariant, for example,

$$(A \,\tilde{\oplus}\, B_z) = (A \,\tilde{\oplus}\, B)_z, \tag{12.5}$$

where z denotes any two-dimensional coordinate. It also has the properties of duality and decomposition as follows.

Property 12.2:

$$(A^c \,\tilde{\oplus}\, \hat{B}) = (A \,\tilde{\ominus}\, B)^c, \tag{12.6}$$

where A^c represents the complement of set A and $\hat{B}$ represents the reflection set of B with respect to the origin.

Property 12.3: For $B = C \cup D$

$$A \,\tilde{\oplus}\, B = (A \,\tilde{\oplus}\, C) \cup (A \,\tilde{\oplus}\, D) \tag{12.7}$$

$$A \,\tilde{\ominus}\, B = (A \,\tilde{\ominus}\, C) \cap (A \,\tilde{\ominus}\, D) \tag{12.8}$$

In finding the collision-free path, Property 12.2 provides a significantly efficient way of finding the free position (FP) set. That is, computing the obstacle regions saves more time than computing the entire FP set. Property 12.1 states that the solution will not be affected by translation, whereas Property 12.3 can be used to reduce computational complexity by decomposing the structuring element into the union of small subsets.

12.4 The Shortest Path–Finding Algorithm

Due to the nonrotating movement of conventional morphological operations, the path determination algorithm of Lin and Chang [1993] cannot really generate the shortest path in practical cases. By using rotational morphological operations, a rotation movement can be modeled to fit real-world applications and the optimal path can be found. The shortest collision-free path-finding algorithm can be divided into two major parts that are described below.

12.4.1 Distance Transform

Distance transform (DT) is an operation that converts a binary image with feature and non-feature pixels into a gray-level image whose pixel values represent the corresponding distance to the nearest feature elements. In many applications of digital image processing such as expanding/shrinking, it is essential to measure the distance between image pixels. DT is a specific example of a more general cost function transformation. Since the contents of a distance map reveals the shortest distance to the referred points, tracing the distance map from a specified point to a referred point will result in the shortest path between the specified and the referred points. In this section, recursive DT with (3, 4)-distance metric is used and the following procedures compute the distance map $D(a, s)$ for all $a \in Z$. In this procedure, $d[a]$ denotes an array representing the current transformation value at pixel a. Definitions corresponding to all variables can be found in Piper and Granum [1987].

1. Set $d[s] =$ REFERENCE where s denotes the starting point; set $d[a] =$ UNTRANS for all $a \in Z - s$ where UNTRANS is the largest positive integer available and REFERENCE = UNTRANS − 1. If a rectangular array representation is used, then set background points (obstacles set) = BACKGROUND, which is an arbitrary negative integer, so that nondomain points may be avoided.
2. Apply procedure propagate(s).
3. When the algorithm terminates, $D(a, s) = d[a]$ for all $a \in Z$.

Procedure propagate() modifies pixel values and, recursively, the pixel neighbor values. It replaces $d[a]$ by the minimum of its original value and the minimum of the value of $d[\ ]$ at a neighbor plus the distance from that neighbor. Whenever $d[a]$ is changed, propagate() is applied recursively to all neighbors of a. The propagate() procedure is expressed in pseudo-C as follows:

```
propagate(s)
pixel a;
```

```
{
  distance nd;
  extern = distance d[];
  if (d[a] == REFERENCE)
        nd = 0;
  else
        nd = min(d[N_i(a)] + d_i | N_i(a) ∈ Z);
  if (nd < d[a]) {
        d[a] = nd;
        for(all neighbors i | N_i(a) ∈ Z)
        propagate(N_i(a))
        }
}
```

12.4.2   Describing the Algorithm

Assume we have a working support Z, an obstacle set O, an arbitrarily shaped car set B, a starting point s, and an end point e, as shown in Figure 12.4a.

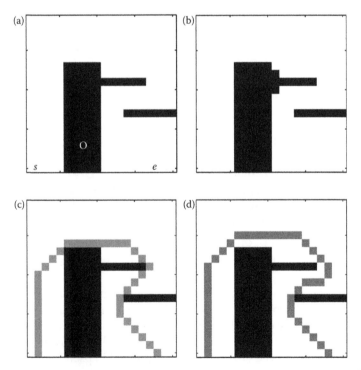

FIGURE 12.4 (a) The original FP set, denoted as the white region, and the obstacle set O, denoted as the black region, (b) result after applying all-directional erosion. Note that the FP set is shrunk. (c) A collision-FP set attaching "s" and "e" points. (d) The practical result of our path-planning algorithm.

The candidate space is denoted as $A = Z \backslash O$. Our goal is to find a collision-free path for B to move from s to e inside the set Z. To find the shortest avoidance-free path among obstacles, the candidate space A first undertakes an all-directional morphological erosion, that is, $A \ominus B$. The resultant set $\check{A}$, shown in Figure 12.4b, is called a "free-position" space and represents all the positions that the car set B can be contained in for at least one orientation. After the free-position space is computed, a propagated DT with $(3, 4)$-distance metric described above is applied on Z with the constraint set $\check{O} = Z \backslash \check{A}$ to compute the distance map. Thus, a shortest path can be found by tracing the distance map with a pre-defined rule, choosing the nearest distance in this example, from s to e as shown in Figure 12.4c.

However, the shortest path obtained above (candidate solution) may be no longer valid if the practical case of a smooth turning transition of a car-like vehicle is considered. For each point the vehicle moves through, if its direction is not contained in the orientation information matrix $\tilde{A} = A \,\tilde{\ominus}\, B$ or its direction violates the smooth turning-angle limitation, the path is said to be impractical and this shortest path is not really a path, but just an FP space. To find the practical collision-free path, those points that disobey the above constraint will be set as the obstacles and the above steps should be repeated until a new path, where all its path points obey this constraint, is found. Figure 12.4d shows a practical path after the continuous turning-angle constraint is applied. The one-pixel upward displacement of the horizontal upper part of the path added to escape the problem of turning 90° is shown in Figure 12.4c. Figure 12.5 shows the sample paths with continuous and discontinuous turning angle, respectively. After a satisfactory path is found, the directional information is recorded by chain-coding and the coordinate of the starting point and the orientation chain codes are sent as the final solution. The shortest-path-finding algorithm is described in the procedures below.

Step 1: Given sets Z, O, B, e, and s, which are defined previously, set a candidate space as $A = Z \backslash O$.

Step 2: For A, perform rotational morphological erosion, $\tilde{A} = A \,\tilde{\ominus}\, B$, and an all-directional morphological erosion, $\check{A} = A \,\check{\ominus}\, B$, to find the

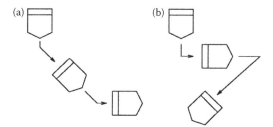

FIGURE 12.5   Two examples for illustrating the conditions of (a) obeying and (b) violating the continuous turning-angle constraint when the moving object is a car-like model.

directional information matrix $\tilde{A}$ and the FP set $\check{A}$, respectively. The new obstacles set now becomes $\check{O} = Z \backslash \check{A}$, where $\backslash$ denotes a set subtraction.

Step 3: Apply a propagation distance transform (PDT) on Z with the obstacles constraint set $\check{O}$ to build a distance map.

Step 4: Route the distance map from s to e with a predefined rule to find one candidate path.

Step 5: For each point in the path, record its directional information by the chain-coding technique and trace the chain code with the directional information matrix $\tilde{A}$. If there exists a point p such that its direction is not contained in $\tilde{A}$ or that its direction does not satisfy the smooth turning-angle constraint, add that point to the obstacles set, that is, $\check{O} = \{\check{O} \cup p\}$, and repeat Steps from 3 to 5.

Step 6: If a path connects s and e such that the shortest path is found, send the coordinates of the start point and the orientation chain codes as the final solution. Otherwise, there is no solution to the current field Z.

The entire process can be illustrated by tracing each step of Figure 12.4. The properties of rotational morphology are used to simplify the implementation of the algorithm. Using the duality property, rotational erosion of A by B can be replaced by rotational dilation of O by B to reduce computational complexity since in most cases the obstacle set is far less than the candidate set. Another way to reduce the complexity is to compute only those boundary points of O denoted as ∂O, but not the entire obstacles set O. The shape decomposition technique described in Equations 12.7 and 12.8 is adopted to further speed up the computation of the FPs of a moving object.

The significant advantage of using morphology tools to solve the pathfinding problem is that complex geometric problems can be solved by set operations. By the way, growing the sizes of obstacles is equivalent to shrinking the vehicle to a single point, thus simplifying the spatial planning problem.

12.5 Experimental Results and Discussions

In this section, several complicated examples based on two vehicle models with a 30×30 field "Z" are examined to validate the system. Two types of vehicle models are used in simulating real-world robot motions. One type of vehicle has its driving and turning wheels located at the center of the vehicle body (center-turning), while the other type, like a bus, has turning wheels located at the front of the vehicle body (front-turning). Considering the motion behavior of these two models, the former permits rotations at a fixed point, while the other leads rotations and translations simultaneously; thus they produce different paths. An important constraint for front-turning vehicles is the rotating angle. That is, if the turning angle is not continuous, as shown in Figure 12.5b, it is not possible for a car-like vehicle to move across obstacles without collision. Figure 12.6 show an example of the resultant paths obtained based on the two models. Readers should note that for the center-turning model, since the moving objects can rotate by arbitrary angles without translation, the continuous turning-angle constraint used in the proposed algorithm can be removed. However, due to limitations of the mechanism of the front-turning model, this constraint must hold. Figures 12.7 through 12.10 show the simulation results for these two models with variant vehicle shapes and target points. We have observed that the proposed algorithm works successfully for many different situations.

Other constraints can be placed within the proposed algorithm to fit into variant situations. For some applications, it is necessary to assign the orientation of the vehicle at the start point as well as at the end point. For example,

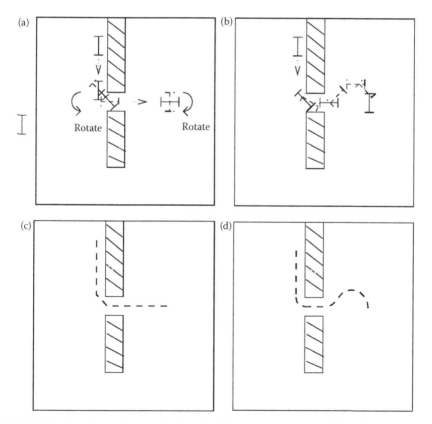

FIGURE 12.6 Resultant paths of adopting (a), (c) the center-turning and (b), (d) front-turning vehicle models in the proposed algorithm.

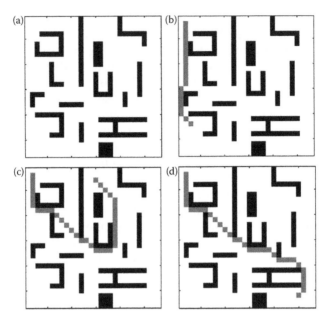

FIGURE 12.7   Simulation results of the proposed algorithm by using the |- shaped front-turning vehicle model. (a) Original work space and resultant paths when end points are located at (b) [22, 3], (c) [3, 15], and (d) [26, 25].

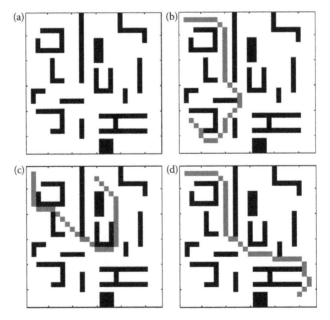

FIGURE 12.8   Simulation results of the proposed algorithm by using the T-shaped front-turning vehicle model. (a) Original work space and resultant paths when end points are located at (b) [22, 3], (c) [3, 15], and (d) [26, 25].

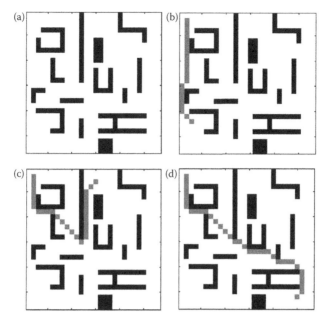

FIGURE 12.9 Simulation results of the proposed algorithm by using the |-shaped center-turning vehicle model. (a) Original work space and resultant paths when end points are located at (b) [22, 3], (c) [3, 15], and (d) [26, 25].

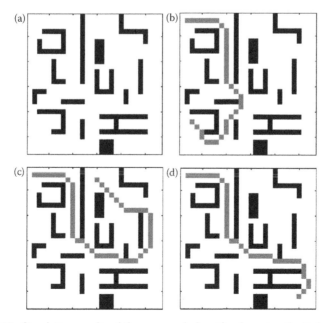

FIGURE 12.10 Simulation results of the proposed algorithm by using the T-shaped center-turning vehicle model. (a) Original work space and resultant paths when end points are located at (b) [22, 3], (c) [3, 15], and (d) [26, 25].

it is necessary to drive a car in an assigned direction as well as park it in a fixed direction. These constraints are trivial to the proposed algorithm since we only need to put an additional directional constraint into Step 5 of the proposed algorithm. Figure 12.11 shows an example of such case.

The computational complexity of the proposed algorithm is somewhat difficult to analyze due to the following reasons: one is that the complexities of applying the PDT are different for different cases. Another comes from the variety in allocating obstacles as well as starting and ending points; thus the resultant path and the number of process iterations cannot be predetermined. However, a coarse analysis based on calculating the

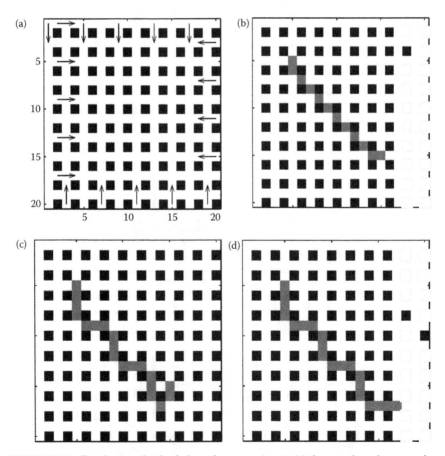

FIGURE 12.11 Resultant path of a |-shaped car running in (a) the tested work space when conditions are imposed (b) no restriction on driving direction, and (c), (d) moving direction is restricted for each road. Note that the arrows denote the allowed moving direction for that row or column and the gray scale represents the moving paths from start position (5, 5) to end positions (15, 15), (15, 15), and (17, 17).

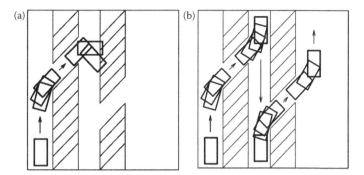

FIGURE 12.12 Resultant path when considering (a) only a forward movement and (b) forward and backward movements, in the vehicle model. Note that there is no solution for the condition in (a).

CPU time, where the computing platform is a SUN SPARC-II workstation running the MATLAB program, is given as a reference for the complexity. In the proposed algorithm, because DT occupies near by 80% to 90% of the computational complexity and because rotational morphology can be implemented in parallel, it is reasonable to estimate the complexity by just considering the repetition of the PDT algorithm. For the cases in Figures 12.7 through 12.10, the CPU time for one iteration of the PDT algorithm is 22 s, and the number of iterations is about 10 to 20 depending on the shape of vehicles and obstacles.

In implementing the proposed algorithm for robotics control applications, the algorithm succeeds if both obstacles and moving objects are modeled as sets and the rotational degree is small enough. However, some other considerations are not included for the sake of simplicity. One is the rule used in tracing the distance map to find a candidate path. For example, different paths are generated horizontal-first or vertical-first rules are used. Sometimes, this results in different solutions. Another consideration, the backward motion of vehicles, helps find more heuristic paths [Shih, Wu, and Chen 2004], as shown in Figure 12.12.

12.6 Dynamic Rotational MM

Since we set the value zero at the destination point in the distance map, we can always find a neighbor with a smaller SED value from any other pixel. Therefore, the shortest path from an arbitrary source to the destination point can be guaranteed to be achieved. Assume that an arbitrarily shaped object moves from a starting point to a destination point in a finite space with arbitrarily shaped obstacles in it. We relate this model to MM using the following strategy:

1. Set the finite space of free regions as the foreground and the obstacles as "forbidden" areas with value –1, and model the moving object as the structuring element.
2. Set the destination point as 0.
3. Apply morphological erosion to the free regions, followed by a DT on the region with the grown obstacles excluded, and then trace the distance map, from any starting point to its neighbor, of the minimum distance until the destination point is reached.

When dealing with shortest path planning using the SEDT, we cannot obtain the correct SEDT by the acquiring approach. Figure 12.13 shows a problematic case in a two-dimensional domain where several obstacles denoted by –1 exist. Note that the SED at the location enclosed by a rectangular box should actually be 29 instead of the 35 obtained by the iterative algorithm. The reason why the iterative algorithm fails is that the structure component, $g(l)$, only records the relations between a pixel and its neighborhood. For example in Figure 12.13, when the pixel enclosed by the rectangular box is processed, the adopted structure component is

$$
g(5) = \begin{bmatrix} -18 & -9 & -18 \\ -9 & 0 & -9 \\ -18 & -9 & -18 \end{bmatrix}
$$

After erosion is applied, the resulting SEDT is 29 from its neighbor, the circled values in Figure 12.13a. When we add obstacles in Figure 12.13b, the pixel enclosed by the rectangle is 35, which is apparently an error. This error could be corrected by the proposed two-scan based algorithm to generate Figure 12.13c. As we can see, the actual SED of this pixel relative to the background pixel 0 is correctly obtained as 29.

(a)

2	1	2	5	10	17	26	37	50
1	0	1	4	9	16	25	36	49
2	1	2	5	10	17	26	37	50
5	4	5	8	13	20	29	40	53
10	9	10	13	18	25	34	45	58
17	16	17	(20)	25	32	41	52	65
26	25	26	29	34	41	50	61	74
37	36	37	40	45	52	61	72	85
50	49	50	53	58	65	74	85	98

Without obstacles

(b)

2	1	2	–1	–1	–1	194	211	230
1	0	1	–1	–1	–1	169	186	205
2	1	2	–1	–1	–1	146	263	182
5	4	5	–1	–1	–1	125	142	161
10	9	10	–1	–1	–1	106	123	142
17	16	17	–1	–1	–1	89	106	125
26	25	26	35	46	59	74	91	110
37	36	37	46	57	70	85	102	121
50	49	50	59	70	83	98	115	134

With obstacles

(c)

2	1	2	–1	–1	–1	178	185	194
1	0	1	–1	–1	–1	153	160	169
2	1	2	–1	–1	–1	130	137	146
5	4	5	–1	–1	–1	109	116	125
10	9	10	–1	–1	–1	90	97	116
17	16	17	–1	–1	–1	73	90	109
26	25	26	29	40	58	68	85	104
37	36	37	40	45	58	73	90	109
50	49	50	53	58	65	80	97	116

Correct SEDT

FIGURE 12.13   A problematic case.

In a digital image, rotational morphology is considered to divide the full 360° neighborhood into eight equal parts. Let P denote an 8-bit codeword with base 2. The value can be interpreted as the degree of the SE orientation that can be fitted into the object region. Directional erosion can be used to indicate the FPs in path finding, so the object can be in the free space for at least one orientation under collision-free conditions.

In Pei, Lai, and Shih [1998], although rotational morphology is used in collision-free situations, the additional space and time complexity in these algorithms is not optimal. The additional record is an 8-bit codeword with base 2, and the processing of MM is eight times for each pixel in the two-dimensional domain. Moreover, the cost increases dramatically with increasing dimensions. For example, in the three-dimensional space, the additional record is a 26-bit codeword and done morphological processing is 26 times for each pixel.

To reduce time-complexity and the need for additional space, we develop the DRMM. We do not calculate morphological erosion for each pixel of the original image until the pixel is selected in the SPP. Figure 12.14 shows an example of deciding the next step by DRMM. Figure 12.14a and b are the original image and the structure element, respectively. The underlined pixel indicates the origin of the structuring element. Figure 12.14c is the structuring element with a rotation of 45°. Note that a counter-clockwise rotation is counted as a positive angle. Let pixel w be the starting point. When deciding the next neighboring pixel in the SPP, we use the following two criteria:

1. The neighbor which has the minimum SED is selected.

2. The neighbor remaining as an object pixel after rotational erosion (i.e., without collision) is selected.

When we decide the next step of pixel w, pixel q is selected because of its minimum SED and satisfactory of morphological erosion. Note that the adopted structuring element is rotated by 45° because pixel q is located 45° of pixel w. Assume that pixels v and f are obstacles. When deciding the next

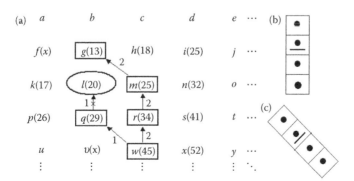

FIGURE 12.14   An example of DRMM.

step of pixel q, pixel l is not selected because the condition of applying the erosion on pixel l by the structuring element in Figure 12.14b is not satisfied (i.e., collision with the obstacle pixel v). Therefore, we go back to w and reconsider the next step of w. Pixel r is selected since it satisfies the above two criteria. Later on, m and g are selected.

Let W, O, E, T, D and S denote a working domain, an obstacle set, an SE set, a stack, and destination and starting points, respectively. The algorithm for the shortest path planning is described below and its flowchart is shown in Figure 12.15.

1. Set the values of the obstacle pixels O and the destination pixel D to be –1 and 0, respectively.

2. Obtain the SEDT by the two-scan based algorithm.

3. Select a starting point S on the SEDT image.

4. Choose pixel m which has the minimum SED in the neighborhood of S, $N(S)$ and decide the direction R_1, which is a vector from S to m to be recorded as a chain code.

5. Apply morphological erosion on pixel m by the structuring element, which is rotated based on the direction obtained from the previous step. If the result is satisfactory, both m and R_1 are pushed into the stack T and S is set to m; otherwise, we pop the preceding pixel u from the stack, remove pixel m from $N(S)$, and set S to u.

6. Obtain the shortest path by repeating Steps 4 and 5 until the destination point D is reached.

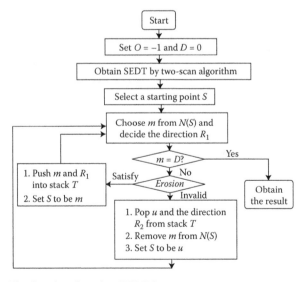

FIGURE 12.15 The flowchart based on DRMM.

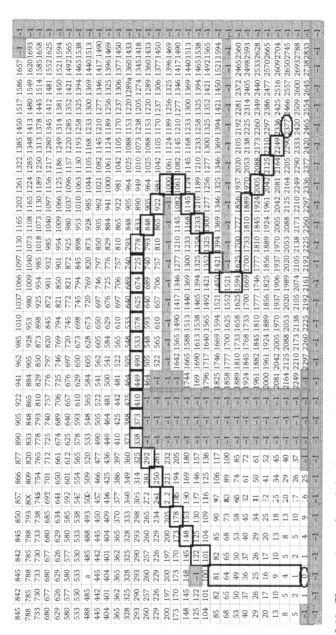

FIGURE 12.16 The experimental result for shortest path planning based on DRMM.

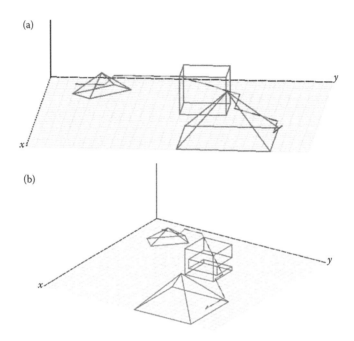

FIGURE 12.17 The results of applying our shortest path planning algorithm on the three-dimensional space (a) with convex obstacles, (b) with convex and concave obstacles.

Figure 12.16 illustrates an example of performing shortest path planning based on DRMM, in which the two pixels encircled with values 0 and 2377 indicate the destination and the starting points, respectively. The dark pixels with value –1 are obstacles and the pixels enclosed by a rectangle constitute the shortest path. Figure 12.17a shows the result of applying our algorithm on the three-dimensional space with convex obstacles, and Figure 12.17b shows the result when there is a concave obstacle. The starting and destination points are located at (3, 3, 1) and (27, 27, 3), respectively.

An example is also shown in Figure 12.18, in which (a) is the original image with a hammer as obstacle and (b) is the distance map obtained by the

FIGURE 12.18 The example of an image for shortest path planning.

two-scan-based EDT. Note that Figure 12.18b is different from Figure 12.12b, since different settings in the object and background are used. In Figure 12.18a, the pixels within the hammer are set to –1 to represent the obstacle, the white pixels are set to 1 to represent the object pixels, and only one pixel is set to 0 as shown in the sign × to represent the destination point. We obtain the shortest path as shown in Figure 12.18c.

12.7 The Rule of Distance Functions in Shortest Path Planning

The distance function applying to shortest path planning may not necessary be ED. As long as a distance function satisfies the following rule, it can be adopted.

Rule: Let $N^{n-1}(p)$, $N^{n-2}(p)$, ... , and $N^{n-n}(p)$ be the n categories of neighbors of p in an n-D space based on a defined distance function. If the values of $d(N^{n-1}(p), p)$, $d(N^{n-2}(p), p)$, $d(N^{n-3}(p), p)$, ... , and $d(N^{n-n}(p), p)$ are all different and are in increasing order, the distance function can be used.

Based on the rule, we can observe that Euclidean, squared Euclidean, and city-block distance functions can be used for shortest path planning; however not the chessboard distance function. Overall, the ED is the one that can preserve the true metric measure.

References

Brooks, R. A., "Solving the find-path problem by good representation of free space," *IEEE Trans. Systems, Man, and Cybernetics*, vol. 13, no. 3, pp. 190–197, Mar./Apr. 1983.

Ghosh, P. K., "A solution of polygon containment, spatial planning, and other related problems using Minkowski operations," *Comput. Vision, Graphics, Image Processing*, vol. 49, no. 1, pp.1–35, 1990.

Hwang, Y. K. and Ahuja, N., "Gross motion planning—a survey," *ACM Computing Surveys*, vol. 24, no. 3, pp. 219–290, 1992.

Kambhampati, S. K. and Davis, L. S., "Multiresolution path planning for mobile robots," *IEEE Trans. Robotics and Automation*, vol. 2, no. 3, pp. 135–145, Sep. 1994.

Latombe, J., *Robot Motion Planning*, Kluwer Academic Publishers, Boston, MA, 1991.

Lin, P. L. and Chang, S., "A shortest path algorithm for a nonrotating object among obstacles of arbitrary shapes," *IEEE Trans. Systems, Man, and Cybernetics*, vol. 23, no. 3, pp. 825–833, 1993.

Lozano-Perez, T., "A simple motion planning algorithm for general robot manipulators," *IEEE Trans. Robotics and Automation*, vol. 3, no. 3, pp. 224–238, 1987.

Pei, S.-C., Lai, C.-L., and Shih, F. Y., "A morphological approach to shortest path planning for rotating objects," *Pattern Recognition*, vol. 31, no. 8, pp. 1127–1138, 1998.

Piper, J. and Granum, E., "Computing distance transformation in convex and non-convex domains," *Pattern Recognition*, vol. 20, no. 6, pp. 599–615, June 1987.

Serra, J., *Image Analysis and Mathematical Morphology*, New York: Academic, 1982.

Shih, F. Y. and Mitchell, O. R., "Threshold decomposition of grayscale morphology into binary morphology," *IEEE Trans. Pattern Analysis and Machine Intelligence*, vol. 11, no. 1, pp. 31–42, Jan. 1989.

Shih, F. Y., Wu, Y., and Chen, B., "Forward and backward chain-code representation for motion planning of cars," *Pattern Recognition and Artificial Intelligence*, vol. 18, no. 8, pp. 1437–1451, Dec. 2004.

Index

Printed and bound by CPI Group (UK) Ltd, Croydon, CR0 4YY

18/10/2024

01776259-0013